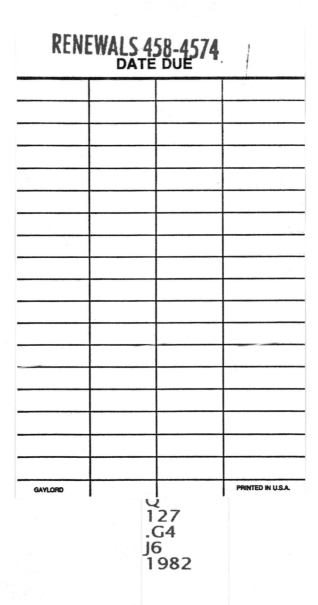

ANCIENTS AND MODERNS

A Study of
the Rise of the Scientific Movement
in Seventeenth-Century England

ANCIENTS AND MODERNS

A Study of
the Rise of the Scientific Movement
in Seventeenth-Century England

Richard Foster Jones

DOVER PUBLICATIONS, INC.
NEW YORK

Copyright © 1961 by the Committee on Publications, Washington University, St. Louis.
All rights reserved under Pan American and International Copyright Conventions.

Published in Canada by General Publishing Company, Ltd., 30 Lesmill Road, Don Mills, Toronto, Ontario.
Published in the United Kingdom by Constable and Company, Ltd., 10 Orange Street, London WC2H 7EG.

This Dover edition, first published in 1982, is an unabridged and unaltered republication of the revised edition published in 1961 by The Washington University Press, St. Louis. The original edition (same publisher) appeared in 1936. The present edition is published by special arrangement with Washington University, St. Louis, Mo. 63130.

Manufactured in the United States of America
Dover Publications, Inc.
180 Varick Street
New York, N.Y. 10014

Library of Congress Cataloging in Publication Data

Jones, Richard Foster, 1886-
    Ancients and moderns.

    Reprint. Originally published: 2nd ed. St. Louis : Washington University Press, 1961. (Washington University studies)
    Includes bibliographical references and index.
    1. Science—Great Britain—History. 2. English literature—Early modern, 1500-1700. 3. Bacon, Francis, Sir, viscount St. Alban, 1561-1626. I. Title. II. Series: Washington University studies.
    [Q127.G4J6   1982]   509'.42   82-7288
    ISBN 0-486-24414-8 (pbk.)   AACR2

TO THE ENGLISH DEPARTMENT
OF STANFORD UNIVERSITY
THIS BOOK IS DEDICATED

PREFACE

No exaggerated estimate of the value of this book is responsible for its second appearance. First published in a limited number of copies in 1936, it has been out of print for some years, and requests for copies have prompted Washington University to ask the author to prepare a second edition. Originally the work was the outgrowth of a previous study of the controversy between the upholders of antiquity and those of modernity in the seventeenth century, evidence of which fact is seen in its first title and subtitle. Since the book is now rather closely identified with that title, it has seemed best to retain it, but the subtitle, together with certain passages in the text, has been changed to conform more closely to the nature of the study, the scope of which in the process of its first composition expanded, almost unconsciously, to include more elements in the scientific movement than were indicated on the title page. The first preface and conclusion recognized this condition and attempted to reconcile book and title, but with less than complete satisfaction. For this reason a more truly descriptive subtitle seemed desirable. Much has been done in the history of science in the seventeenth century since the book was first composed, and it is, perhaps, to be regretted that the present edition does not, with rare exceptions, take note of material that may bear upon it. Since, however, these additions to knowledge lie rather near the periphery of the present author's conception of the scientific movement in the way of amplifying and clarifying matters subservient to it, and since, as far as he knows, his general ideas have not been attacked, he has left the text almost entirely in its original form.

The subject of this book is that thought movement in the seventeenth century to which modern science in England traces its source. Therefore, the author is interested more in what scientists and fellow travellers say than in what they do, more in the complex of thought and feeling out of which scientific achievement sprang than in the achievements, more in the attitude expressed toward discoveries than in the discoveries themselves. This movement was thoroughly, though not exclusively, Baconian. The present study strives to be objectively historical, but if a thesis can be discovered in it, it is that Sir Francis Bacon

vii

bears the same relationship to the movement under discussion as, for example, Karl Marx bears to the development of Communism, but, of course, to much better purpose. It is that type of relationship in which a leader becomes the human center of a movement, to whom men can feel and pay loyal allegiance in the spirit of discipleship. The encouragement and stimulation derived from this attitude is by no means a negligible factor in the development of science during the mid-seventeenth century, though obscured by more tangible evidences of Bacon's influence.

Sir Francis both expressed and molded his age. Before his scientific writings appeared, not only had significant discoveries been made, but also important elements of the movement itself had taken form, such as hostility to the authority of the ancients, and the importance of experiments, yet in England it was indeed moving slowly until he rang his bell, and as the century advances it assumes more and more a Baconian complexion. Some of the most influential ideas in the current of scientific thought are too uniquely his to render any other source credible. An outstanding example is his idea of the necessity of an universal natural history drawn from authentic records of the past as well as from contemporary observations and experiments, one wide enough indeed to cover all nature. Upon this material his inductive method was to be employed for the purpose of discovering the fundamental laws of nature, knowledge of which would enable men to use them, singly or in combination, to secure all the effects of which nature is capable. Bacon cheerily announced that this happy event would follow soon after the ages required for the compilation of his history. The unquestioning acceptance of this impossible idea by the best scientific thinkers of the day may be termed one of the marvels of history, which only Bacon was capable of bringing about. He inspired in his followers such complete confidence in his ideas that they were stimulated and incited to carry on experiments with an intensity hitherto unknown, so much so, in fact, that Sprat was moved to call his times "this age of experiments." It is hardly an exaggeration to say that the compilation of this history was the conscious goal of most of the scientific activities of the day.

Another unique element is the emphasis which Bacon placed upon experimentation, for the latter, though by no means new, had never been accorded the absolute authority and exclusive role upon which he insisted. This attitude is to be associated with his attack on the authority of antiquity, to which men had for ages largely turned for knowledge of nature. In experiments he had an alternative to offer men whom he would turn from the ancients, and whom he thus brought to nature herself. He and the age he influenced never tired of contrasting the two sources of information, as the antitheses which the contrast suggested testify: nature versus books, works versus words, laboratories versus libraries and closets, industry versus idleness. At times it seems as if men were charmed by the magic word "experiment" itself, and it must be confessed that in general no great progress was made in penetrating deeply into the true nature of experimentation. A few like Boyle and Hooke, it is true, were inquiring into the characteristics of truly scientific experiments and the proper method of making them, but some let the word rather than the proper conception serve their purpose. Even the physical exertion required by experimental research assumes an aspect of virtue, and the physical inactivity of reading becomes nothing more than sloth. Sometimes the contrast comes perilously near that between doing and thinking. The science of the day was most frequently called "the experimental philosophy," though other adjectives, "new," "free," "solid," "useful," "real," find expression. And finally faith in experimentation influenced the Baconians in their attitudes toward other types of science. It softened their criticism of the bragging chemists, and rendered suspect, to a certain extent, the mechanical philosophy of matter, motion, and mathematics, which under the influence of Descartes and the atomic philosophers was slowly establishing science on the firmest foundation it had ever possessed, but which at the time was criticized as being a theory only and not based on a sufficient number of experiments.

There was one other characteristic of seventeenth-century science for which Bacon was largely, if not entirely, responsible, namely, its pronounced utilitarianism. The purpose of his *Magna Instauratio* was to enable man to know nature in order to command her. It is true, Bacon divides experiments into two groups, those that give light and those that bear fruit, but the purpose of the first was to enlarge man's knowledge so that he might thereby enjoy more fruit. Bacon feared that in observing and experimenting, men would gain bits of knowledge which they could immediately turn into material benefits, and thus be tempted to abandon further investigation; so he cautioned them not to be diverted by such opportunities from a goal larger but not different in kind. The utilitarian spirit is everywhere evident in the twenty years preceding and following the Restoration, that is, among the Baconians, not among the atomic scientists. The effects of it are seen in the attitude of the age toward what we call pure science. The satisfying of intellectual curiosity through scientific inquiry is regarded as a legitimate activity, but it is associated with the proud speculative man in contrast to the humble experimenter working for the good of mankind. The utilitarian spirit is also revealed in the interest shown by scientists in agriculture and in the "mechanical" arts and manual trades. Both Bacon and his followers looked upon these as able to furnish much valuable information to investigators, but they also stressed the great contribution which experimental science could make to them, and through them to the world.

A number of other characteristic values and attitudes of the age find in Verulam their most likely source, such as the need to develop a critical mind, which would be slow and cautious in accepting any proposition not thoroughly proved by experiments. This attitude led directly to a distrust of reason and an antipathy to systems, which were regarded as the product of the mind and not of experiments and observations. The attack on the ancients was partly inspired by the conviction that reason played a more important part than sensuous observations in their philosophy, which for this reason comprised abstract generalities derived from the mind and not from nature. The relegation of reason and learning to a position subordinate to the senses, together with the large number of men requisite for the compilation of the natural history, was responsible for lowering the intellectual qualifications required for scientific work to such extent that anyone who had hands and eyes was thought capable of performing experiments. Bacon, however, found another reason for the low qualifications he established for scientific brains, and that was the need to remove the despair which the moderns felt of ever equalling the ancients in mental powers because of the widespread belief in the decay of nature, an hallucination which rendered modern inferiority inevitable. In England the great lord chancellor came to be regarded as the champion of modernity against the ancients, and he had much to do with the intellectual independence of his followers, but he never tried to equate modern and ancient genius. Instead he offered the world his method as requiring a minimum of intellectual power, and yet as more than offsetting the advantage of better brains enjoyed by antiquity. It is hardly too much to say that during this period reason and mind were somewhat suspect in the domain of science.

There were also certain social correlatives of the scientific ideas mentioned above that, especially in the puritan era, receive vigorous expression because congenial with attitudes arising from other sources. These sprang almost entirely from the utilitarian element in scientific thinking and from the lowered intellectual gifts demanded of experimenters. The farmer, artisan, and mechanic, because they were in contact with natural things and were unhindered by any intellectual principle of exclusion, rise in importance and receive considerable recognition. The democratic implication in this situation can hardly escape notice, and indeed becomes conspicuous in what Sprat says about the liberal membership policy adopted by the Royal Society. Certainly science contributed something to that stream of democratic thought and feeling which, arising in the Reformation, was gathering force among the Puritans. The emphasis placed by Bacon upon the material benefits which his philosophy would bestow upon man could not but introduce or re-enforce the social motive of the public good, for this philosophy made such a goal seem quite feasible. A specialized version of this social conception is revealed in the appearance of a humanitarian spirit which found in the prospective blessings of science the hope and the means of bettering the condition of the poor. Bacon speaks of his philosophy as capable of relieving the miseries of man, and he describes the Father of Solomon's House, a scientific organization in the *New Atlantis,* as looking as if he pitied men. During the mid-seventeenth century the social utility of science is recognized; the beneficent results of scientific activity are contrasted with the evils of war; and the heroes of science supplant those robed in martial glory.

But Bacon did more than furnish ideas; he offered encouragement and hope to men oppressed by the subtlety of nature, for which the science handed down from antiquity was no match, and by the prevailing opinion of nature's decay. He eloquently and convincingly (for fortunately he had the power of convincing) held out his historical and inductive method as quite adequate to overcome those bugbears. Furthermore, the very title of his first scientific works displayed the promise of progress, and was adopted as a slogan by others. Today his experiments do not arouse much enthusiasm, but the numerous editions of the *Sylva Sylvarum* which appeared before the Restoration, and the frequent and approving references to it suggest the comfort which his example, as well as his philosophy, must have given his followers. In fact, he offered himself as an example of what one man, though beset by many duties, could achieve in the field of experimentation. The scientific spirits of that day vividly realized that they were carrying out his instructions, and this realization, supported by the hope of ultimate success, made them feel that they were living at a momentous time in history. "A universal light," says Sprat, "seems to overspread this Age." They were also aware of the character which their own active pursuit of knowledge was imposing on their generation, as the terms they use to describe it make clear: "a learned and inquisitive age"; "a prying and laborious age"; "an age of industry and inquiry" marked by a "searching spirit" and an "affection to sensible knowledge." Bacon had made it plain that the "far-off divine event" toward which they were making their way was indeed far-off, yet occasionally we hear a confident voice proclaiming it just around the corner. Whereas Glanvill and Boyle had only seen in visions a future technological paradise, Sprat speaks of the "wonderful perfection" already achieved by the "mechanical Arts." The age needed no future historian to apprise it of its significance.

The purpose of the foregoing sketch is only to suggest Bacon's significance in the history of science. It is hoped that the following pages will support and supplement it.

London   *R. F. J.*
October 1, 1959

CONTENTS

ANCIENTS AND MODERNS

A Study of
the Rise of the Scientific Movement
in Seventeenth-Century England

THE RENAISSANCE

CHAPTER I

THE SCIENTIFIC ATTITUDE OF THE ELIZABETHANS

*Where as Galen with other auncient and approbat doctours doth
prayse porke I dare nat say the contrarye agaynst them, but thys I
am sure of, I dyd neuer loue it*
Andrew Boorde, *A Compendyous Regyment*, 1547.

To speak of the scientific attitude of the Elizabethans is, perhaps, to do violence to words, for certainly, with a few notable
exceptions, they manifest no attitude which could by modern
standards be so described. Yet we may, with the indulgence of
the reader, employ the expression to mean their idea of the
sources from which knowledge of nature was to be drawn and of
the method to be followed in gaining such knowledge. This chapter can hardly hope to add anything to what is already known
about the subject. Yet sometimes generalizations become so far
removed from the particular cases or data upon which they are
based, that they become vague and tend to lose force. So a fresh
description of representative sources of evidence may serve a
useful purpose in making ideas more vivid and in sharpening
their significance. In the present instance this procedure seems
all the more desirable, for an accurate understanding of the rise
of modern science in England requires a clear knowledge of the
circumstances in which it arose. Bacon's task and achievement
can be justly appreciated only when cast into relief against the
the background of his age, instead, as seems more frequently to
be the case now, of being subjected to standards drawn from the
fully developed scientific method of today. I have tried, therefore, to reconstruct the background by examining typical treatises of the period, which examination, though not based upon
exhaustive investigation, may indicate clearly the scientific temper of the period.

One of the results of the revival of learning was the establishing of direct contact with the extant scientific works of antiquity.
To a certain degree ancient science had reached the Middle Ages,
largely through the Arabs, "but not till the period of the Renaissance did the western world recover an adequate and direct
knowledge of ancient science, learn to examine it critically, or

endeavour to find a path of its own in the new experimental method."[1] It was, then, with something of the sense of new discovery that the English of the sixteenth century explored the recently published stores of learning which proved so much more satisfactory than traditional science. It is not strange that having followed the muddied stream of their knowledge to its unsullied fountain, they should have been content to rest by its side in the belief that this was the source of all learning. So men looked back many centuries for explanations of the surrounding universe, concerning which they had come to be curious. They faced backward rather than forward in the quest for truth, and prepared themselves for scientific investigation by mastering the Greek and Latin languages. The natural result was an ardent worship of the great classical minds, and a submission to their authority which it is hard for us to appreciate. Then it was considered as strange that an "auncient and approbat doctour" should be wrong as it is now considered that he should be right.

Of all the ancients who exercised despotic sway over human opinions Aristotle stands out conspicuously. So broad was the scope of his intellectual interests that few phases of man's thought escaped him. His syllogistic logic became the main instrument of scientific reasoning. He had dominated medieval theology, and in the Renaissance he continued to dominate science. The absoluteness of his authority is revealed by one of the statutes of Oxford University, which decreed "that Bachelors and Masters who did not follow Aristotle faithfully were liable to a fine of five shillings for every point of divergence, and for every fault committed against the Logic of the Organon,"[2] and at least one scholar was expelled for daring to attack the Grecian. In fact, the "Prince of Philosophers" became so important that his name was used as representative of the whole of ancient authority. Two other ancients, however, attracted much attention, Galen in medicine and Ptolemy in astronomy. In his own field Galen's authority was as pronounced as that of Aristotle in other sciences; when the ancients came to be driven from their strongholds, he was as bitterly attacked as the Stagirite himself. But, as we shall soon see, antiquity did not have it all its own way. Entrenched as it was in the universities and some of the professions, there were those who on the outskirts of respectable learning, sometimes in most despised precincts, were lifting vigorous voices of protest. On the continent, various thinkers from various directions had moved to dethrone Aristotle, and Bruno in 1583 had even extended the campaign to England, though with no results. The spirit of curiosity regarding nature and man, a distinctive trait of the Renaissance, was too strong to permit an entirely unbroken acquiescence in ancient learning. Yet the fact remains that with the exception of William Gilbert and a few others no respected English name was arrayed against it.

Nowhere is ancient authority more clearly revealed than in Elizabethan medicine. Thomas Linacre, who received his medical training in Italy toward the close of the fifteenth century, and who was physician to Henry VIII, had in the first quarter of the sixteenth century firmly established Galen as the basis of respectable medicine. He introduced lectureships in medicine at Oxford and Cambridge, and was largely instrumental in the founding of the London College of Physicians, which remained a Galenical stronghold for more than a century. In a lesser degree other ancient writers on medicine were regarded as authorities. Thus it came about that a library of old writers was the only requisite for the training of a doctor, and one could even become an author of medical books without having ever attended a patient. Sir Thomas Elyot, now known for his *Boke named the Gouernour*, did not scruple, even in the midst of a life busied with other activities, to enlighten the world on medicine, and, to the objection raised by certain members of the College of Physicians that he was not a physician, he offered the irrefutable answer that a noted physician had read the works of Galen to him, and that later he had read Celsus, Pliny, Dioscorides, and others.[3] The author of the first English treatise on anatomy, published four years after Vesalius had laid the foundations for modern anatomy in a book not drawn from Galen but based upon actual dissection and observation, stated that it was written "Not for them that be expertly seene in the Anatomie: for to them Galen the Lanterne of all Chirurgions, hath set it foorth in his Canons, to the high glory of God, and too the erudition and knowledge of al those that be expertly seene and learned in the noble Science of Chirurgerie."[4] Again and again this extreme worship of classical authorities appears. The author of a book with the whimsical title *Bulleins Bulwarke of defence againste all Sicknes*, 1562, calls Hippocrates, Galen, and Dioscorides the captains of his

"Bulwarke." Galen bore the title of "Prince of Phisitions";[5] he
and Hippocrates were considered the "very foundation of our
profession." These two had made the art of medicine perfect;
they appear more like gods than men; and all who have written
since "haue dronke of the water, that flowed out of their two
welles, and doe greatly reioyce therein."[6] They would have had
good reason to rejoice if medicine had been perfect, but unfor-
tunately it was not, and their attitude was a great obstacle to its
advancement, and reveals how necessary Bacon's revolt was.
Sometimes we find experience and observation placed on the
same plane as authority and proposed as an alternative or
supplementary route to knowledge. John Banister included in
his *Needefull, new, and necessarie treatise of Chyrurgerie*, 1575,
"certain, experimentes of mine owne inuention, truely tried, and
daily of me practised," and in the dedication he claimed his work
was compiled "partly by the study of other good authors, partly
also by my own experience." The innumerable references to
Galen sufficiently explain the "other good authors." John Wood,
in the preface to a work strewn with references to Hippocrates,
Galen, and others, says, "quicquid in hoc libello continetur,
aut per propriam nostram experientiam confirmatur, aut ex
Hipocrate, Gal. et reliquis principibus vtriusque medicinae,
summa cum industria, et fide colligitur."[7] There was, however,
with the exception of the chemical doctors, no disposition to oppose experience to authority, and the former was admitted only
when it did not run counter to the latter, or supplied some
unaccountable oversight on the part of the heroes of old.

The same attitude is revealed in the herbals of the times,
which really constituted a part of medicine, since interest in
botany had not developed beyond the use of plants in the preparation of medicines. Just as medieval philosophy in general
considered the whole universe constructed for man's own private
interest, so it thought every herb was a specific for some ailment,
and possessed no other use or significance. Here Dioscorides was
the prime authority though he by no means stood alone. One
medico-botanist declaims violently against "the upstart physicians" who spend no time on the study of Dioscorides, Galen, and
Pliny, assuring his readers that the ancient trio together with
Theophrastus were his chief sources.[8] The pages of the herbals
are strewn with references to the ancients.[9] Frequently we do

find descriptions of plants drawn from observation, and the discovery of others unknown to the ancients, but such refreshing variations occupy a very inferior place. Thus we see in what is perhaps the most noted herbal of the period, John Gerard's *Herball or Generall Historie of Plantes,* 1597, descriptions of plants grown in his own garden, though most of his material is "gathered" from books. Prefixed to this book is a letter from a St. Bredwell physician which advocates chemical medicines as well as simples, and declares that physic would approximate perfection if the two were joined. "The Phisicke reader by their meanes, shall not onely come furnished with authorities of the Ancients, and sensible probabilities for that he teacheth, but with reall demonstrations also in many things, which the reason of man, without the light of the fornace, would neuer have reached vnto." It is unusual to find a Galenist advocating chemistry, for the Galenical and chemical doctors were mortal enemies. Here, of course, experiment and observation are raised to the level of authority.

The chemical physicians traced their lineage to Paracelsus, that remarkable Swiss genius who burned his Galen and turned to the study of nature and chemical experiments. In doing so he began what later became the most pronounced anti-authoritarian current in the seventeenth century. The Paracelsians did perhaps discover a few drugs, and they certainly made experiments, however blindly, but their greatest interest to us lies in their attacks on ancient authority and in their praise of experimentation and sense-observations as the criterion of truth. Most assuredly the chemists of the sixteenth and early seventeenth century in general produce no favorable impression upon the modern reader. Much of what they write is pure jargon, obscure and imposing terms being used to impress the people, while such an air of mystery is thrown about their theories, so much is said and so little revealed—if indeed known—that we can understand their being branded as quacks by the outraged Galenists. Furthermore, the extravagant claims made for their remedies, and especially for the *elixir vitae,* or universal remedy, justly provoked the established physicians to great wrath. Yet they were the most complete and consistent opponents of authority, and the loudest shouters for experiment and observation.

The attitude of the Paracelsians is well represented in a book written by one who signs himself "R. B."[10] It is true that the author, suffering from a strong religious bias and incensed by the "heathnish Philosophie" of Aristotle and Galen, praises chemistry because it attributes all phenomena to God and thus satisfies his religious nature. Yet in numerous passages he evinces an unmistakable faith in experiment, or, as he says, "palpable and visible experience." By experimenting with fire the chemical physician shall "knowe all things by visible and palpable experience, so that the true proofe and tryal shal appeare to his eyes and [be] touched with his hands . . . . For them [salt, sulphur, and mercury] shall he know, not by his own braines, nor by reading, or by report, or hearesay of others, but by experience, by dissolution of Nature, and by examyning and search of causes, beginnings and foundations of the properties and vertues of thinges."[11] He inveighs in no uncertain manner against the Galenists for their reliance upon authority and for their implicit following of Aristotle, Galen, and Avicenna. Prophetic of mid-seventeenth-century attacks on Peripateticism are his strictures upon the mere speculation, sophistry, arguments, reasoning without demonstrations, and "dunsicall wrangling" of the established physicians. Naturally he is indignant at the privileges and protection bestowed by law upon them. Full of the faith of observation and experiment, he maintains relentless opposition to the tyranny of classical writers and of modern laws which would hinder the free search for truth. He finds in the Reformation the same hostility to authority and the same insistence upon freedom of thought for which he is pleading in medicine, a relationship between science and Protestantism which will be discussed in later chapters.

But as long as the vnskilfull and sluggish Phisition may enjoy that immanitie [immunity] and freedome, and as long as it shalbe allowed in the Scholes to be heresie and foule ignoraunce to speake against any part of *Aristotle, Galen Auicen,* or other like heathens doctrine: as long as the *Galenists* may shrowde themselues vnder the Wings and protection of Princes, Priuiledges and Charters, the cause of the Chimycall Phisition must needes lye in a desperate state. And no man almost shalbe able to attayne to the perfection in true Phisicke. As long as *Scotus* or *Thomas Aquinas,* and such other were so priuiledged in the scholes, that no interpretation of Gods worde was allowed, but such as was brought out of them, or agreed with them, the cause of true Religion, and seruing of God was in desperate state, and it lay oppressed and hidden. And as long as those that were noseled in such puddle, were mainteyned, defended and priuiledged by princes and potentates, it was hard for trueth to shewe his face abroade openly. Wherefore if the Chymicall doctrine agreeing with Gods worde, experience and nature may come into the Scholes and Cities in steade of *Aristotle, Gallen,* and other heathen and their followers. And if it were lawfull and commendable for euery honest student to labour in the Philosophicall searching out of the trueth, by the fire or otherwise, and thereby either confirme and make manifest the trueth by this Arte taught, eather to add newe things wel tried to the old that be good, and then to reiect the other bastard, adulterat, sophisticat stuffe, and so ioyne words and deedes together, then should there be no time spent in vayne, and vaine-glorious bable and sophisticall disputations, without due triall by labor and worke of fire, and other requisite experiments, then should it easely be seen whether *Gallen* and other heathen or the Chimests were most to be folowed and allowed. And whose writings and trauailes were more auaileable for mans health, either conseruing or restoring, & who seeketh more paynefully, faythfully, sincerely, charitably and Christianlike, for the certeine helpe of his neighbour, and not lucre or veine glory and pompe, the auncient Chimical Phisition or *Gallen* and his folowers.[12]

Far as the Paracelsians were from the truth, with their salt, sulphur, and mercury, they were more nearly right than the Galenists with their qualities and humours, and though with them experiments are generally restricted to those performed with fire, we find them frequently basing their claims for the truth of their theories squarely upon the virtue that inheres in experiment and sense-observation in striking contrast to reliance upon books and ancient authority.

We see the authors of other supposedly scientific works ransacking the stores of the printed page, and justifying their work by the authorities from whom their material was "gathered."[13] When Thomas Hill, writing on agriculture, says that his ideas were "gathered out of the treasures of the beste wryters (of the Greekes and Latines) in this arte, and experienced by the often practises, of sundrie skilfull Gardiners, in diuerse Realmes and Countries, vnto our great vse commoditie,"[14] he is really basing the truth of his theories, not upon an empirical foundation, but upon the authority of their origin. The second part of the sentence is only a minor and expected appendage to the significant first part. Dazzled by the recovered light of the past, the Elizabethans so invested the ancients with the robes of authority that the latter became oracles, to question which bordered upon sacrilege. They restricted the study of nature to the narrow confines of a library, and tested the truth of scientific theory by the names on the title page. Aristotle, Galen, Dioscorides, Pliny, Ptolemy, and numerous others exercised a powerful dominion over the minds of men, one result of which was to make the study of nature an intellectually aristocratic pursuit, to clothe it in the gowns of pedantic learning, and remove it from the grasp of those who live in closest contact with natural phenomena.

Mathematics received slight consideration in Elizabethan England, the significant work of Napier and Briggs belonging to the next century. In 1590 Thomas Hood says that "it is but late since the Mathematicall Sciences began to be in request within this Citie." Robert Recorde is the only name that creeps into most histories of mathematics, and his performances were slight, being largely elementary treatises to teach the uninformed. Indeed, he finds it necessary to defend the study of mathematics as well as that of astronomy, which is closely associated with it.[15] His defence rests largely on an appeal to authority, particularly that of Hippocrates and Galen (for astronomy), after the mention of whom he says, "But omittinge the testimonies of famous wryters (which would make a wonderfull volume of themselues, if they were written only together) I wyll vse a simple plaine proofe manifest to all men." Authority evidently was the more impressive argument but not suited to simple folk. The great discovery in astronomy which was most likely to lead to a questioning of ancient theories was, of course, that of Copernicus, first published in 1543. But in England a profound silence regarding it was for the most part maintained throughout the sixteenth century. We may be sure that it was discussed,[16] but the slight progress made in mathematics may have had something to do with its not being treated more seriously. Recorde refers to the theory, which he seems both to condemn and approve,[17] but his own astronomical writings are for the most part only popular presentations of Johannes de Sacro Bosco's (John Holywood's) *Tractatus de Sphaera,* itself a somewhat elementary and very popular exposition of the Ptolemaic system.[18] The old geocentric theory of the universe held practically undisputed sway over Renaissance England, and Bruno, who had sworn allegiance to the new system, in his visit to England seems not to have dented English conservative thought. Ancient authority and "the consent of all learned multitude of writers" were sufficient grounds for the accepted astronomical views.[19]

Two men, however, in the Elizabethan period clearly embraced the Copernican theory, William Gilbert, who will soon be discussed, and Thomas Digges, who because of his mathematical training had a more intelligent conception of it.[20] Not only was the latter the first to support the new astronomical system, and he did it with enthusiasm and without reservation, he also introduced into his discussion translations from Copernicus' *De Revolutionibus Orbium Coelestium*. Furthermore, he added to the theory the conception of an infinite universe with numberless stars located at varying distances from its center.[21] His espousal of the theory inspired him to attack the authority of antiquity, especially that of Aristotle, who in his eyes was largely responsible for the fact that all universities subscribed to the Ptolemaic system.[22] He realized the necessity of testing hypotheses by reason and observations, and gives some evidence of having employed the experimental method himself.[23] Digges shares with Gilbert the honor of being one of very few Englishmen to express the true scientific attitude at this time.

If there was one branch of knowledge calculated at this time to reveal ancient deficiencies and modern superiority, it was geography. The sixteenth century witnessed innumerable voyages which continually widened the knowledge of lands unknown to the ancients. This enormous expansion in geographical learning was not without effect in militating against the standing of antiquity, especially in the next century, and nowhere did it show its influence more significantly than in the numerous geographical figures of speech with which the moderns urged their claims upon mankind. And yet so firm a hold had ancient authority obtained upon the Elizabethan mind that it was with difficulty and apology that any deviation from the path of antiquity was ventured. "Perchaunce," says Cuningham, "ther mai arise sundry sicophants (reading this my labour folowing) which will not let to accuse me of arrogancie, in that I take this part in hand [description of countries, cities, inhabitants, etc.] sence Ptolomæus that famous Kinge hath of that argument left to vs his Geographie. Vnto whome I answer, confessing me not worthy to kisse his fotesteppes, aswell for the maiestie of his person, whan he liued, as also for his excellent Learning, Science, & diligence," that the work was too large to be completed by any one man. He finds it necessary to apologize for even attempting to contribute to a field pre-empted by an ancient. As regards another geographical matter Cuningham says, "although Ptolomæus and Auicenna were of sufficient authoritie, to make you credit this thing . . . . yet I wil vse other reasons then testimonies in this behalfe."[24]

It was impossible, of course, for English geographers, in the light of modern discoveries, not to notice the errors and deficiencies of ancient geography. It is not strange to find them calling attention to such defects and elaborating upon more recently acquired knowledge. What is remarkable is the respect and reverence still accorded the ancients, the tender excuses offered for their ignorance, and the innumerable citations of them whenever opportunity offered. They were never to be treated lightly, much less condemned. Sometimes it was absolutely necessary to differ from antiquity, as, for instance, in the cherished proposal for finding a northwest passage to the Far East. Relative to the latter, Richard Willes says that though the maps of "Ptolome the father of geographie, and his eldest children" show land all the way to the pole, Ptolemy was ignorant of the world north of latitude 63, and in refutation of him Willes offers not only recent Portuguese voyages but also the authority of Aristotle, Pliny, Plato, and others. For the most part he is pitting ancient against ancient, and the authority of the past stands on the same level with modern discoveries.[25] Likewise Richard Eden, in upholding the possibility of a northwest passage, states that though Ptolemy was "an excellent man," he was ignorant of America and other parts of the world recently discovered,[26] and therefore not in a position to speak authoritatively on the matter. Eden was more independent than most of his contemporaries. He instances printing and gunpowder (but omits the compass, strange to say) as evidence that "this our age maye seeme not onely to contende with the Auncientes, but also in many goodly inuentions of Art and wyt, farre to exceede them";[27] and he repeats with approval Fernelius' condemnation of those who contend that "the auncientes haue inuented and comprehended al thynges." Purchas, if we may step beyond the limits of our period, holds that modern opinions regarding the size of the earth are superior to ancient by virture of the discovery of the Americas and Australia by the Spaniards, English, and Dutch, yet he thinks it nothing short of miraculous that such knowledge should have been revealed to

"the old and decrepit Age of the World." Purchas also feels constrained to part company with classical writers and early Christian fathers in their categorical denial of the habitability of the
tropics and antipodes—an error which later is frequently urged
against antiquity—though he does it with all reverence.

> But Poets, Philosophers, Fathers, (in other things worthy our loue, for their
> delightful Poems; our admiration for their profound Science; our awefull
> respect and reuerence for their holy learning, and learned holinesse) herein
> wee bid you farewell: . . . . our America, subiect to that supposed burning
> Zone, with clouds and armies of witnesses in her well-peopled Regions can
> auerre, that the parts betwixt the Tropickes are both habitable and inhab-
> ited.[28]

Modern geographical observations added to and frequently corrected traditional knowledge, but they certainly did not undermine respect for the ancients, nor, in academic treatises at any
rate, very much lessen their authority in many things.

Closely associated with both mathematics and geography was
the compass, and consequently the phenomenon of magnetism.
Of the three inventions that impressed upon the Renaissance
mind the character of the period in which it lived—gunpowder,
printing, and the compass—the last appeared by far the most
significant, not only because it was largely responsible for the
discoveries that amazed and thrilled the age, but also because its
mystery defied explanation and invited attention and study.[29]
Magnetic phenomena were known to the ancients, but their explanations of them were indeed weird, though faithfully adopted
by medieval and Renaissance writers. To the sailor in the mists
and the cold of the north Atlantic or on strange southern waters,
the needle's action was a matter of life and death, and was by
necessity an object of observation and induced interest.[30] Other
matters might indeed be left to speculation in libraries, but magnetism as revealed by the compass was an ever-present phenomenon, in which centered the reality and necessity of actual
observation. Certainly interesting accounts of the variation of
the needle observed in quarters the ancients never reached were
recorded and brought back to England.[31] And when magnetism
came to be more thoroughly explored, mariners were urged to
study the action of the compass, and become the agents of scientific observations. Thus it is not surprising to find in treatises
on magnetism the first elaborate manifestations of the true ex-

perimenting spirit and of a genuine perception of the limitations
of the ancients and of the evil results of subserviency to their
authority.

William Gilbert is the outstanding figure in magnetic studies
during this period, but long before he published his work, and
probably before he had even turned his mind to the subject, the
attitude later revealed by him is definitely expressed. In 1581
Robert Norman, a maker of mathematical instruments who had
spent some eighteen years at sea, published a work[32] which, content to establish in convincing fashion only one fact about terrestrial magnetism, illustrated the method and evinced the spirit of
inductive science. He opposed ancient authority, but he felt constrained to handle with kid gloves that object of adoration.
After rightly complaining that many ancient philosophers had
left their opinions on the loadstone to posterity as infallible
truths, he feels that he must show them due reverence, though
not to the extent of sacrificing the evidence of his eyes and mind.
"And as I maye not, nor meane not herein willyngly to condemne
the learned or auncient writers, that haue with greate diligence
laboured to discouer the secrets of Nature in sundrie thynges
with their operations and causes, yet I meane God willyng, without derogatyng from them, or exaltyng my self, to set doune a
late experimented truth found in this Stone, contrary to the
opinions of all them that haue heretofore written thereof." Nor
does he hesitate to describe in most approving tones the true and
effective nature of his method in contrast to the traditional one:
"Wherein I meane not to vse barely tedious coniectures or imaginations, but briefly as I maie to passe it ouer, foundyng my
argumentes onely vpon experience, reason, and demonstration,
whiche are the groundes of Artes." He next offers a sturdy defence against the charge which he felt might be preferred against
him by the high priests of traditional science, namely, that he
was not sufficiently learned to express any ideas.

And albeit it maie bee saied by the learned in the Mathematicalls, as hath
been alreadie written by some, that this is no question or matter for a
Mechanician or Mariner to meddle with. . . . But I doe verely thinke, that
notwithstandyng the learned in those sciences beeyng in their studies
amongest their bookes, can imagine greate matters, and sette doune their
farfetcht conceiptes in faire shewe, and with plausible woordes, wishyng
that all Mechanicians, were such as for want of utteraunce should bee

forced to deliuer vnto them their knowledge and conceiptes, that they might florishe vpon them, and applie them at their pleasures.

The contrast here between observation and reading is evident enough. The honest mariner heartily condemns those learned men who speak disdainfully of the searchers out of the secrets of their arts because the latter lack knowledge of Latin and Greek. In turn, he accuses the pedants of promising much and performing little.[33]

Norman insists on experiment in opposition to reliance upon books as distinctly as Gilbert. In fact, he engages in a stout defence of those who with little learning place their faith in humble experiment and observation. In so doing he is this early striking a note which resounds much louder in the seventeenth century, for one of the most interesting effects of the experimental philosophy was the development of a democratic attitude. The method made all men equal; intellectual aristocracy received many a rude blow. From whatever source Norman may have derived his conception of experimental induction, he always manifests utmost confidence in it. He introduces illustrations in a manner later characteristic of Gilbert, and he describes experiments remarkably prophetic of his successor. Especially good are his experiments showing that the needle dips below the horizon and that the point toward which it is directed is not the source of the attractive power. Keenly aware of the dangers inherent in following ancient theories and stories, he perceived as clearly as Bacon did later that the only escape lay through experimentation. After giving a few of the more remarkable accounts of magnetism from the past, he remarks, "Not these onely, but many other Fables haue been written by those of auncient tyme, that haue as it were set downe their owne imaginations for vndoubted truthes, and this moste of all in *Geographie,* and *Hydrographie,* or *Nauigation.* Therefore I wishe experience to bee the leader of Writers in those artes, and reason their rule in settyng it downe, that the followers bee not ledde by them into errours, as oftentymes it is seen."[34] It is experiments, or, as he calls them, "reason's finger," that will direct men to the truth; it is upon experiments that he bases the validity of his theories.

Interest in the compass and magnetism continues unabated throughout the rest of the century. Sailors were instructed to

observe the action of the compass in various parts of the world
for the purpose of gathering data for theories, and of testing
theories by direct observation, the importance of the latter being
thus convincingly demonstrated.[35] This fact is made clear by a
translation of a Dutch treatise under the title *The Haven-Find-
ing Art,* 1599. The translator, Edward Wright, published in the
same year *Certaine Errors in Navigation* which, together with
other treatises of his, exerted considerable influence on the science of navigation. In his second dedication, Wright urges that
instructions be issued to the sailors in the navy to make careful
observations on their various voyages "that out of diuerse experiments some certaine reason and rule of the variation might
be gathered." (The problem most students of the compass
were interested in at this time was the ascertaining of longitude
by the variation of the compass, an idea founded upon the fallacious belief that the variation remained constant for any one
place.) He also recommends his translation and *Certaine Errors*
"to be dayly tried and examined by the touchstone of your long
and skilfull experience at sea: nothing doubting but as they
haue endured the more exquisite triall of exact obseruation, and
Geometricall demonstration both by seamen and landmen on
shore, so they shall be found agreeable to the heedeful experiments of all skilfull Nauigators at sea." In Wright's works all
discussion is based on observation and experiment.[36] The ancients are completely ignored, either because through ignorance
of the compass they could furnish no information, or because in
their comments on magnetism they presented only fabulous accounts and stories.

The climax, of course, in this rising interest in magnetism and
the compass was reached in the epoch-marking *De Magnete,*
1600, of William Gilbert, president of the Royal College of Physicians and physician in ordinary to Queen Elizabeth. Though
Gilbert had been carrying on experiments for some eighteen
years, we should not regard his work as a startling phenomenon
in a country entirely blind to the scientific method, but as the finest achievement in a current of activity which flowed through
Norman, Wright, and others.[37] If anyone has claims of priority
in this field, it is not Gilbert but Norman, whose work truly
prophesies Gilbert's greater achievement. But the latter's experiments were so much more varied and far reaching, his discoveries

so much more numerous and important, and his break with antiquity so much more pronounced that his obscuring his predecessor is understandable.

At the very outset Gilbert proclaims freedom of thought and investigation, asserting that "it is permitted us to philosophize freely and with the same liberty which the Egyptians, Greeks, and Latins formerly used in publishing their dogmas: whereof," he adds, unable to forego a thrust at their errors and their worshippers, "very many errors have been handed down in turn to later authors: and in which smatterers still persist, and wander as though in perpetual darkness." Though he makes the ancients a polite bow, he soon recovers his proud and independent posture: "To these early forefathers of philosophy, Aristotle, Theophrastus, Ptolemy, Hippocrates, and Galen, let due honour be ever paid: for by them wisdom hath been diffused to posterity; but," and he is never silent about modern claims, "our age hath detected and brought to light very many facts which they, were they now alive, would gladly have accepted. Wherefore we also have not hesitated to expound in demonstrable hypotheses those things which we have discovered by long experience."[38] As he warms to his subject, however, he is anything but polite to the ancients and their theories. Of Galen's and others' divergent views regarding iron he says, "Thus do the smatterers cross swords together, and puzzle inquiring minds by their vague conjectures, and wrangle for trifles as for goats' wool." At a time when the Ptolemaic astronomy ruled supreme in England, he denounced it with the utmost assurance as a superstition and fable "now only believed by idiots, ... yet in former ages, that motion . . . . was actually accepted as a basis of computations and of motions, by mathematicians." Again and again he castigates the innumerable errors discovered in ancient science, and he condemns the authority it commanded, citing the silly theorists who "stubbornly ground their opinions on the sentiments of the ancients." Apropos of the absurdities in the Ptolemaic system, he calls upon philosophers to witness "how unfruitful and vain a thing it becomes . . . . to take one's stand on the principles and unproved opinions of certain ancients." He even goes so far as to insist that nature was practically unknown to the ancients; that knowledge of it belongs essentially to modernity. He is unmerciful in his denunciation of ancient fallacies, realizing, as he

did, how these errors continued to be propagated by the blind
followers of antiquity in the protected seats of learning.[39]

In fact, it was the propagation of ancient mistakes that constituted his chief charge against the philosophy derived from
antiquity. He holds up to his age the horrible example of those
who "not being practised in the subjects of nature, and being
misled by certain false physical systems . . . . adopted as theirs,
from books only, without magnetical experiments, certain in-
ferences based on vain opinions, and many things that are not,
dreaming old wives' tales." He feels such a healthy scorn for the
traditional ideas concerning magnetism that he refuses "to overthrow by argument those errors and impotent reasonings of
theirs, nor many other fables told about the loadstone, nor the
superstitions of impostors and fabulists." The perpetuation of
error through the blind following of ancient records and the
acceptance of theories without reference to observation or experiment find emphatic condemnation in vigorous language.

Very often the herd of philosophizers and plagiarists repeat from the
records of others in natural philosophy opinions and errors about the attrac-
tions of various bodies. . . . Why the sucking fish Echineis or the Remora
should stay ships has been variously treated by Philosophers, who are often
accustomed to fit this fable (as many others) to their theories, before they
find out whether the thing is so in nature. Therefore, in order that they
may support and agree with the fatuities of the ancients, they put forward
even the most fatuous ratiocinations and ridiculous problems, cliffs that
attract, where the sucking fish tarry, and the necessity of some vacuum, I
know not what, or how produced. Pliny and Julius Solinus make mention of
a stone Chatochitis. They say that it attracts flesh. . . .[40]

Gilbert proposes but one avenue to truth, that of observation
and experiment: "Wherefore we but seldom quote antient Greek
authors in our support, because neither by using greek arguments nor greek words can the truth be demonstrated." He seems
to relish ascribing the mistakes of others to lack of proper experiments. Had Fracastoro, he says, "observed by a large number of experiments that all bodies are drawn to electricks except
those which are aglow and aflame, and highly rarified, [he]
would never have given a thought to" certain theories. He
broadens his comment to say, "It is easy for men of acute intellect, apart from experiments and practice, to slip and err."
Gilbert frequently notes the reason for errors in matters beyond
his immediate interest, and generalizes—almost moralizes—on

the necessity of experimentation. He cites an error of Cardan's as an example of the danger of falling into mistakes in the absence of genuine experiments. To him experiments, which he considers as certain as mathematical proofs, are the true foundation, not only of the science of terrestrial magnetism, but of science in general. In contrast to the guesses and opinions of "professors of philosophy" he praises experimental induction and the labor spent therein in no half-hearted manner:

Nor have we found this our labour idle or unfruitful; since daily during our experimenting, new and unexpected properties come to light; and our Philosophy hath grown so much from the things diligently observed, that we have attempted to expound the interior parts of the terrene globe, and its native substance, upon magnetick principles; and to reveal to men the earth (our common mother), and to point it out as with the finger, by real demonstrations and by experiments manifestly apparent to the senses.[41]

Gilbert sensed the tendency in those who drew their information only from books, without verification through observation or experiment, to lose themselves in a nebula of fruitless words which produce only the sensation of knowledge and not the reality. In noting the injurious effect upon the advancement of learning of indulging in language which has no counterpart in the material world, Gilbert anticipates Bacon and Hobbes, and at this early date expresses a linguistic attitude which constituted an integral part of the scientific movement in the seventeenth century, and which was responsible for the stylistic program adopted by the Royal Society.[42] "There are many modern authors," he says,

who have written and copied from others about amber and jet attracting chaff, and about other substances generally unknown; with whose labours the shops of booksellers are crammed. Our own age has produced many books about hidden, abstruse, and occult causes and wonders, in all of which amber and jet are set forth as enticing chaff; but they treat the subject in words alone, without finding any reasons or proofs from experiments, their very statements obscuring the thing in a greater fog, forsooth in a cryptic, marvellous, abstruse, secret, occult, way. Wherefore also such philosophy produces no fruit, because very many philosophers, making no investigation themselves, unsupported by any practical experience, idle and inert, make no progress by their records, and do not see what light they can bring to their theories; but their philosophy rests simply on the use of certain Greek words, or uncommon ones; after the manner of our gossips and barbers nowadays, who make show of certain Latin words to an ignorant populace as the insignia of their craft, and snatch at the popular favour.[43]

Perhaps the most significant ideas destined to be expressed some years later by Bacon are stated clearly and convincingly by him: the attack on the authority of the ancients; impatience with those who copy books in lieu of observing nature; and the importance of inductive experiments. His pages are strewn with refutations of the ancients and with innumerable experiments, intelligently conducted and skilfully reasoned upon, in which there are testing and verification of hypotheses. In short, the *De Magnete* represents a fine piece of scientific research. He perceived that the trouble with the world of science lay in the fables authoritatively handed down by the ancients, for which reason he is severe in his condemnation of medieval and modern writers who blindly followed them: "Deplorable is man's ignorance in natural science, and modern philosophers, like those who dream in darkness, need to be aroused, and taught the uses of things and how to deal with them, and to be induced to leave the learning sought at leisure from books alone, and that is supported only by unrealities of arguments and by conjectures."[44]

Gilbert established the science of terrestrial magnetism. His experimental discoveries in electricity alone have been estimated at a score. He was the second in England to espouse openly the Copernican theory, to which the sixth book of the *De Magnete* is devoted.[45] It has been thought that he went only so far as to accept the rotation of the earth, but there is sufficient evidence to show that he also believed in the revolution of the planets about the sun.[46] He stressed the rotation of the earth, because, believing that it was the result of terrestrial magnetism, he felt more interest in it. It is true, he believed in animism and judicial astrology, a link connecting him with his age, but such relics of fading doctrines did not in the least impair his assured faith in the experimental method nor invalidate his discoveries. His reputation was high with those who were qualified to appreciate his work, such as Edward Wright, William Barlow, and Nathaniel Carpenter, and on the continent no less a mind than Galileo's was stirred to the highest pitch of enthusiasm and praise.[47] Throughout the first half of the seventeeth century his book, though never widely noticed, continued to touch minds; he had more to do with the acceptance of the Copernican doctrine than any one man in England, and in all investigation of magnetic phenomena his name is sure to appear. To the scientists who joined together to form the Royal Society he was one of the heroes of the new science. Subsequent generations seemed somewhat to slight him, but he stands today immortalized in the eyes of students of electricity and magnetism.

So the ancients were not entirely without opposition in this age. The Paracelsians in dark and discounted ways were opposing experiments to ancient authority, and paving the way for the large part which chemists later played in toppling over the citadels of antiquity. Norman, Gilbert, and a few others were applying induction to the principles underlying the compass, and were achieving results which justified their claims for the experimental method. Yet as a whole, the Elizabethans were committed to a reverenced antiquity, a fact Bruno was made to realize in no uncertain fashion. The opinion of the learned world was based solidly upon classical philosophy; the universities and the professions drew their prestige from it. Isolated and, in the case of Gilbert, important revolts were in evidence, but little progress could be hoped for until these scattered forces could be marshalled into solid ranks, until the face of the learned world could be turned from the past to the future. What was needed was the infusion of new values and new attitudes in thinking minds, and the unifying and solidifying of these ideas in a definite and widely accepted philosophy. The achievement of this desideratum was the destiny of one man—Sir Francis Bacon.

CHAPTER II

THE DECAY OF NATURE

Whereunto I will adde the weaknesse of the elements, decay of
the heauens and a generall imperfection in al things now, in this
last old and cold age of the world.

Godfrey Goodman, *The Fall of Man*, 1616.

Probably no single factor was so responsible for the feeling of
modern inferiority as the belief that all nature was decaying in
its old age. It lay, indeed, at the bottom of most manifestations
of the worship of antiquity, though frequently not finding expression in words. And just as ancient authority persisted longer in
the esthetic than in the scientific world, so the last significant
attack on this debilitating idea is discovered in Edward Young's
Conjectures on Original Composition, 1759, an arraignment of
the literary doctrine of imitation. Opposition to the theory near
the beginning of the second quarter of the seventeenth century
produced the first definite controversy in England over the relative merits of the ancients and moderns, a controversy which
recurs in the Restoration and again near the close of the century.
The two issues involved in the quarrel were the question of
genius and the question of learning; that is, which age possessed
greater mental powers, and which enjoyed greater knowledge.
The idea of the decay of nature, of course, bore directly upon the
first. Although the early phase of the dispute over nature's decay
was in one way not closely associated with the rise of the modern
scientific spirit, in another it was near allied to the latter, for
the most vexing problem which confronted the youth of modern
science was that of reliance upon classical authority, and, therefore, any conception which weakened or stengthened such
authority was of tremendous importance. Thus it is not strange
to find in the later development of experimental science the belief
in universal decay receiving definite attention from the Baconians.

The question naturally arises, Did admiration for the classics,
inspired by the revival of learning, give rise to the theory of
decay as an explanation or cause of the superiority of the ancients, or did the theory help to create the sense of modern

inferiority? Although the question defies a definitive answer, there are some considerations which would lead us to infer that the latter is more probable. The first is that the theory comprehended much more than man. In fact, in the early discussions of the matter the comparison between ancient and modern men occupies an important but secondary place. The whole universe— stars, sun, planets, the air, earth, water, plants, animals, and minerals, as well as the purely physical qualities of man—constitutes the larger field of discussion. It seems hardly possible that the question of intellectual superiority alone could have given rise to a problem of such cosmic dimensions. Secondly, the belief that humanity was approaching its old age appears long before the revival of learning, especially in Cyprian and the apocryphal *Esdras*.[1] Furthermore, the most important expression of the belief in England devotes only a dozen out of some four hundred and fifty pages to the "wits" of men, and draws much of its material from the Bible.[2] In fact, the whole question assumes a decidedly religious cast, and probably grew out of the religious thought of the past.

The first mention of the theory in England which I have discovered appears in a treatise entitled *A blazyng Starre*, 1580, written by Francis Shakelton, an obscure divine. Approaching his subject from a purely religious point of view, he is especially interested in showing that the "heathenish" idea of the world's being eternal, set forth by Aristotle and Galen, can be refuted by the Bible. Though holding to the Biblical prediction of the destruction of the world by fire, he joins with it a belief in the old age of the universe and its attendant debility. He is confident that

it shall manifestly be proued that this worlde shall perishe and passe awaie, if wee doe but consider the partes whereof it doeth consist, for doe we not see the yearth to be changed and corrupted? Sometymes by the inundation of waters? Sometymes by fiers? And by the heate of the Sunne? And doe we not see that some partes of the same doe waxe old, and weare awaie euen for verie age? Doe wee not in some places also read, that mountaines haue falne doune, by reason of earth quakes? And Rockes haue been cracked, and broken so in peeces, that by the meanes thereof, certaine Riuers haue been (as it were) dronke vp, or els, haue had recourse an other waie? Also haue ye not read, that the seas haue rebounded backe, ouerwhelmed whole Cities, and vtterly drouned whole Prouinces? And what are these strange alterations els, but euident argumentes that the worlde shall one daie haue an ende?

Naie I dooe saie moreouer, that if wee will giue credite to the Mathematiques, the constitution of the celestiall worlde, is not the same that it hath been in tymes paste, for so much as the Sunne, is not so farre distant from vs now, as it hath been heretofore. So that some dooe affirme, that the distance betweene the Sonne and the earth (which is the centre of the worlde) is lesse now, by the fourth part almost, then it was in the tyme when *Ptolomeus* liued. Where hence thei dooe coniecture and gather, that the Sune (at this daie) is nerer vnto vs by 9976. German miles then it was before.

Wherefore, if there be so greate alteration in the superior worlde, what shall wee saie of the inferiour? Let this therefore be a forcible argument to proue, that the world shall haue an ende: for so muche as it doeth waxe old, and euery part thereof doeth feele some debilitie and weakenesse. For there is lesse vertue in Plantes hearbes then euer before. And more feeble strength in euery liuing creature then euer was before. And less age in men than euer was before. It remaineth therefore (of necessitie) that shortly there shall be an ende and consumation of the Worlde, because it is (as it were) subiecte to olde age, and therefore feeble in euery parte.[3]

These paragraphs contain the germ of much that was said a generation later in the most ambitious attempt to prove inevitable decay. They would seem to indicate that at this early date the theory had become well established in England. Probably it enjoyed even wider currency on the continent. Bruno, who visited England in 1583, touches upon the matter in several of his works.[4] Were it not for the passage quoted above, one might be tempted to attribute to him the introduction of the idea into England, for his visit created considerable stir, especially in Oxford.

To those who maintained a critical attitude toward their own age, and they were many, the idea proved useful in affording a convincing explanation of the many evils which met their eyes. "Multum dubitaui (lector humanissime)," says a writer on medicine, "hoc exiguum opusculum in hoc senescenti mundo in publicum emittere, in quo tanta viget inuidia, et in hac arte opinionum dissensio, vt nemo fere aliorum scripta legat, nisi vt calumnietur, sed qui studet ab omnibus approbari (si *Gal.* credamus) nihilo plus agit, quam si det operam, vt omnium opes possideat."[5] Men seemed to be blind to any possible virtues in their own times. Convinced as they were that nature had about run its course, and was tottering on senile legs to a final dissolution, their eyes were all the keener to discern evidences of decay, and to read in such signs confirmation of their belief. If some virtue did appear, it seemed little less than miraculous. In speaking of the great expansion of geographical knowledge in the Renaissance, Purchas exclaims, "It cannot be without some great worke of God, thus in the old and decrepit Age of the World, to let it haue more perfect knowledge of it selfe."⁶ All this in an age which pro- duced Shakespeare, Spenser, and Bacon!

The idea of nature's decay was just the kind to appeal to Donne's able and fantastic mind. He makes extravagant use of it in *An Anatomie of the World: The First Anniversary*, 1611, weaving it in and out of his metaphysical conceits. Heaven and earth decay, and man's shortened stature and abbreviated life, as well as his weakened mind, reveal the old age of the species. Again, in a sermon preached in 1625, two years before Hakewill published his elaborate denial of the theory, he says:

As the world is the whole frame of the world, God hath put into it a reproofe, a rebuke, lest it should seem eternall, which is, a sensible decay and age in the whole frame of the world, and every piece thereof. The seasons of the year irregular and distempered; the Sun fainter, and lan- guishing; men lesse in stature, and shorter-lived. No addition, but only every yeare, new sorts, new species of wormes, and flies, and sicknesses, which argue more and more putrefaction of which they are engendered. . . . *S. Cyprian* observed this in his time, when writing to *Demetrianus*, who im- puted all those calamities which afflicted the world then, to the impiety of the Christians who would not joyne with them in the worship of their gods, *Cyprian* went no farther for the cause of these calamities, but *Ad senescen- tem mundum*, To the age and impotency of the whole world.⁷

In this last passage Donne was obviously indebted to the most complete expression which the gloomy doctrine received: *The Fall of Man*, 1616,⁸ by Godfrey Goodman, chaplain to the Queen and according to his own statement, "a poor country parson." Goodman's approach to the subject is altogether religious, and the thesis which he attempts to support is that corruption entered into man through his fall, and through him into nature. He is obsessed with the idea of corruption: "me thinkes whatsoeuer I see, whatsoeuer I heare, all things seeme to sound corruption."⁹ Evidence of corruption he finds in the evils of his times, in the sickness, suffering, vices, passions, and unhappiness of man, in the warring of the elements, in flies, worms, and monsters, in the decay of beauty and withering of fruit. Nature gives man in- finite desires, as if she had an infinite treasure, but in truth she is barren and defective. Man himself is vicious and corrupted in every state and course of life, a corruption which finally appears in his death. As man through his fall brought death upon himself, so he imposed death upon all nature. In general, Goodman's idea is that the course of man and nature has been one of continual decline from a perfect state to the decay of old age. He holds to the theory that the farther anything proceeds from its source, the more corrupt it becomes, as water becomes more impure the farther it runs from the fountain. And as we see decline and decay in individual parts of nature, as for instance in man, so the universe itself must partake of the nature of its parts and pass through the cycle of youth, old age, and death.

He finds proof for his thesis in a comparison between men of past and present time. The clothing of the ancients was heavier, coarser, and more fit for exercise; their food was coarser and simpler, fit for strong stomachs. There were none of the dainty foods and spices of the modern era, and tobacco was not needed as an aid to digestion! There is a dash of primitivism in Goodman's theory, a belief in the simple life as opposed to civilization. Naturally, if from the first, man has pursued a downward path, he was in a happier condition when there was less civilization to separate him from the natural state. It is only one step further to assert that those who are less civilized, in other words, savages, represent a more desirable type of man. Goodman did not take the step, for it obviously would have made against his idea of universal decay, but many of his arguments could be used by primitivists. But whatever his idea of savages may have been, he argues that men in the past were stronger, larger, and longer-lived (the antediluvians, for example), adducing as proof the discovery of great bones and large weapons, great castles and cathedrals compared with "the paper-buildings of today." Finally, there were in the days of yore few diseases and little medicine, whereas in modern times physicians and apothecaries are kept busy repairing man's deteriorating body, "for thus the old age of the world, as it is a weaknes in it selfe, so is it accompanied with many infirmities."[10]

Only three out of the four hundred and fifty pages which the book contains touch upon the differences in mental powers between ancients and moderns, and even those are not confined entirely to the Greeks and Romans.[11] His own age, he asserts, did in its pride and arrogance "vniustly claime vnto it selfe the name and title of the learned age," and elsewhere he speaks sarcastically of "this great learned age." The mental superiority of the ancients is revealed in their being the inventors of the arts. "Was it a worke of small difficultie," he asks, "to hew and square out nature, allotting to euery Science her proper subiect, her due limitation? to reduce all the seuerall starres into constellations, to obserue their motion, their qualitie, their influence?" After speaking of the excellent invention of grammar, he makes the general statement that

For all Arts whatsoeuer, the best authors are the most ancient, euen vnto this day: I could instance in euery one in particular, though wee building vpon their foundations, haue added some ornaments, yet such as are not necessarie to perfit the Art: and generally for the Ancients, whatsoeuer you shal obserue in practise amongst them, you shall finde that it stood with great wisdome and prouidence, if you please to haue relation to the times and occasions. And wherein they seeme to be defectiue, you may ascribe it to the happinesse of their times; for their plentie was such, as that they were not inforced to trie all conclusions in husbandrie, whereas our wants seeme to require our best inuentions.

Goodman believed that "Art serues like a cobbler, or tinker, to peece vp the walles, and to repaire the ruines of nature," and so he held, somewhat inconsistently, that the superiority of the ancients was evidenced by their not finding it necessary to develop certain industries such as agriculture, since nature furnished an abundance to satisfy commendably simple needs.

In certain passages Goodman clearly has classical antiquity in mind. He claims that poetry, oratory, philosophy, and history were "neuer more excellent" than in the Augustan age. He finds occasion to discuss at some length the simile of the dwarf on a giant's shoulders, which frequently appears in the controversy between the ancients and moderns, and which was a product of the theory of decay.[12] One would suppose that Goodman would have embraced the figure as showing deterioration in mankind, but he will not allow the race even the small crumb of comfort found in such an elevated position.

But this great learned age hath found out a comparison, wherein we might seeme to magnifie the Ancients, but indeed very cunningly do presse them downe, making them our foot-stooles; preferring our selues before them, extolling and exalting our selues aboue measure; for thus it is said, that we are like dwarfes set vpon the shoulders of Gyants, discerning little of our selues, but supposing the learning and ground-worke of the Ancients, wee see much further then they, (which in effect is as much, as that we prefer our own iudgments, before theirs).

He heartily believed that the moderns were dwarfs, but he would deny them the point of vantage indicated by the simile for two reasons. First, the ancients were much greater than is revealed in their writings, and second, modern men did not have sufficient intelligence to understand what the ancients had bequeathed to them.

But evidence of decay is not discovered in man alone; it is manifested in the elements. The seas have grown fruitless and barren, for proof of which fact he refers to Camden as authority that fish are not so plentiful as in the past. The earth also has lost its fertility and does not produce the crops which our forefathers raised. Decay is to be discovered even in the heavens themselves.

Now since the fruitfulnesse, or barrennesse of the earth, proceedes from the influence and disposition of the heauens; in the last place I dare accuse the materiall heauens, as being guiltie, conspiring, and together ioyntly tending to corruption; Scripture shall warrant me, *the heauens shall waxe old as doth a garment*, Psalme 102, vers. 26. Reason and all humane learning shall backe me, for certaine it is, that the Sunne hath descended much lower by many degrees, then he was in the time of King *Ptolemie;* the same Mathematicall instruments, which agree together in all other dimensions doe vndoubtedly proue the diuersitie; by vertue of perspectiue glasses, we haue lately discerned spots and shadowes in the Moone, and within our memorie, in the yeere 1572, a true Comet did appeare in the eighth Heauen, which as it had a time of beginning, so had it a period, and time of dissoluing. And thus being mortall of our selues, wee dwell in houses of clay, the roofe of this world, as well as the foundations shall together be mooued; for wherefore serues the diuersitie of seasons, the day and the night succeeding each other, Summer and Winter, the rising and setting of Starres, the different and contrarie motions, the various aspects and oppositions? but that in some sort they partake of our nature, and shall haue their part and portion with ours.[13]

This theory is, of course, the direct antithesis of the idea of progress, which has profoundly influenced modern thought. And just as to those who are imbued with the latter idea any change seems a change for the better, so in the Renaissance those who were oppressed by the belief in a declining universe considered every change a change for the worse. Thus we find Goodman indignantly assailing Ramus and his followers for discarding "the grounds, principles, and rules of that most illustrious & thrice renowned *Aristotle,* whom all ages doe reuerence and acknowledge to haue bin natures chiefe secretary, or

best counseller of estate, the father of all humane knowledge,"
and "for reducing all things, (euen the very scope & foundation
of arts), to their own fond inuention, and barbarous innoua-
tion."[14] Indeed, Goodman is so conservative that he defends
the scholasticism of the medieval ages against his own times,
calling Lombard, Aquinas, and Duns Scotus the "lights and
lampes of all true learning." In his belief in the general
superiority of antiquity, so well revealed in the giant-and-dwarf
simile, and especially in his utter subserviency to Aristotle, he
manifests an attitude which constituted the chief object of attack
on the part of the new scientists. Though only fifty pages of his
book are devoted to a comparison of ancients and moderns, and of
these only three are concerned with intellectual superiority,
he clearly foreshadows the later upholders of antiquity. Furthermore, his treatise provoked the first significant defence of modernity in England.

This work was George Hakewill's *Apologie,* 1627, a learned
treatise, which reached a third and enlarged edition in 1635, and
the long title[15] of which sufficiently describes the wide scope and
ambitious purpose of the author. He declares that the work was
motivated by his desire to redeem captivated truth, to vindicate
the Creator's honor, and to encourage men oppressed by belief in
necessary degeneration. The spirit evinced throughout the whole
volume marks him as a man of high courage and sincere faith.
He perceived the real cause of the hallucination as well as the
evil effects necessarily produced by it. Vindicating the dignity of
mankind, he calls upon his contemporaries to arouse themselves
and make use of the faculties which they possess.

That he had to take such a stand is indicated by what he says
regarding the wide currency of the idea of nature's old age: "The
opinion of the Worlds Decay is so generally received, not onely
among the Vulgar, but of the Learned, both Divines and others,
that the very commonnesse of it makes it current with many,
without any further examination: That which is held, not onely
by the multitude, but by the Learned, passing smoothly for the
most part without any checke or controll." He also emphasizes
the fact that he was the first to undertake to combat the idea.
"I have walked (I confesse) in an untroden path, neither can I
trace the prints of any footsteppes that have gone before mee, but
onely as it led them to some other way, thwarting, and upon the

by, not directly: some parts belonging to this discourse, some
have slightly handled, none thoroughly considered of the whole."[16]
He did not trust his own powers unaided to accomplish the ambitious task of proving the modern universe undecayed in all its
parts, but, as he informs us, he called in aid from philosophers,
historians, mathematicians, grammarians, logicians, poets, ora-
tors, soldiers, travellers, lawyers, and physicians. Philosophy
and the arts were his own profession, but for physic and law he
consulted his chief acquaintances both in the university and out
of it. In his use of history and mathematics he not only had the
approbation of professors of the subject, but he also inserted in
his treatise some pieces from them. In short, he collected his
material from as wide sources as possible.

Hakewill attributed the glorification of antiquity to various
causes. Even ancient poets, he thought, had been guilty of prais-
.ing past times at the expense of their own: Homer, Virgil,
Juvenal, and Horace were *laudatores temporis acti*. The "pretty"
invention of the four ages, which Goodman had used to good
effect, he considered instrumental in making men yearn toward
the past, and he maintained that they had produced such an impression upon men's minds that they could hardly be rooted out.
Other causes he ascribed to the morose and crooked disposition
of old men, who imagined everything sour because their taste
was, and to the excessive admiration of antiquity, which after
all sprang not so much from a due respect for the ancients as
from envy of the living. Such detractors praised the old writers
to disparage the new.

The author is careful to note the depressing influence of the
belief in the decay of mankind:

For being once thorughly perswaded in themselves, that by a fatall
kinde of necessitie and course of times, they are cast into those straits,
that notwithstanding all their striving and industrie, it is impossible
they should rise to the pitch of their noble and renowned Predecessours,
they begin to yeeld to the times and to necessity, being resolved that their
endeavours are all in vaine, and that they strive against the streame.[17]

In opposition to the view that men's minds had undergone a sad
decline since ancient times he proposes his theory of circular
progress: "There is (it seemes) both in *wits* and *Arts,* as in all
things besides, a kinde of *circular progresse*: they have their
birth, their *growth*, their *flourishing*, their *failing*, their *fading*,

and within a while after their *resurrection,* and *reflourishing*
againe."[18] He stoutly maintains that there is no such thing as
natural superiority or inferiority determined by the time in
which men live. If the moderns are inferior to the ancients, it is
simply because the former are slothful, indolent, and negligent.
He indignantly turns from the small comfort which some of the
subscribers to the theory of decay found in the thought that
though men had dwindled from giants to dwarfs, yet by possessing the knowledge of the ancients, they were in a position to
know more, though adding little themselves. Both Goodman and
Hakewill attack this simile, but how differently! The latter says:

> But if wee conceive them [the ancients] to be *Gyants,* and ourselves
> *Dwarfes,* if we imagine all Sciences already to have received their utmost
> perfection, so as wee need not but translate and comment upon that which
> they have done, if we so admire and dote upon Antiquity as wee emulate
> and envy, nay scorne and trample under foot whatsoever the present age
> affords, if we spend our best time and thoughts in clyming to honour, in
> gathering of riches, in following our pleasures, and in turning the edge
> of our wits one against another, surely there is little hope that we should
> ever come neere them, much lesse match them.[19]

But to the idea of necessary inferiority he returns a ringing
denial. Men of all times are of one stature, neither giants nor
dwarfs, with the difference, however, that if the moderns possessed the studiousness, watchfulness, and love of truth which
characterized the ancients, they would be lifted up higher by
means of the latter. He who claims to be a dwarf sitting on a
giant's shoulders is nothing but a man of competent stature
grovelling on the earth. If the moderns are inferior to the ancients,

> The fault, dear Brutus, is not in our stars,
> But in ourselves, that we are underlings.

"Forth, beste, out of thy stal" is the call that Hakewill makes
to his contemporaries. Just as Bacon, as we shall see later, stood
on the calm assurance of an intellectual faith, so Hakewill stands
upon an unshaken faith in a beneficent Providence, which makes
him reject without hesitation the idea of predestined decay.
Without depreciating the ancients, he sounds a rallying call for
man to use the powers given him, and he will find that neither
time nor necessity has sway over him. The spirit of Bacon,
Sprat, and, to a less degree, Wotton, is splendid, but the sight of

this courageous man standing at the close of the Renaissance and delivering a message that will suit all ages is indeed inspiring.

Hakewill divides his book into four parts. The first is introductory, treating the question of decay in general. The second deals with the supposed degeneration of the heavens, elements, and elementary bodies. The third attempts to show that in regard to length of life, strength, stature, arts, and wits man has suffered no decline. The fourth upholds modern improvements in manners, especially as regards virtue, and tries to prove "a future consummation of the world." Comprehensively as Goodman considered the question, Hakewill far surpasses him in wide and exhaustive treatment. No other work touching on the subject approaches his in this respect. In England the controversy was narrowed to its scientific aspects, largely owing to the influence of Bacon, and to the attack of science on the authority of antiquity, though other elements are not entirely forgotten. Since, however, the progress of the controversy confined the arena chiefly to man's mental powers and accomplishments, we are for the most part concerned only with that portion of the third book which deals with "arts and wits." Yet mention may be made of the way in which Hakewill, by citing a multitude of authorities, wherein fiction mingles freely with fact, and by much ingenious reasoning, shows that the celestial bodies, the earth, the air, water, animals, and plants are not suffering from any universal decrepitude.

The first five chapters of the third book maintain that in longevity, strength, and stature the moderns are not inferior to the ancients. The sixth upholds the modern in imagination, memory, and judgment, giving account after accounts, some authentic, some smacking of the fabulous, of moderns renowned for one or more of these mental powers, with whom ancient examples are compared. In this chapter he also speaks of modern aids to learning, the most important of which was printing, and of hindrances, among which he mentions the giant-and-dwarf figure and the bitter controversies among Christians.

The next four chapters present the problem as it was generally presented throughout the century. The field of man's achievements is divided into four large areas, which are again divided into smaller sections. Chapter 7 deals with the "three principall professions," divinity, law, and medicine, in which superiority

is granted the moderns, especially in medicine because of Paracelsian chemistry and increased knowledge of anatomy. These
last were to figure later in the attack on the ancients which pre-
ceded and followed the founding of the Royal Society. The eighth
chapter compares ancient and modern history, poetry, and the
art military, a peculiar combination. The moderns excel in history by virtue of their improved knowledge of chronology and
geography, are not much inferior in poetry, and equal, if they do
not surpass, the Romans, who were far ahead of the Greeks, in
arms, fortifications, and strategy. The ninth chapter is a verita-
ble *potpourri*, the first division embracing grammar, rhetoric,
logic, mathematics, and philosophy, and the second, painting,
agriculture, architecture, navigation, horsemanship, and heraldry. The moderns excel in grammar but are surpassed in
rhetoric. In logic also they are inferior because of the wrangling
of the schoolmen who desired victory rather than truth; but they
have added method without disturbing Aristotle's text. In support of modern mathematics, which is divided into astronomy
and geometry, Hakewill inserts an essay of his friend, Henry
Briggs, Savilian professor of astronomy at Oxford, in which special attention is called to the Copernican theory, Galileo's telescope, and certain mathematical discoveries. These also were to
figure largely in the future development of the controversy. He
divides philosophy into metaphysics, physics, ethics, and politics,
but postpones discussion of the last two to his fourth book. After
granting superiority in metaphysics to the moderns, he says that
physics or natural history was enriched by Aristotle but is defective in the historical part, perhaps a Baconian echo. Then
follows a passage which is significant in showing the only direct
influence which Bacon seems to have exerted upon the book:

And for the speculative, both himselfe [Aristotle] and his followers seeme
to referre it rather to *profession* and disputation, matter of wit and
credit, then use and practice: It is therefore a noble and worthy endeavour
of my Lord of S. *Albanes*, so to mix and temper practice and speculation
together, that they may march hand in hand, and mutually embrace and
assist each other. *Speculation* by precepts and infallible conclusions preparing a way to *Practice*, and *Practice* againe perfecting *Speculation*.

In agriculture, horsemanship, and heraldry the moderns, where
they have not created or made discoveries, have made more perfect the practice of the ancients. For modern champions in paint-

ing he mentions Angelo, Raphael, Vasari, the English Hilliard, and, whom he thinks the greatest, Dürer of Nuremberg. In architecture the best of the moderns follow the rules of the ancients, but at times vary from them with good discretion. In navigation the moderns far surpass their competitors, since, owing to the compass, they have widened their travels to include the whole globe, in which connection Drake, Magellan, and Hakluyt are mentioned.

The last chapter is interesting by virtue of the emphasis placed upon discoveries and "many singular artificial inventions, for the use, ease, delight or ornament of mankind." Some of these are mere fiction, such as the wooden flying eagle of Nuremberg, and the invention of a perpetual motion machine, but these, Hakewill says, "are in truth but toyes and trifles in regard of those three most usefull inventions, which these *latter ages* challenge as due and proper to themselves, *Printing, Gunnes*, and the *Mariners* Compasse; of which *Cardane* comparatively speaks in high tearmes—All antiquity can boast of nothing equall to these three. Upon these then will I insist, and with these conclude this comparison of *Arts* and *Wits*."[20] Discussion of the three inventions occupies more than one-fourth of the chapter, in the course of which much attention is paid to their effects upon the world, and to upholding modern claims to the honor of inventing them. Previously men had not been wanting who attributed everything new to the ancients, so that Hakewill found it necessary not only to point to inventions as proof of modern superiority but also to prove them the product of modernity.

Hakewill is a true child of the Renaissance. Unlike Bacon he does not look forward in proving his points. He stands in the afterglow of the Renaissance, Bacon in the dawn of modernity. The latter seems to have exerted only the slightest influence upon him, though he doubtless was familiar with the Chancellor's writings. In the idea of circular progress he deliberately embraced a theory which, as we shall see, Lord Verulam rejected, and refused an analogy which the latter accepted though giving it an interpretation different from the usual one. It is true that Bacon had frequently mentioned the three inventions which Hakewill emphasizes, but for a purpose directly opposite to that of Hakewill's. In the *Novum Organum* they are distinctly attributed to chance, while in the *Apologie* they are cited as examples of the superiority of modern wit. Finally, Hakewill with one slight exception seems oblivious to the method which, as will be revealed in the next chapter, was to establish the superiority of modernity over antiquity. The passage, less than a paragraph, quoted on a previous page indicates the extent to which he had embraced the faith of experimental philosophy. He sought tangible evidences of modern superiority, and did not rely upon dreams and visions of the future.

As is true of the controversy through the rest of the century, Hakewill in his discussion of ancient and modern intellectual achievements is interested chiefly in philosophical and quasi-scientific matters. He expresses little concern in the literary or esthetic world. He considers, it is true, as the very nature of his task necessitated, painting, architecture, and poetry, but his defence of modern painting and architecture is brief and faint-hearted. In discussing poetry he is even more superficial. Most of his argument is devoted to the attempt to prove Latin poetry superior to Greek, a point that is pertinent to his purpose, but irrelevant to the controversy between the ancients and moderns. He considers Sidney's *Arcadia* in its "inventive parts" not inferior to the choicest ancient work, and asserts, "if I should match Virgill himself with *Ariosto* or *Torquato Tasso* in Italian, *Bartas* in French, or *Spenser* in English, I think I should not much wrong him." He also quotes extensively from Scaliger's praise of Buchanan and Ronsard. Ten years after Shakespeare's death he advances against the poetry of the ancients the "inventive parts" of the *Arcadia*. It is a sad commentary on the state of poetic appreciation in that day, and well justifies Sidney's *Apologie* and other works calculated to enhance the value of poetry. Yet we should be thankful for the mention of Spenser, even though he is ranked beneath Virgil. Hakewill is not interested in poetry and the arts; philosophy, learning, and inventions hold the center of the stage. These represent the point of conflict; the arts are introduced only to preserve the completeness of the design.

Hakewill's influence upon subsequent phases of the conflict was second only to that of Bacon. We find his work used in the clash over the Royal Society and in the conflict precipitated by Temple. His most important contribution to the quarrel lay in furnishing ammunition to the moderns and in showing them the

way to use it. As we shall see later, Bacon determined the bone
of contention; he decided the issue that was at stake, but he did
not hand on to his disciples any means of upholding the new
philosophy, besides faith in the idea and the compelling appeal
to reason. Hakewill showed how a tangible basis of comparison
might be obtained by observing discoveries and inventions. His
references to anatomical discoveries, presumably those of Vesalius and Harvey, and to chemistry, together with the mention of
the Copernican theory and Galileo's telescope in Briggs's essay
and the one reference to Bacon's method, represent the first
appearance in the controversy of the new science which was destined later to play an overwhelming part in it. Hakewill introduced concrete evidence. Emphasizing the human element in the
general problem of nature's decay, he established for the whole
quarrel over the relative merits of ancients and moderns the
method of comparing men and accomplishments in particular
fields, thus analyzing the problem and making the issue more
clear cut. In this he was followed by Glanvill, who in turn was
imitated by Wotton.

Hakewill's book almost immediately had repercussions at home
and abroad. Discussion of the question became so popular that in
the year following the book's appearance it was made the subject
of a philosophical disputation at Cambridge. The respondent,
who argued against the theory of decay, called upon Milton for
aid, and the poet complied with a Latin poem entitled *Naturam
non pati senium*, which emphatically denies that the heavenly
bodies are diminishing, or that the earth and earthly things are
losing their vigor. Especially does Milton reject the idea that the
sun is losing its heat and the earth is incapable of producing
metals, two points that were hotly debated. He does not, however, touch upon man. In 1632 John Jonston, a naturalist born
in Poland, who at the time was studying medicine at Leyden,
published a treatise on the old age of the universe,[21] derived in
large part from Hakewill. Jonston probably first developed interest in the problem when he was a student at St. Andrews,
Cambridge. Though the continent lies outside the province of
this volume, his work is worthy of notice because of its great
indebtedness to Hakewill, and because a quarter of a century
later, when antagonism between antiquity and the modern world
was growing more intense, it was translated into English to serve

the purpose of the new thinkers. In the points discussed and arguments advanced he is greatly indebted to his predecessor. He adopts unchanged the two main theories of Hakewill, compensation, that is, the gaining by one part of the universe of what is lost by another part, and circular progress. He also resembles the Englishman in spending little time upon the literary and artistic aspect of the controversy. He grants superiority to the ancients in painting, sculpture, architecture, and oratory, but thinks the poetical invention revealed in Barclay's *Argenis* and Sidney's *Arcadia* as good as any the ancients could boast of.

Jonston's chief advance over Hakewill lies in the greater emphasis placed upon modern discoveries, inventions, and science in general. He, like his predecessor, dwells upon printing, gunpowder, and the compass, and points to the great increase in geographical knowledge resulting from the voyages of Americus, Drake, and Cavendish. The ancients, he says, wrote *ne plus ultra* on the Pillars of Hercules, an expression frequently encountered in the rise of the scientific spirit. But he has much more to say than Hakewill regarding the remarkable extension of astronomical knowledge due to the invention of the telescope and the observations of Copernicus, Brahe, and Galileo, by which new stars had been discovered unknown to the ancients. He also lays stress upon recent developments in mathematics, especially Napier's logarithms. He enlarges upon modern improvements in medicine, due to dissection, and styles Vesalius the restorer of anatomy. "If Hippocrates, Aristotle *or* Galen, *himself, were now alive, he would be amased* to see this art [medicine] augmented and adorned with so many Ornaments, enriched with so many new Inventions, and confirmed with so many wonderfull operations."[22] He is much more alive than the earlier writer to the importance and greatness of Bacon, to whom he refers a number of times: "But the practick part of Philosophy was, till now, in the greatest darknessse; at last in our age the way to it was opened by the famous *Verulam, Viscount of S. Albanes, Chancellour of England* in his New Organum, his Sylva Silvarum, his Historie of Life and Death, and of Windes. And those that have afforded any thing notable therein, were either of the age newly past, or of our times."[23] Hakewill was a cleric; Jonston was a naturalist, whose numerous scientific works bespeak his interest in, and knowledge of, science past and present. Thus he was in a better position to appreciate the remarkable advances which science had already made and to urge them as arguments against the theory of senile degeneration. The religious element which hitherto has been so pronounced in the controversy gives way to the scientific and indicates what actually became the main arena of the battle between the ancients and moderns.[24]

The idea of the decay of nature, frequently symbolized by the figure of giant and dwarf, traverses the whole of the seventeenth century, and crops out whenever ancients and moderns clash. For, as has been said before, in England the two important issues in the controversy were genius and knowledge, and the theory of decay bore in a vital manner upon the former. Hakewill's influence in combating the theory is felt throughout the seventeenth and even into the eighteenth century.[25] John Hall, in arguing for a reformation of the universities by way of substituting the new experimental science for the traditional philosophy, maintains that such a reformation

would finde it easie to bring it [knowledge] into a just and beautifull body, and make an happy inversion of that common saying, That our Ancients were Gyants, and we are Dwarfs. And whereas some of the Heathen wise men could say, That those were the best personages that liv'd nearer, and lesse remoter from the Age of the gods: we might contrarily by experience finde, that we had made up the decayes of Humanity, and inforced backe time into its first happy and lusty circle.[26]

Another in defending the search for truth against reliance on authority asserts that *"Nature* certainly is not grown old to feebleness and sterility; or if it produce now as great ingenuities as it hath formerly, can it possibly be, that she will not finde employment for them? otherwise a *Traditional* acquiescence were as commendable as all *Disquisitional* recherches after *Truth."*[27] At the time when the Royal Society was founded Glanvill expressed the opinion that such men as Descartes, Galileo, Harvey, and the members of the Royal Society would "strike dead the opinion of the worlds decay, and conclude it, in its *Prime."*[28]

Nowhere do we find more convincing evidence of the baleful influence this wide-spread and deep-rooted idea exerted upon men than in a vigorous defence of the new science written by Henry Power, an ardent admirer of the Royal Society. In speaking of the hindrances to learning, he declares there is one more general Impediment, which is an Authentick discouragement to the promotion of the Arts and Sciences, and that is, The Universal Exclamation of the World's decay and approximation to its period; That both the great and little World have long since pass'd the Meridian, and, That the Faculties of the one doe fade and decay, as well as the Fabricks and Materials of the other.

Though the idea, he says, is to be found in all past ages, "yet the Clamour was never so high as it is now." To combat this view of the universe he attempts to prove the "World's Duration, from the slow motion of the Sun's *Apogaeum,* or the Earth's *Aphelion,*" and by assuming that the Biblical age of the world is correct. For proof that the mental faculties of the race have not declined he relies upon the discoveries of the new science and upon Hakewill's arguments: "how far our Modern Wits have outdone the Ancient Sages, the parallel twixt the few inventions of the one, and the rare Discoveries of the other, will easily determine. But the Learned *Hackwell's* Apology shall be mine at present, for not treating any further of this Subject; he having long since perform'd that Task, to the conviction of Prejudice it self." He makes interesting use of some of the scientific instruments invented in his century to show how man has remedied a few of the effects of Adam's fall. The "Aged world," he says, "stands now in need of Spectacles, more than it did in its primitive Strength and Lustre," but though Adam's senses were superior to sinful man's, the former "could never discern those distant, or minute objects by Natural Vision, as we do by the Artificial advantages of the *Telescope* and *Microscope.*"[29]

Hakewill's influence upon the supporters of modernity continues undiminished throughout the century. Francis Osborne, in maintaining that a vast undiscovered world lies beyond the bounds of knowledge established by the ancients, refers to the *Apologie* for confirmation,[30] and Marchamont Nedham makes similar use of it in supporting the new thinkers of his age against their detractors.[31] But in Joseph Glanvill's *Plus Ultra,* 1668, we discover the clearest evidence of Hakewill's influence upon modern science's struggle for recognition, for in it the field is divided in the same way and into some of the same sections as in the *Apologie.* In each subject modern is matched with ancient as had been done more than two score years before; the same emphasis is placed upon inventions, in the account of which Hakewill's treatise was evidently used. From the latter also, the author of which he calls "a Learned Man of our own," Glanvill derived his discussion of ancient anatomy.[32] Even in the last outbreak of the quarrel between ancients and moderns[33] Hakewill again appears. Sir Thomas Pope Blount, an unoriginal critic of the day, was incited by Temple's *Essay upon the Ancient and Modern Learning,* 1690, to take up cudgels in behalf of the moderns, and to combat what he calls the "universal opinion that the world daily declines." Practically all his arguments, including some passages copied *verbatim,* are taken from both Hakewill and Glanvill without acknowledgment.[34]

Perhaps, more than we discover in print, this lingering conception of universal decay was at the bottom of the worship of antiquity and the regard for Latin and Greek writers which characterized the criticism of the neo-classical period.[35] Somewhere underlying the doctrine of imitation, which oppressed the age, must have been the feeling that modern minds were by necessity inferior to ancient. Pope's "Hail, Bards triumphant, born in happier days" could hardly have been inspired by any other idea but this.[36] Though the critics of the day seldom express the theory directly, it is not improbable that in their subconscious minds resided the conviction of inevitable inferiority through the decline of nature.[37] How else could a critical theory like that of close imitation secure such a strong hold upon intelligent people, or how could there have been such extravagant and servile worship of men who had lived many ages before? At any rate, when the first conspicuous onslaught on imitation appeared after the middle of the next century, the author found it necessary to attack this very idea of nature's decay, and that, too, from the scientific point of view. In fact, Edward Young's *Conjectures on Original Composition,* which in places sounds startlingly like the mid-seventeenth century, proves that in order to overthrow the doctrine of imitation Young felt compelled to combat the idea of universal decay.

CHAPTER III

THE BACON OF THE SEVENTEENTH CENTURY

> Wherefore since I have only taken upon me to ring a bell to
> call other wits together, it cannot but be consonant to my desire
> to have the bell heard as far as can be. And since they are but
> sparks which can work but upon matter prepared, I have more
> reason to wish that those sparks may fly abroad, that they may
> the better find and light upon those minds and spirits that are
> apt to be kindled.
>
> Sir Francis Bacon's letter to Dr. Playfere.

It would hardly be possible to give in a book, much less in a
chapter, a satisfactory discussion of the philosophy of one who
took all learning for his province, and did much to encompass
it. There were, however, certain elements in Bacon's philosophical conceptions and attitudes which were especially influential in
the development of the scientific movement in seventeenth-century England. Some of these have received due notice from historians of ideas; others have been too slightly treated. In fact,
the true measure of the part that Bacon played in the early
development of scientific thought has never been taken. It is the
purpose of this chapter to throw into relief those ideas of his
which inspired and stimulated the progressive minds of the second and third quarters of the century, which indeed determined
the scientific complexion of the century. A knowledge of them is
essential to a satisfactory understanding of the subsequent chapters of this book, by no means the least purpose of which is to
make clear the remarkable domination which Bacon maintained
over human thought.

Bacon's ideas concerning the reformation of learning fall
roughly into three groups: the recognition of the inadeqacy of
existing knowledge, and of the need and possibility of its advancement; the hindrances that prevented this advancement;
and the means by which it might be secured. The dissatisfaction
which Bacon felt over the state of learning in his own day was
of great importance. Insisting that men overestimated their
store of knowledge, he points out the evil of their believing that
all truth had been discovered, an attitude which was as "pillars

of fate" set in the way of science.[1] This discontent he continually manifests in pointing out those branches of learning
which were incomplete. The inventory of knowledge contained
in the *Advancement of Learning* revealed such deficiencies that
it could not but tend to change a complacent into a critical attitude toward it. Down through the century we hear this indict-
ment of undisturbed satisfaction with imperfect or erroneous
ideas preferred by those eager to supply the deficiencies. But
even more important than the realization of the inadequacy
of science was the view implied in the title, *Advancement of
Learning,* and necessitated by the very fact of his exposing the
deficiencies in knowledge; namely, the appreciation of the need
of progress and faith in its possibility. All Bacon's writings are,
in fact, animated by the desire and confident hope of pushing
forward the frontiers of learning. The influence of this spirit
was to turn men's faces from the past to the future. It was as if
a bar had been taken down and a door opened on a new vista.
In this view was born the idea of progress, not yet, it is true, the
vision of a future limitless in time and improvement; but certainly the foundation of such a conception. In every stage of the
subsequent development of science, this progressive attitude, emphasized by the frequency with which the word "advancement"
is used, appears, and at no time more conspicuously than during
the Commonwealth.

The importance of inspiring men with faith in the possibility
of progress is disclosed in Bacon's recognition of despair as the
greatest of all the hindrances to the progress of science.[2] The
obscurity of nature, the vastness and difficulty of the task of
discovering truth, the weakness of the judgment, the deceitfulness of the senses, all conspired in Bacon's eyes to impress men
with the hopelessness of their condition. Likewise, the belief that
there was nothing new to be discovered, that the best had come
down to us from the past, a view which finds frequently condemnation later in the century, discouraged man from attempting
anything new. The received philosophy, also, because of the
erroneous nature of many of its tenets, imposed upon men a de-
liberate and factitious despair, and effected an unfair circumscription of human power.[3] For the most part Bacon attributed
this feeling of hopelessness to the fact that men underrated their
power and were ignorant of their own strength. He felt espe-

cially called upon to give them a higher opinion of their ability and resources, and to encourage them in their conflict with nature. In fact, he lays modesty aside sufficiently to offer himself as an example of what a single man, though occupied with multifarious duties, could do toward advancing science, in order to suggest how much could be achieved by the joint labors of many men continued through many ages.[4] Like Hakewill, he issues a rallying call to mankind to know their own abilities and bestir themselves from the slough of despond. His call echoes throughout the century.

Perhaps the most significant obstacle to the advancement of science was reverence for antiquity, and in combating this evil Bacon helped to establish an essential attitude of the new science, for scientific progress depended primarily upon a favorable outcome in the controversy between the moderns and the upholders of antiquity, in which science was and continued to be the central issue until the end of the century. Bacon's condemnation of antiquity, though not extreme, is quite apparent. His chief objection to the Greeks was that they relied more on reason than on direct observation of nature. The ancients, he says, took experiments upon credit, and built great matters upon them; knowledge derived from them is talkative, not generative, full of controversies rather than works, and of contentions rather than fruit. Had there really been any life in ancient philosophy, learning would not have stagnated for such a long time, but some profitable results would have come from it. He contrasts the progress of the mechanical arts, in which philosophy and authority did not figure, with the corruption of ancient philosophy; one develops with time and the other degenerates.[5] He praises Aristotle's *Problems*, from which he took much himself, and admits that the Stagirite possessed wonderful acuteness of mind, but his desire to conquer made his philosophy dogmatic and magisterial, and destroyed the more significant atomic philosophy of the ancients. For the most part he speaks disrespectfully of Aristotle, whose philosophy he considered "contentious and thorny."[6]

Upon subserviency to the authority of the ancients Bacon engaged in repeated attacks, perceiving, as he did, how it militated against all intellectual advancement. How long, he exclaims, will men let a few accepted authors stand up like Hercules' Pillars, beyond which there can be no sailing or discovery! More than once he employed this figure of speech, the popularity of which in the seventeenth century is striking, and which he did much to make popular.[7] He urges men not to be dazzled by the great names of ancient philosophers, nor the huge volumes of their modern followers, for though they may be large in bulk and manifest the great diligence of the authors, they shrink to pitiful proportions when they are stripped of fables, antiquities, quotations, idle controversies, philology, and ornaments which are fitted for table-talk rather than for the advancement of learning. He laments that the sciences are confined to a few prescribed authors who are imposed upon the old and instilled in the young, teaching them to affirm or deny, but not to explain or satisfy themselves, with the result that all progress is precluded and dominion over nature forfeited. Since new discoveries, he points out, must be sought from the light of nature and not brought back from the darkness of antiquity, little hope can be entertained of men who are enslaved to established opinion, itself full of verbal contentions and barren of works.[8]

Bacon took due notice of the influence which the idea of nature's decay exerted upon the worship of antiquity, to which idea, however, he does not return an emphatic denial. Men, he says, fear that time is past children and generation, a fear which he seeks to allay with his paradox, *Antiquitas sæculi, juventus mundi:*

As for antiquity, the opinion touching it which men entertain is quite a negligent one, and scarely consonant with the word itself. For the old age of the world is to be accounted the true antiquity; and this is the attribute of our own times, not of that earlier age of the world in which the ancients lived; and which, though in respect of us it was the elder, yet in respect of the world it was the younger. And truly as we look for greater knowledge of human things and a riper judgment in the old man than in the young, because of his experience and of the number and variety of the things which he has seen and heard and thought of; so in like manner from our age, if it but knew its own strength and chose to essay and exert it, much more might fairly be expected than from the ancient times, inasmuch as it is a more advanced age of the world, and stored and stocked with infinite experiments and observations.[9]

Bacon seems not to deny the superiority of the ancients in genius, but finds the same comfort in his paradox which others derived from the figure of the giant and dwarf. In knowledge the moderns should be superior, for they are the heirs of all that has been discovered, and because knowledge grows by increments, those who come last are more fortunate than their predecessors. In speaking of the great extent to which reverence for antiquity had prevented progress he repeats his paradox and stresses the greater stock of information possessed by the moderns owing to geographical discoveries.[10] Whether he really believed that the ancients were superior to the moderns in genius, or whether he thought it best to make such a concession to his age, it is true that he repeatedly insists that he is not matching ancient and modern wits, but is pitting his method against ancient wits. The ancients, he grants, were wonderful in matters of wit and meditation; he does not disagree with the estimate put upon their genius but with their method. Again and again he insists that he is opposing to the ancients only a new method, "by them untried and unknown."[11] It is also possible that Bacon in emphasizing the fact that no unusual mental qualifications were requisite for his method was partly making a concession to the wide-spread belief in modern deterioration in order to encourage his age. The idea of circular progress, or what Bacon calls the ebb and flow of sciences, which Hakewill later was to advance against the theory of decay, Bacon deliberately rejects as causing men to despair of progress beyond a certain point, and also, perhaps, because it contradicted his idea of the incremental advancement of learning.[12]

In spite of his aversion to the philosophy of the ancients and his emphatic condemnation of servility to their authority, Bacon goes as far as he can toward conciliating the supporters of antiquity. The ancients, he says, deserve reverence; modernity should take its stand upon them before determining the best way to proceed. He does not wish to eradicate the received philosophy, but concedes its use for disputation and ornament. There is no doubt, however, that he considered the moderns superior to the ancients in learning, partly because of the addition to knowledge made by the geographical discoveries of the preceding century. He bases his faith that his era will surpass any period of antiquity on the number of modern wits, the monuments of ancient writers, printing, and the multitude of experiments unknown to the ancients which have been revealed by the discovery of new lands.[13] Although he staunchly upholds the moderns in the controversy with antiquity, it is the possibilities of the future which arouse his keenest enthusiasm. The circumnavigation of the globe, he says, the privilege and honor of his age, has brought it about "that these times may justly bear in their motto .... *plus ultra*— further yet— in precedence of the ancient *non ultra*—no further."[14]

In characterizing ancient philosophy and its later versions as contentious, erroneous, and barren of works, in denouncing the enslaving of minds to authority as derogatory to the dignity of the human mind and as an effectual bar to intellectual progress, in exalting his own times over antiquity, as well as in his optimistic view of the future, and in opposing his inductive method to whatever the past could boast, Bacon was outlining the campaign which the moderns relentlessly carried on against the ancients throughout the next half-century. Gilbert had previously opened the campaign, and his influence is pronounced during the first generation of the seventeenth century; but with the ascendency of the Puritans into power, Bacon rapidly becomes the chief inspiration in the revolt from the past. His followers, however, express fainter praise for antiquity and denounce with more virulence the ancients and their upholders, probably because it became increasingly evident that subserviency to authority presented the greatest obstacle to the growth of science, and because the growing number of discoveries and the wider popularity of the experimental method rendered progressive minds more certain of their faith and more confident that the conservatives in science were hopelessly wrong.

Some of the hindrances to the advancement of learning which Bacon pointed out seem to have made little impression upon the century. In fact, one of these impediments, specialization in research, constitutes a virtue in modern eyes. The basis of his own method was a true, universal natural history, and he looked with suspicion upon any endeavor which sought to encompass only a small field. He finds fault with investigators' confining themselves to some one subject as an object of research, such as the magnet, the tides, and the heavens (he undoubtedly had Copernicus and Gilbert in mind), because he thinks it is unskilful to investigate the nature of anything in itself, since the same nature is also manifested in other things. He maintains the same attitude toward the arts and professions, which, he insists, draw their strength from universal philosophy, and which, when studied individually, retard the progress of learning.[15]

Bacon makes a great deal of the injury to learning which derives from men's relying more on mind than on nature, so that they withdraw from the observation of nature and seek truth through the operations of the intellect. They desert the senses for reason, and are more eager to build systems than to understand the physical universe.[16] This antithesis between mind and the senses, which is more fully developed in the discussion of his method, is one of the most characteristic scientific attitudes of the seventeenth century. The same may be said of the injury to science which Bacon finds in men's not proposing a utilitarian goal for knowledge, which failure, indeed, he styles the greatest error of all.[17] Though he condemns the seeking for experiments of fruit only and the unseasonable eagerness to grasp benefits, he does so because such a procedure interferes with the greater material good for man, which will follow a patient and uninterrupted searching out of the laws of nature and the proper utilization of them. Another hindrance, which later Baconians constantly combat, Bacon describes as the deceitful and hurtful opinion of long standing, that the dignity of the human mind is impaired by long and close intercourse with experiments and particulars, with material things subject to sense.[18] It is amusing to listen to the ardent advocates of experimentation in the Commonwealth and Restoration jeering their opponents for withholding their hands from material things, and at the same time glorying in their own commerce with low and common objects.

Bacon was particularly concerned over those psychological factors which, as he expresses it, render the understanding a false mirror of the external world, and to which he gave the name of Idols. He seems to call attention to them in the hope that men being warned of them will be on their guard, or at least can reckon with them. The Idols of the Tribe are those mental characteristics common to all men: proneness to suppose more order and regularity in the world than exist; the tendency of mental prepossessions to bend all things into conformity to themselves; and the influence which the will and affections, such as pride, hope, impatience, and the like have upon the mind, but particularly the part which desire plays in determining thought. Finally, the greatest hindrance to learning associated with the mind is the dullness, incompetency, and deception of the senses, which without the proper aids are infirm and erring, and instruments can do little to assist them. The defects inherent in the senses can be remedied only by the senses themselves assisted by experiments. Of these mental imperfections, it was the last which later scientists stressed.

The Idols of the Cave, or of individual men, refer to the bias which education and environment impose on the variable and erratic spirit of man, and later are frequently cited as the reason for the obstinacy with which many men still cling to the past. The Idols of the Theater represent the circumscription of the mind by philosophical dogma and received systems, which represent nothing real in nature because they are based upon insufficient observations and experiments, as with Aristotle's philosophy, which he calls sophistical, or are based upon too highly specialized fields of experimentation, as with the theories of the alchemists and Gilbert, which he calls empirical.

The Idols which Bacon considered the most troublesome of all, and which in his eyes had rendered philosophy sophistical and inactive were those of the Market-place. These refer to the reaction of the understanding to words, and Bacon's recognition of them marks the beginning of a movement that is one of the characteristic features of early English science. Bacon believed that words filled the understanding with misapprehensions, because, being invented to accommodate vulgar minds, they were defective. He cites as evidence the fact that when superior wits engage in argument the discussion often becomes a mere dispute over words. He did not have the faith in definitions which Hobbes later professed, because in natural and material matters, definitions consist only of words, which beget other words, and so the problem is not solved. The express fault which he finds with words is that they either represent things which do not exist in nature, or else convey very confused ideas of things. In short, language does not impart to the mind a true or accurate picture of material reality, but fills it with more or less fantastic ideas of nature. That he particularly had traditional philosophy in mind is revealed in the remedy proposed for this evil, namely, that all theories be dismissed and recourse had to individual instances, or direct observations. He reveals the same linguistic attitude in his discussion of syllogistic logic. Syllogisms consist

of propositions, and propositions of words, which in turn are only
"marks of popular notions of things" (cf. Hobbes's definition of
words as the marks of things), so that if these notions have no,
or only a confused, relation to the material world, the whole logical edifice crumbles.[19] This linguistic view had two marked effects upon later scientists. First, it determined the most consistent charge which they brought against the philosophy of the
past, namely, that it was merely verbal, with no counterpart in
reality; and second, it influenced them to consider a reformation
in language and style as essential as one in science, to formulate
their linguistic and stylistic ideals, and to try to make them operative. The effort to construct a more satisfactory language
received its most complete expression in John Wilkins' *An Essay
towards a Real Character and Philosophical Language,* 1668, and
the stylistic views, which played a great part in the simplification
of English prose style in the seventeenth century, are clearly expressed in Sprat's *History of the Royal Society.*[20]

Bacon firmly believed that progress in knowledge was possible
only when the ground had been completely cleared and a new
foundation laid. Men must begin, he says, from the very foundations and effect a total reconstruction of all sciences. Especially
must the mind be freed from every opinion or theory, and proceed straight to nature. He exhorts men "with unpossessed
minds" and "with minds washed clean from opinion to study it
[nature] in purity and integrity." Much, he claims, may be expected of men of well-purged minds, who apply themselves directly to experience and particulars.[21] This insistence upon
sweeping the mind clear of all opinions is very suggestive of the
procedure which Descartes in the *Discours de la Méthode,* 1637,
tells us he adopted, but there is a significant difference between
the two methods. The French philosopher started with the simple
principle, *Cogito, ergo sum,* upon which, by means of reasoning
and clear ideas, he sought to construct a sound edifice. The
English philosopher, on the other hand, though insisting just as
emphatically upon the necessity of purging the mind of all notions, proposed a sensuous and material basis for man to build
his ideas of nature upon. Reason, or logic, says Bacon, has
for so long been divorced from facts that it has fixed errors
rather than discovered truth, and, therefore, the important step
is to return to a purely sensuous knowledge of natural things,

and from that foundation to work slowly upward, constantly guiding and controlling the mind by observations and experiments. As we shall see later, this difference is reflected in the distinction drawn in the Restoration between the experimental and mechanical philosophies, between Bacon's experimental learning and Descartes's rationalism.

Bacon's realization of the nature of the received philosophies, in which logic played a much greater part than fact, and his emphasis upon direct contact with nature made him depreciate the understanding and exalt the senses, suspect reason and trust direct observation. The intellect, he says, is far more subject to error than the sense, which at once apprehends the appearance of an object and consents to its truth. By itself the mind can effect little and is not to be trusted. Thus it stands in constant need of being directed and assisted by the senses in order to cope with the subtlety of nature.[22] More than anything else Bacon stresses the material object, the concrete fact, and the necessity of using one's eyes. Careful and severe examination and visual evidence are the fundamental tests of truth.[23] But the senses themselves, though more trustworthy than the mind, are deceptive. "To the immediate and proper perception of the sense therefore I do not give much weight," he says in apparent contradiction of other opinions which he has expressed.[24] Therefore, it is necessary to examine the evidence of the senses and rectify it where necessary. The remedy for the defects of the senses, as for the defects of the understanding, lies in sense itself assisted by experiments which produce and urge things too subtle for the sense to some manifestation which may be comprehended by it. As he expresses it, "the office of the sense shall be only to judge of the experiment, and . . . the experiment itself shall judge of the thing." He sums up the whole matter by saying that he does not take away the authority of the senses, but supplies them with help; does not slight the understanding, but governs it.[25] The importance which this idea accords experimentation does much to explain the all-embracing value ascribed to it later in the century.

Perhaps the most important service which Bacon contributed to the cause of science lies in this very fact, that he forcefully called men's attention back to the physical world, made them distrust the sheer operations of the mind divorced from material reality, and taught them the value and necessity of sensuous observation of nature as a *sine qua non* of scientific knowledge. More than once he calls attention to the fact that he himself dwelt constantly among the data of nature, withdrawing his intellect from them no farther than was necessary to perceive them properly, and this continual conversancy with the external world, he claims, possessed more value than his wit, for truth was to be sought not in the mind but in the world.[26] This high valuation of the importance of observation and experiment, together with a sceptical attitude toward the more ambitious operations of the reason, became the chief value of seventeenth-century science in England.

A corollary of this emphasis upon the importance of sense-observation and the corresponding distrust of the understanding unless securely tied to the physical world is disclosed in his attitude toward systems, to which he maintained an abiding aversion, because they represented to him the unlawful operation of reason upon an insufficient physical basis. He mentions a dozen ancient philosophers, including Aristotle and Plato, and a half-dozen modern writers, including Gilbert, who invented systems after their own fancy to no good effect.[27] This opinion explains his view of the great discoveries of his age. The Copernican system he regarded as only a contrivance by which celestial phenomena could be explained mathematically,[28] and not as necessarily being true in nature. In this case, as in others, he objects to a system which depends more upon reason than upon physical data. The Ptolemaic and Copernican systems, he held, were equally supported by the phenomena, which other theories might be invented to explain. In short, they are only hypotheses devised for calculations and the construction of tables.[29] One of the investigations which he proposes is to find out whether the Copernican theory is to be found in nature, or whether it was merely invented for the convenience of calculation.[30] He admits that the theory of the earth's rotation, which, he says, was prevalent in his own day, cannot be refuted by astronomical principles, for it explains the phenomena, but believing that it is essentially false, he holds that it may be refuted by the principles of natural philosophy.[31] In astronomy, just as in science in general, hope for discovery does not lie in invented systems, but in a deeper and wider investigation into the nature of things.

... it is not merely calculations or predictions that I aim at, but philosophy: such a philosophy I mean as may inform the human understanding, not only of the motion of the heavenly bodies and the period of that motion, but likewise of their substance, various qualities, powers, and influences, according to natural and certain reasons, free from superstition and frivolity of traditions; and again such as may discover and explain in the motion itself, not what is accordant with the phenomena, but what is found in nature herself, and is actually and really true.[32]

While he believes that astronomical theories, invented to explain phenomena, have corrupted natural philosophy, he at least concedes that the findings of the latter should be reconcilable with the phenomena. He raises the rather shrewd question, fully in keeping with his belief that the mind of man is prone to see more order in nature than exists, as to whether there is any system at all, or whether man has merely imposed his own invented scheme of things upon nature. In short, in astronomy, as in other sciences, his injunction is to abjure theories and find out by observation and experiment what is actually in nature. His attitude toward the Copernican theory is in full accord with his attitude toward science in general.[33]

The same is true of his opinion of Gilbert, for whom he had a high regard. He appreciated his countryman's laborious employment in the study of the loadstone, speaks of his collecting observations "with great sagacity and industry," and approves of a question raised "by Gilbert, who has written upon the magnet most laboriously and after the experimental method."[34] But he objects to his making a philosophy out of his observations on the magnet and to his constructing an entire system in accordance with his favorite subject. In other words, Gilbert let his mind go beyond his data, inasmuch as he confined his experiments to a field too narrow for his philosophy. Furthermore, Bacon believed that to understand a subject, a larger field of inquiry than the subject itself was necessary.[35] He maintained the same attitude toward the chemists: their theories, which only confuse their experiments, he rejects, but he grants that a number of discoveries and useful inventions have come from their experiments.[36] This was exactly the opinion of them later entertained by Boyle and other members of the Royal Society.

Associated with Bacon's hostility to systems was his insistence upon the necessity of suspended judgment, or a critical attitude.

The mind, he explains, is constitutionally hasty in framing and asserting notions, much to the hindrance of learning. One should always maintain a sceptical attitude towards systems and writers, taking nothing on faith and authority and without examination, and one should also be very slow in forming ideas or reaching conclusions about things. The understanding must not be allowed to fly from sense and particulars to the most general conclusions, the practice prevalent in his day and the chief fault he finds with syllogistic logic, which is based upon general principles hastily devised from the scantiest data.[37] The raising of doubts guards philosophy against errors and attracts an increase in knowledge, so that "he who makes too great haste to grasp at certainties shall end in doubts, while he who seasonably restrains his judgment shall end in certainties."[38] He distinguishes, however, between his suspension of judgment and philosophical scepticism, which denies the certainty of the senses and understanding, on the ground that his view maintains only that nothing can be known except in a certain way and course which bring the proper aid to the mind.[39]

The method which Bacon stresses so heavily, and which he claimed was new and untried, falls into two parts. The first was "a Natural and Experimental History," such as would serve for the foundation of a true philosophy, and perhaps was considered by him the more important of the two. He insists that such a history is essential to an interpretation of nature, is, indeed, the foundation of all, "for we are not to imagine or suppose, but to discover, what nature does or may be made to do." Without it, he says, his Organon, or inductive method, would be of little use, though the history by itself would much advance science. He even goes so far as to claim that when this history has been satisfactorily established, the investigation of nature and all the sciences will be the work of but a few years.[40] To its neglect he attributes the unreal philosophies, barren of works, against which he never tires of inveighing. The purpose of this natural history, as he defines it, is twofold: to furnish knowledge of particular things; and to supply the primary materials of philosophy. He was especially interested in the second, which, he says, had never been taken in hand. Thus in his eyes, the noblest end of natural history is not pleasure or profit, but to be, as it were, the nursing mother of philosophy, to furnish the stuff and matter of true lawful induction, and thus become a solid and eternal basis of true and active philosophy.[41] The history was to be a most comprehensive collection of experiments and observations, gathered over the whole field of nature, to furnish data from which the mind could construct a universal philosophy by means of Bacon's inductive method.

Bacon stresses the importance to his history of observations collected from the widest sources. The experiments of natural magic were to be diligently sifted for any authentic material. Even superstitious stories of sorceries, marvels, witchcraft, charms, dreams, and divinations were not to be altogether excluded, if they could show any clear evidence of fact.[42] But he was particularly impressed with the contribution which a mechanical and experimental history could make to his natural history, material which had been scorned as low by other natural historians, and which was to be gathered from the mechanical arts, the operative parts of the liberal arts, and experiments not yet grown into an art. The greatest diligence, he says, must be bestowed upon this history, "mechanical and illiberal as it may seem, (all fineness and daintiness set aside)," and he enumerates, as examples of the arts which might contribute to the history, agriculture, cookery, chemistry, dyeing, glass, enamel, gunpowder, and paper.[43] The importance of experimentation loomed large in his eyes. As we have already seen, it provided the means of correcting the defects of the senses, and so of governing the reason. Furthermore, he believed that experiments take off the mask from natural objects, and disclose the struggles of matter, for nature reveals herself under vexations. Mechanical experiments also do not vanish in speculations but lead directly to operations to relieve the inconveniences of man's estate.[44]

Bacon a second time divides his natural history into the speculative, or the search for causes, and operative, or the production of effects. It is the second in which he is interested, for all true fruitful natural history ascends from experiments to axioms, or principles, and descends again to new experiments, at each stage rising to more general principles.[45] Speculation is justified on the ground that it makes possible other effects. Though Bacon was more concerned with the final philosophy to be drawn from his natural history, he was not blind to the benefits to be derived more directly from observations and experiments. He denies that he despises the benefits which will accrue to the various arts from experiments, and calls attention to the immediate good which will be gained in the discovery of new commodities through the transfer of the observations of one art to the use of others.[46]

His conception of natural history necessitated two other ideas: the need of co-operation, and the fact that no unusual intellectual ability is required of those who would compile it. Certainly the immense amount of data required by his conception was so great that men could very well despair of ever achieving it.[47] He himself realized that the proposed goal necessitated much labor and expense, was, in fact, a "royal work," but he insists that all difficult works can be overcome by amplitude of reward (which he hoped the King would furnish), soundness of direction (which he would supply), and conjunction of labors.[48] "For the last," he says, "touching impossibility, I take it that all these things are to be held possible and performable, which may be done by some persons, though not by one alone; and which may be done in the succession of ages, though not in one man's life; and lastly, which may be done by public designation and expense, though not by private means and endeavour."[49] As another move toward co-operation, Bacon proposed the establishment of a college of natural history.[50] Since his plan required a great number of men, the question of ability naturally arose, for geniuses are rare. Bacon solves the problem by insisting that the compilation of his history does not require men of unusual abilities: "But the course I propose for the discovery of sciences is such as leaves but little to the acuteness and strength of wits, but places all wits and understandings nearly on a level." And again, "For my way of discovering sciences goes far to level men's wits, and leaves but little to individual excellence; because it performs everything by surest rules and demonstrations."[51] In one case he compares his method to a compass, and in another to a machine, both of which require little of their operators. In stressing the fact that his method would more than compensate for the limited abilities of his followers, he was really seeking to remove the despair effected by the theory of nature's decay, which bestowed upon the moderns a necessary intellectual inferiority to the ancients.

Bacon lays down some general rules for his history. Particulars should be carefully entered, and in doubtful cases the reader should be warned by some such expression as "It is reported."

Sometimes the name of the observer should be given. In all cases, the manner of conducting the experiment should be clearly described so that men may judge and verify it, and especially that others may be stimulated to improve it, since natural history is open to every man's industry.[52]

More than any other element in his philosophy Bacon's conception of natural history and its importance influenced the seventeenth century. As will be seen, the most prevalent motive animating scientists in the third quarter of the century was the desire to contribute data to such a history. Boyle repeatedly claims this as his motive; Sprat says that the aim of the Royal Society was the accumulation of a vast pile of experiments, complains that the ancients did nothing toward this end, and prefers the joint labors of many men to those of individuals no matter how significant. The multifarious activities of individual scientists and of the Royal Society as a whole can best be explained on this basis. Later the need of co-operation is not only emphasized in words, it is also met in practice by the various groups of experimenters out of which developed the Royal Society, itself the final embodiment of the co-operative plan in Bacon's philosophy. The faith in, and hope of, a comprehensive natural history, based on experiments and observations, were responsible for much of the enthusiasm with which later English scientists entered into their work. But they were not, any more than Bacon, indifferent to the immediate good which might arise from the materials of this history, especially as regards the so-called mechanical arts, which received from them the same emphasis accorded by Bacon.

Bacon did not conceive of his natural history as a haphazard collection of unorganized observations and experiments. In fact, he says that little is to be expected from the intellect unless particulars are apt, well arranged, drawn up, and marshalled in some order. He also states that a greater abundance of experiments is to be sought according to a fixed law and in regular order, and he elaborates upon this order and direction in conducting experiments, which he explains under the terms of Variation, Production, Translation, Inversion, Compulsion, and Chances of experiments.[53] When, however, experiments and observations of all arts have been collected, digested, and brought into one man's knowledge,[54] his own inductive method is to be applied. There are three distinctive features of the latter. The first has to do with the various stages of the process of establishing general principles, in which he contrasts his logic with the syllogistic logic of the day. Whereas the latter merely glances at experiments and particulars, and at once establishes the highest generalities, abstract and useless, his induction dwells duly and orderly among particulars, and rises gradually to general principles.[55] In fact, this gradual and unbroken ascent from sense and particulars to the most general principles Bacon stresses as the true but untried way. The understanding, he asserts, must not be permitted to jump from particulars to remote axioms, but must rise by successive steps from particulars to lesser, to middle, to greater axioms. The lowest axioms, he explains, differ little from bare experience, and the highest, are notional and abstract; the middle, are the true, solid, living axioms upon which depend the affairs and fortunes of men.[56] In another passage he says that one is not to extract works from works, or experiments from experiments, but to extract principles from them, and from these principles new works and experiments.[57] In short, the interpretation of one experiment furnishes a principle which suggests other experiments, and these in turn lead to higher principles.

The second unique feature of his induction is found in his conception of form, which is more clearly understood in connection with his attitude toward atomism. Bacon based his science upon a purely physical foundation, though he frequently found it necessary to employ current terminology, which may sometimes be misleading. Spirits, for instance, which figure frequently in his writings, are nothing but natural rarified bodies. What he styles the original passions or desires of bodies, or the primary elements of nature, are only the qualities of dense and rare, solid and fluid, hot and cold, heavy and light.[58] He used the word "form," because it was familiar, though he considered the forms of the received philosophy mere figments. His dictum that bodies are not acted upon but by bodies suggests a purely materialistic basis for his philosophy.[59] There is also little doubt that in a way he accepted the atomic philosophy.[60] He asserts that his purpose is not to resolve nature into abstractions, but "to dissect her into parts; as did the school of Democritus, which went further into nature than the rest." Matter rather than forms should be the object of our attention, "its configuration and changes of configuration, and simple action, and law of action or motion."[61] Force, or motion, was implanted by God in the first particles, from the multiplication of which all variety proceeds. He declares that the motion and virtue of the atom are the beginning of all motions and virtues. In one passage he compares solid bodies to machines, which from their configuration admit innumerable variations, and in another maintains that the diversity of things is due to the magnitude, configurations, and positions of one fixed and invariable substance.[62] All this, though rather far from the mathematically elaborated mechanical philosophy of Descartes, indicates that Bacon leaned to the theory of atoms.

There is nothing in nature, Bacon says, but individual bodies performing individual acts according to fixed laws, the discovery and explanation of which are the foundation of knowledge and operation. This law he calls form. And again he defines forms as those laws and determinations of absolute actuality which govern and constitute any simple nature such as heat, light, and weight in every kind of matter that is susceptible of them. His own observations and experiments led him to the conclusion that the nature of which heat is a particular case appears to be motion.[63] If we combine these ideas, we arrive at the conclusion that the form of heat is the law governing the particular motions of individual bodies (certainly his particles), which produce the nature of heat. Bacon believed that the forms, or laws, of simple, or abstract, natures are few in number, but that they may be combined in innumerable ways to produce the infinite variety seen in nature.[64] They are like the letters of the alphabet, which, though not numerous, can be variously combined to form any number of words. If, then, one understands these forms, he will be able to combine them at will to achieve any result of which nature is capable, and thus command her.

His induction, the explanation of which he did not complete, was his method of arriving at a knowledge of these fundamental forms. He is careful to distinguish his particular inductive logic from the common one of simple enumeration, which he calls childish, because being based on too small a number of such facts as lie at hand its results are precarious and likely to be contradicted by a single instance.[65] What chiefly distinguishes his

method is his conception of rejection or exclusion. In the process
of exclusion, he declares, are laid the foundations of true induction.[66] He further explains that the principle of exclusion oper-
ates in cases where several natures are not found when the
nature under investigation is present, or are found when the
given nature is absent.[67] Thus his process is essentially one of
elimination, though he says that conclusions can be reached only
on affirmative instances, which, after proper exclusion, are solid,
true, and well defined. Perhaps a simple example may clarify
this explanation. Flames produce both heat and light, but the
moon, according to Bacon, gives forth only light. Therefore, in
investigating the form, or law, of heat, exclude light. Bacon
looked upon his method as insuring absolutely certain results, as
finally making the understanding of man a match for nature.[68]

His ideas of forms and exclusion did not influence the seventeenth century very much. In fact, his influence on the scientific
movement was general rather than specific. He impressed upon
his followers the importance of experimentation and inspired
them with confidence in the certainty of its results, but he cast
little light upon the proper method of experimentation. This
fact, together with his failure to appreciate the great discoveries
of his day and to make any himself, is responsible for the present
depreciation of his importance in the history of science.[69]

Bacon proposed for the most part a utilitarian end for science.
He believed that the greatest error men had made in learning was
in mistaking the end of knowledge, which to him was the relief
of man's estate.[70] He proposed his method to enable men to
secure dominion over nature, "to endow the condition and life of
man with new powers and works."[71] In his eyes, human learning
and human power were one and the same. It is true that he
warns men not to seize too soon the fruits of their experiments
and observations, but only because he believed that by pursuing
their investigations to a knowledge of more general principles,
they could so command nature that fruits would come "in clus-
ters."[72] He classifies experiments as "experimenta fructifera"
and "experimenta lucifera," "either of use or of discovery,"[73]
terms which are constantly employed throughout the century, but
the experiments of light are of value because ultimately they lead
to the securing of greater benefits. One justification of this ultili-
tarianism is seen in his humanitarian spirit. He seems to have

been deeply impressed with the wretched condition of the human race, and cherished the hope that out of the marriage of mind and things there might spring "a line and race of inventions that may in some degree subdue and overcome the necessities and miseries of humanity."[74] Certainly he is one source of the humanitarianism and social-mindedness which are so conspicuous in the period of Puritan domination. Bacon firmly imposed the utilitarian spirit upon seventeenth-century science in England.

A few other views of Bacon's played their part in the later development of science. His emphatic separation of science and religion[75] certainly made scientific progress easier and enabled the Puritans to embrace his philosophy. The great value he placed upon inventions[76] long survived him, and he is frequently cited in the praises of them. His criticism of the schools and universities and the reforms he would introduce are sufficiently close to those later published by the Puritans as to suggest that he influenced the latter either through Comenius or directly. The more important are: that children come too early to the study of logic and rhetoric, when they should be taught matter rather than ornaments; that the universities are too much committed to certain authors as authorities; that no liberty of thought is permitted; that readings, disputations, and other scholastic exercises should be abolished; and that correspondence with European universities should be established.[77]

Bacon impressed upon his age the need of advancement and held out to it the hope of scientific progress. His own unfailing assurance in his philosophy supported the encouragement which his optimistic spirit offered men. He did more than anyone else to break the fetters which bound his age to servile submission to the authority of the ancients, and he inspired his followers to face the future rather than the past. His repeated insistence upon experiment and observation as indispensable for the discovery of scientific truth, his reiterated injunction to learn the appearance of things and find out what really is in nature called men from mind to matter, and from libraries to laboratories. His conception of a universal natural history and all it entailed was mostly responsible for the comprehensive and varied experimentation carried on by individuals or by groups with the assiduity and indefatigability which characterized the scientific temper of the third quarter of the century. In the material welfare of man he found for his age an impressive sanction for his scientific program, and thus furnished a social motive for a materialistic and utilitarian enterprise. His was a stimulating and vitalizing influence hardly to be overestimated. Without it science would have moved forward, but in England scientific activities would have been retarded to no inconsiderable extent. The bell he touched rang with increasing intensity throughout the years.

CHAPTER IV

THE GILBERT TRADITION, 1600-1640

The rest of *Magneticall* proprieties I find in ancient Writers, as litle knowne as their causes; & if any matter herein were broached, it was merely coniectural, and depending on no certain demonstration; neither had we any certain or satisfactory knowledge of this thing, vntill such time as it pleased God to raise vp one of our Countrymen D. *Gilbert*, who to his euerlasting praise hath troden out a new path to *Philosophie*, and on the Loadstone erected a large *Trophie* to commend him to posterity.

Nathaniel Carpenter, *Geographie Delineated*, 1625.

Bacon's influence was slow in getting under way. It grows more noticeable as we approach the period of Puritan supremacy, when it was to be revealed in all its potency, but during the years covered by this chapter manifestations of his spirit are rather vague and indirect. Yet there do appear some signs that his ideas had not fallen upon barren ground. Here and there, even though he may not be mentioned, are revealed attitudes and values that are so nearly his as to furnish presumptive evidence that he is the source. The disposition to rely upon antiquity for ideas about nature begins to weaken, and the word "experiment" appears more and more frequently. The hostility to Aristotle, though traceable to various sources, finds a possible explanation in his writings. The spirit of utility and the conception of the common good, which were not unknown to the Elizabethans, now show signs of detaching themselves from the nationalistic spirit, and of embracing mankind in a universal view which soon became, and is now, characteristic of science, the most international of man's intellectual activities. Yet the line of scientific development in which the modern spirit is most clearly revealed traces its source not to Bacon but to William Gilbert.

The *De Magnete,* though not receiving the notice later accorded other important scientific discoveries, furnished for a long time the basis for most works on magnetism, electricity, and, to some extent, navigation. As has been stated, Gilbert was more closely associated with the acceptance of the Copernican theory in England than any other man, in spite of the fact that he himself understood only dimly the mathematical demonstrations upon which it rested. The completeness with which he established the basic principles of terrestrial magnetism gave much weight to his mere approval of the other great theory. In works on magnetism his influence is predominant. The first treatises which followed Gilbert's volume are largely derived from it and do not add very much to knowledge of the subject. Such was *A Short Treatise of Magneticall Bodies and Motions*, 1613, by Mark Ridley, who is described in interesting fashion on the title page as "Dr. in phisicke and Philosophie. Latly Physition to the Emperour of Russia, and one of ye eight principals or Elects of the Colledge of Physitions in London." Ridley experiments in the same manner as Gilbert, making similar use of the terrella, and though in one or two instances he attempts to prove his predecessor wrong, he considered the latter one "whose labours are the greatest, and best in *Magneticall* Philosophie." But though dependent on Gilbert, he takes nothing from him on trust. Employing the same kind of apparatus as his predecessor, he carries on his own "Magneticall experiments," in which, if we may believe his own words, he experienced the greatest delight. He accepted Gilbert's version of the Copernican theory *in toto*, strengthened, however, by the discoveries of Kepler and Galileo.[1]

In another work[2] Ridley revealed his indebtedness to Gilbert by giving expression to attitudes which grow in strength as the century advances. He deprecates the force of tradition and traditional knowledge, saying that he dares and hopes to persuade those "who are to much adicted to that they sucked from their youth out of Philosophers, which all Schollers read after they haue beene two yeares in the Vniuersities, especially from their old friend *Aristotle*," to surrender their errors. To this end he attacks Aristotle's astronomical views, especially that of the *primum mobile*. Conscious of the new spirit which was beginning to stir in his day, he holds out the hope that though Aristotle's opinions had been embraced by almost all men,

yet now in these daies, when *dies Diem docet*, by the multitude of the writings and labours of late most learned Writers, it is come to bee an vsuall assertion in the Schooles that, *in Anatomicis & Astronomies errauit Aristoteles;* and certainely if wee shall obserue and ponder what a thing the wonderfull vastnesse of expansion, the fine and thin Firmament hath, and what a great distance there is betweene the earth and the farthest limits

(for Spheres are but fained) of the aduancement of *Saturne* from the
earth, but especially of the fixed Starres, which are held to be remoued as
farre againe and much mooue [more] from *Saturne*, it might be said, that
the earth firmed in her position might mooue her whole bodie many thousand miles from place to place, and yet there would bee found no excen-
tritie of the fixed starres by the earth, and all those arguments which are
vsed by *Aristotle, Ptolomie, Regiomontanus, Iohannes de Sacrobosco* and
others to proue, that the earth is *centrum vniuersi* would bee no otherwise,
then now they are, for the earth is not *centrum vniuersi*, but is as it were
punctum insensibile respectu vniuersi & stellarum fixarum, and no doubt
but that *Copernicus* and *Tichobrahe* most perfect and exact Astronomers,
who make the Sunne to be *centrum vniuersi & planetarum* accompted the
said arguments and proofes to be of force, and as truely verified of the
Sunne, as of the earth if a man might passe thether to trie the same.[3]

The realization of the advancement of knowledge beyond the
mistakes of the ancients, rendered keen by the great discoveries
of the Renaissance, together with the consciousness of the vastness of the universe and of the consequent insignificance of the
earth, conceptions that were slowly beginning to seize men's
minds, is here clearly manifested. Ridley's perception of the possibility of scientific progress, of the limitless future of increasing
knowledge again finds expression in the hope that "many new
points in Philosophy and Astronomy will blossome and spring
out of this kind of learning hereafter."[4] The door that was closed
and barred by classical antiquity is beginning to open, not only
on the infinitude of the universe, but also on the infinite advancement of science.

Perhaps even more indebted to Gilbert than the doctor under
discussion was William Barlow,[5] archdeacon of Salisbury, whom
Ridley accused of deriving all his ideas from his great predecessor. In reply the cleric asserted that he had begun his experiments forty years earlier, and for proof of his originality he
pointed out five discoveries of his, the most important being the
difference in magnetic powers between iron and steel.[6] If his
words are true, he would appear to have antedated both Norman
and Gilbert. With the latter he must have been closely associated,
for he published a letter from Gilbert to the effect that Barlow
had been of more assistance to him than anyone else, and elsewhere Barlow states that Gilbert had urged him to publish his
work. The archdeacon evinces true faith in the experimental
method; he would not even accept Gilbert's experiments before he
had tried them with his own hands; and he quotes the latter to

the effect that truth must be learned from things themselves and not from books. Influenced largely by Biblical considerations, he would not consent to Gilbert's interpretation of the Copernican theory, though he subscribed whole-heartedly to his ideas of terrestrial magnetism. The defence he offers, however, for his rejection of the modern astronomical discovery is one which, true to the spirit of science, occurs with increasing frequency as the century advanced, and which was expressed in a quotation destined to become the motto of the Royal Society: *Nullius addictus iurare in verba magistri.* Though not a traveller himself, he maintains that he secured much of his data from "great Trauellers" and "our chiefest Nauigators."[7] Barlow, like Ridley, is a clear example of the influence which Gilbert exerted in turning men from traditional knowledge to observation and experiment, and in introducing a critical attitude of mind which insists on being satisfied only by the test of sense-observations and rational experiments.

In 1621 there appeared a work by Nathaniel Carpenter entitled *Philosophia Libera,* which reached a third edition in 1636. This volume deserves much more attention than it has received, for it is the first work in England to be primarily devoted to an earnest plea for liberty in scientific investigation and discussion. Carpenter was educated at Oxford, where he was made scholar by the mandate of James I. Later Archbishop Ussher prevailed upon him to come to Ireland, where he became schoolmaster of the King's wards in Dublin. The first edition of his treatise was published at Frankfurt, and in the following year the second, under a slightly different title, was issued at Oxford. This possessed an entirely new preface addressed to the students of the university, and some additional sections given over to a discussion of the Copernican theory.[8] This edition, which primarily interests us, is divided into three "Decades" composed of ten "Exercitationes" each, most of which attack some Peripatetic idea. It is the preface, however, which is most significant, for it contains the germ of an attitude later characteristic of the Royal Society. It attacks servility to the ancients, mainly Aristotle, and makes a strenuous plea for a critical attitude of mind and complete freedom of thought. In describing himself as one who *in nullius iuratus verba* does not expect certainty in anything but the Scriptures, he employs a phrase containing in brief a declaration of intellectual independence, which, as has been said before, was to be emblazoned on the standard of the Royal Society. He commiserates those who, being born into an intellectual servitude, accept the ideas of antiquity before understanding them and condemn as vanities anything which they think new.

The remainder of the preface is largely taken up with an attack upon the followers of Aristotle, who in swallowing his principles prove unworthy of his example, for if they possessed the spirit of the Stagirite they would grant liberty in philosophizing to all. Were he alive, he would disown them as spurious and degenerate, yet so completely has his authority prevailed that his name alone has power to unite those who hold his views in common and to hurl his adversaries on the prejudices of the majority. Because of poor luck rather than lack of merit, the older philosophy of the Greeks gave way so entirely to the Aristotelian philosophy that only Peripateticism holds sway in the schools, turning, like Medusa's head, all beholders to stone. Aristotle, being a man, was subject to mistakes, yet the Peripatetic philosophy, admitting no flaws in his works, has grown more imperfect the farther it has moved from its source, until these multiplied phalanxes of errors, like insects bred in the carcass of a dead horse, have reached posterity. To this superstitious Peripateticism he opposes his free philosophy "not adorned with the splendid colours of great names, but naked and simple, conscious rather of sincerity than of authority, which neither avoids the judgment of the learned nor fears the prejudices of sciolists." In contrast to extreme dogmatism he praises that ingenuous candor which dictates a middle way to the more prudent so that they will neither swallow without salt nor blindly reject any theories. He justifies this critical attitude of mind on the ground that many things lie in the bountiful lap of nature, which are beyond the bounds of the accepted philosophy, and to a discussion of which Aristotle can contribute nothing. Such things, he holds, are the special province of free philosophy. Finally, he claims that he is motivated only by the desire to turn others to more profitable pursuits and to stir up more learned men for the promotion of science.

He next passes to a conception characteristic of Bacon and destined to become a distinguishing feature of the scientific spirit in the Restoration. Employing the enervating theory of nature's

decay to his own purpose, he maintains that if in the decrepit age
of the world[9] the moderns are but boys sitting on the shoulders
of the gigantic ancients, yet by the aid of the latter the former
are enabled to grasp and profit by those secrets which antiquity
only guessed at. For certainly science is not the work of one man
or of one age; a beneficent nature is ours, which is not so prodi-
gal as to pour into one age what was meant for all, nor so
niggardly as to withhold her glory from any particular period.
If his own age has been more productive of discoveries than the
past, as he believes, it is not because the moderns are superior
in genius but because they come later in time. If the youth of
the world was richer in genius, its old age is richer in experience,
a compensation for which he expresses gratitude to God. He
finishes with the compromise: "Let the philosophy of the ancients
flourish, provided, however, it does not disdain the civilization of
the moderns; let the fervor and industry of the moderns wax
strong, provided, however, they are strengthened by the aid of
the ancients." He seems to concede deterioration in genius, but
argues superiority in knowledge due to the incremental nature
of the growth of learning, a conception which had been elaborated upon by Bacon.

Carpenter emphasizes reason, evidence, and the need of explaining phenomena by means of investigation and not by
authority or idle speculations. The fact that the ancients were
ignorant of, or opposed to, any theory was of no moment in his
eyes. His attack on the various tenets of Peripateticism was
incited by a desire to shake the blind worship of Aristotle and
to reveal to his followers the falsity of the principles which they
viewed with superstitious awe.[10] In this way he hoped to effect
a sceptical attitude of mind and a willingness to consider ideas
outside the narrow circle of their allegiance. Fully aware of the
fruitlessness of disputatious Peripateticism he perceived that the
root of the evil lay in submission to classical authority instead
of reliance on observation and experiment. "Ego sum," he says,
repeating the significant expression used in the preface, "in
nullius iuratus . . . . verba,"[11] and it is this attitude which he
wishes to create in his readers. In order to show the need of
suspended judgment, he harps upon the bar which prejudice
offers to the reception of new ideas. He himself reveals a singularly open mind in his treatise, attacking Peripatetic ideas but

also giving what evidence he can in their favor, and though he definitely leans toward the Copernican theory and bestows unstinted praise upon the new astronomers, he admits into the discussion all arguments against them. Many of these arguments are of the traditional deductive kind; he does not stress experiments, though he recognizes the necessity of investigating and rationally explaining phenomena; and since his knowledge of mathematics was slight, his astronomical arguments are rather obvious. His great virtue lies in his recognition of the inadequacy of the old philosophy and of the evils resulting from servile subscription to it, in freeing the mind from prejudices and opening it to new ideas, and in sensing the significance of the new thought and discoveries which were beginning to stir the world.

There is no definite evidence that Bacon influenced Carpenter. The latter's attitude toward Aristotle and the Peripatetics, his recognition of the obstacle to the advancement of learning presented by the blind worship of antiquity, his insistence upon suspension of judgment and a critical attitude of mind can all find a source in the writings of Verulam, though we would naturally expect, were this so, that experiments would have figured more in his treatise.[12] One might plausibly infer that he was indebted to Bruno. The preface to the first edition, upholding ignorance against learning, may have been inspired by Bruno's *Cabala Del Cavalo Pegaseo*, though it may be indebted to Cusanus' *Learned Ignorance*, Cornelius Agrippa's *On the Uncertainty and Vanity of all Knowledge*, or Erasmus' *Praise of Folly*, all of which, McIntyre thinks, influenced Bruno.[13] Carpenter makes the same use of the Golden Age as Bruno in his *Spaccio*. Besides the general attitude toward Aristotle and the ancients and the specific attacks upon Peripatetic fallacies, an activity dear to Bruno's heart, Carpenter's desire to show that ignorance and superstition are not requisite to true religion and that the Scriptures have been used to glorify ignorance allies him rather closely to the Italian. Furthermore, Carpenter's insistence upon "free philosophizing" is much that of his predecessor. Neither Bacon nor Bruno finds a mention in his works, but there is another who does, and for that reason I have placed Carpenter in the Gilbert tradition.

In 1625 Carpenter published a geographical treatise,[14] which in a comprehensive and learned manner covers a variety of things: the relationship of the earth to the heavens, the magnetic qualities of the earth, longitude and latitude, descriptions of seas, lands, rivers, mountains, animals, plants, human races, and so on. That Gilbert more than anyone else is responsible for the author's scientific outlook is revealed in a passage which, though long, is worth quoting in full, for it bears full testimony to the strength of the Gilbertian tradition, and reveals the author's faith in experiments more fully than his previous work. After praising Gilbert in a passage quoted at the beginning of this chapter, he continues:

This famous Doctor being as pregnant in witty apprehension, as diligent in curious search of naturall causes: after many experiments, and long inquiry, found the causes of most magneticall motions and proprieties hid in the magneticall *temper* and constitution of the *Earth*, & that the Earth it selfe was a meere *Magneticall* body challenging all those proprieties, and more then haue expressed themselues in the Load-stone. Which opinion of his was no sooner broached, then it was embraced and wel-commed by many prime wits, as well *English* as *Forraine*. Insomuch that it hath of late taken large root, and gotten much ground of our *vulgar Philosophie*: Not that in the maine scope and drift of it, it contradicts or crosses all *Peripateticall* principles, or the most part of such grounds as haue hitherto borne the stampe as well of *Antiquity*, as of *Authority* [Carpenter is playing safe]: But that it hath brought to light matters of no small moment, which neuer found any ground or footsteppes in our ordinary Philosophie. This new Philosophie I dare not commend as euery-where perfect and absolute, being but of late yeares inuented, and not yet brought to mature perfection: yet would it sauour of litle ingenuity or iudgment in any man, peruersely to deny all such Magneticall affections in the Earth as are grounded on plaine experiments and obseruation: sith no *Philosophie* was euery-way so exact, but required experience dayly to correct it. I intend not here an absolute discourse of *Magneticall* Bodies and Motions, but leaue it to their search whose experimentall industrie is more suteable to such a subject.[15]

Certainly this sincere praise of Gilbert indicates the chief source of Carpenter's philosophy, which, as here revealed, embraced the experimental method, for he ascribes his predecessor's success to his experiments and makes "experimentall industrie" the prime requisite of those who would speak with authority on magnetism.[16] Sense-observation, investigation of the phenomenon itself is what he insists upon. In reply to an objection on the part of Joseph Scaliger that Copernicus' demonstration was drawn from phenomena only, he asks,

I would know how *Joseph Scaliger* by any other meanes came to know it? I alwayes supposed it a principle amongst *Mathematicians*, that the τὰ φαινόμενα had bin the surest ground of Mathematicall Demonstration: for euery reason which can be alleadged, must of necessity be grounded on meere coniecture, as forged in a mans braine without any obseruation of Nature; or else suggested vnto vs from the things themselues. How little dependency is on the *Former*, let euery man iudge: where it is as easie for euery man to deny, as affirme; and such fancies are better reserued in the braine, wherein they were first hatched, then bee suffered to proceed further.[17]

The sections which treat the Copernican theory are largely taken from the *Philosophia Libera*. Carpenter was too much influenced by Biblical arguments to accept the theory *in toto*. The rotation of the earth, the feature of the system emphasized by Gilbert, he did subscribe to, but he denied its other motions and considered it the center of the universe. Yet in accordance with his just and open-minded nature, he stated the theory as fully as he could, "iustifying it no farther than will stand with Astronomicall obseruation," and advanced his own ideas candidly but modestly to the view of the judicious, "who herein may vse their Philosophicall liberty, to imbrace or reiect what they please."[18] It was dogmatic and unreasoning allegiance to antiquity and hostility to new ideas more than enthusiasm for the discovery which aroused his wrath. "I would not here be mistaken," he says,

as though I strongly apprehend these grounds [Copernicus' arguments], & reiect all the principles of our *Peripateticke* Philosophie: I only inueigh against their preiudicate ignorance, which ready to licke vp the dust vnder *Aristotles* feet, with a supercilious look contemne all other learning, as though no flowers of science could growe in another garden. I confesse this opinion of the Earths circular motion to be subject to many and great exceptions, and opposed by strong and waighty arguments drawne probably from the booke of God, the touchstone of sincere verity; yet I hold it too strongly fortified to be inuaded by popular arguments drawne from seeming sense, and bolstered vp with names and authorities.[19]

Carpenter is an independent thinker, refuting with much gusto Peripatetic ideas but not always unreservedly accepting new discoveries, constantly insisting upon freedom of thought and discussion and consistently revealing it in his practice. With no respect for authority or the ancients in themselves, he emphasizes the need of demonstration, experiment, and observation of nature. He is more frequently on the side of revolutionary theories than against them. He embraced part of the Copernican system, and was sympathetically interested in the rest; he accepted Gilbert's ideas of terrestrial magnetism entire; and he entertained the highest regard for Copernicus, Brahe, Kepler, Galileo, Gilbert, and others. His emphasis upon experiment, his attack on the ancients and their blind followers, the zest with which he ridicules ancient fables and supporters of them, his acceptance only of the rotation of the earth (which alone Gilbert had emphasized), and his eulogy of, and agreement with, Gilbert suggest, if they do not prove, that the Elizabethan doctor was the chief source of his ideas.

The next work[20] to be considered does not mention Gilbert, but falls so squarely in line with much of the material which entered into his work and so obviously springs out of the scientific interest attached to the compass, with which Gilbert's name was inseparably associated, that it finds a natural place in his tradition. One Thomas James, backed by some merchants of Bristol, set out on a voyage to find the Northwest Passage, and after a winter of indescribable hardships in the northeastern part of North America returned a disillusioned and a sadder man. But just as the navigators in the Elizabethan period had observed the action of the compass and carried their observations back to England to assist Gilbert and the others in their study of magnetism, so James had kept a weather eye open not only for the action of the compass but for other things as well which might be of scientific interest, such as marine readings, topography, fauna, and flora. He evidently had been instructed to keep records of latitude, longitude, and variations of the compass, and when his observations did not agree with certain mathematical conclusions, he says conclusively, *"but thus we found it by practice"* (his italics).[21] He studied the action of the tides by observing marks on the land and by noting the drift of the ship: "By all these experiments I found exactly, that the tide was no stronger there, then that betwixt *England* and *France*."[22] He makes observations on ice in a bay, and studies the refraction of the air, all in a scientific spirit. James is only representative of that scientific activity on the part of voyagers, which began in the Elizabethan period, when, however, the wonderful and strange prevailed over the scientific, and continued with increasing intensity until in the second half of the seventeenth century the scientific attitude became characteristic of them, and they were called upon to furnish data to the Royal Society, the members of which, following Bacon, were trying to spread their energies over the whole of nature.

We have already noted in a previous chapter that in the sixteenth century students of geography were timid in dissenting from the limited knowledge of antiquity, in spite of the remarkable expansion of geographical learning. In the seventeenth century a bolder spirit becomes apparent, and the vast amount of data regarding natural phenomena in parts of the world entirely unknown to the ancients brought out in increasingly sharp relief the frequently false or inadequate theories of antiquity.[23] Nowhere is this fact better illustrated than in the remarks on James's journal written by William Watts,[24] M.A. of Caius College, Cambridge, rector of St. Alban, Wood Street, London, and army chaplain to Lord Arundel and later to Prince Rupert. His words are addressed to "the venerable Artists and younger Students" in Cambridge. James's observations, he says, furnish

So much, and such variety: such variety, and that so various, (I thinke) from what is receiued in the *Schooles*: that it were well worth the disquisition of an *Vniuersity*, (and I wish you the first honour of it) either to find out, how these *Obseruations* may bee reduced to *Aristotles Philosophy*: or whether they need any other enquiry, and ought to be examined by some other *Rules*, then *Aristotle* hath yet light vpon. This is my purpose of inscribing it vnto you.

Sensing the fact that obedience to authority stood in the way of accepting any new truth, and that the avenue to increased knowledge could be prepared only by undermining this authority, he continues:

Of this one thing am I confident: that you are all so *rationall*, and *ingenuous*, as to preferre *Truth*, before *Authority*: *Amicus Plato, amicus Aristoteles*, but *magis amica veritas*.[25] Your *Sciences*, then, being *liberall*; your *Studies*, I know, haue so farre passed into your *maners*, that your minds are so too, and that such as haue already profited beyond the credulity required in a yong learner, and are themselues promoted to be *Masters of the Arts*; though they still reuerence their *old Greeke Tutor*, yet they will not suffer that of *Pythagoras Schoole*, so to domineere in *Aristotle*, as to let an *Ipse dixit*, goe away with it: much lesse allow it the authority of a *Mayors hammer*, with one *knocke* to silence all arguments.

He next asks and proceeds to answer two questions. The first, *"Whether those Rules of Aristotle's Philosophy be to be allowed so Vniuersall, that they hold all the world ouer,"* he counters by asking if Aristotle knew the Indies or the southern hemisphere. "Plainely, those that are conuersant in the *nauigations* and *bookes of voyages* into those parts; haue found so many contrarieties to obserue; that it were rather *tedious*, then difficult, to fill vp a *Notebooke* with them." He makes plain the error of the ancients in limiting habitation of the earth to certain zones by pointing out that Cambridge and the uninhabitable place in which James wintered lie in the same latitude. He easily disposes of Aristotle's theory of thunder being caused by dry exhalations by noting that thunder is constant over the snow-covered Andes. Next follows a general estimate of the value of voyage books to scientific investigation:

I could, (in my smal reading) instance in many many other particulars: which I had rather should be found out by some industrious searchers after *Nature*, in the *Modern Relations* of our *Discoverers*, then in this my short *Proposition*. Tis not to be doubted, but that the carefull reading of our *Books of Voyages*, would more elucidate the *History of Nature*, and more conduce to the improuement of *Philosophy*, then any thing that hath beene lately thought upon. These Nauigations haue in·part fulfilled that of the Prophet, *Many shall passe to and fro, and knowledge shall be encreased*. This, I suppose, might be obserued from this study, That the great and infinite *Creator* hath so disposed and varied euery thing, that it is impossible for mans reason and obseruation to conclude him: and therefore, though vulgar and receiued *Philosophie*, may giue a man a *generall* hint, all the world ouer; yet no Vniuersall and vnfayling certainty.

The second question he proposes is whether any human dictate should be so authoritative as to exclude examination, the answer to which he begins with Bacon's assertion that a slight knowledge of nature leads to atheism, but a deeper knowledge to belief in God. In his attack upon the principle of authority he touches upon a relationship which is significant in the history of scientific thought, and which will frequently appear in this study, namely, the relation of science to Protestantism. No one, he declares, will think it fit for Englishmen to have a pope in philosophy any more than a pope in religion, "one, that no body shall presume to censure of: but all be bound to aduance his *Decretalls*, above the *Holy Scriptures*." He points out as a horrible example the authority bestowed upon Aristotle in medieval divinity: "Philosophy hath taken its turne in the *Schooles*: and the *holy Texts* by the *Schoolemen*, haue euen been submitted vnto *Aristotles*: yea, to the great corruption of *Theologie* .... let Schollers be so bold with *Aristotle,* as to examine him vpon good assurance, by what is *Truths Touch-stone."* As we shall see later, one of the chief considerations which made the Puritans sympathetic with the new science was the fact that both possessed an enemy in Aristotle, who was looked upon by the one as the corrupter of divinity, and by the other as an obstacle to the progress of science. Thus we see that Watts in denying the validity of Aristotle's authority in science arraigns his domination over theology.

In his further attack on the Stagirite, Watts touches indirectly on the idea of nature's decay by declaring that the same God who endowed Aristotle with genius has also raised up many excellent men in later days. The fact that their inventions and observations find no credit in the world because ancient dictates have preoccupied all good opinion offers great discouragement to all attempts to advance knowledge. The only remedy for the situation lies in unqualified freedom to examine the old philosophy, so that the means that have advanced geography, mathematics, and mechanics may also promote physics. He finds justification for a critical attitude toward Aristotelianism in the fact that it had been challenged by various men, among whom he mentions *"Basson* and *Gassendus,* (two braue men) [who] haue newly written point-blanck against him: nor haue they taken away all liberty, from those that are to follow them." Echoing Bacon's attitude, he claims that his words are intended "out of good will to promote learning; to encourage and countenance future vndertakings." The few echoes of Bacon's ideas indicate that he had something to do with Watts's general view of the past and the present, and furnish a sign that we are approaching a period in which his influence will be paramount.

Watts was not the only one to make use of James's narrative. Henry Gellibrand, Gresham professor of astronomy, added a few pages to it entitled "An Appendix touching Longitude," in which he denied that longitude could be determined by the variation of the compass (a fallacious idea which had held ground since it was first proposed by Gilbert), because such variation did not remain constant for any one place. For proof he relied heavily on the mariner's observations "which for Iudgment, Circumspection and Exactness may compare with most." The following year he elaborated upon his idea in a pamphlet entitled *A Discourse Mathematical on the Variation of the Magneticall Needle. To-*

gether with Its admirable Diminution lately discovered, 1635, which supports the variation of the variation by careful experiments and observations, both his own and those drawn from various observers of the preceding half century or more. The chief inspiration of the treatise is revealed in a passage on the very first page, which states that those who wish to know the truth about magnetic qualities should consult "that most learned worke (admired by all Foreiners) long agoe penned by *Mr. Dr. Gilbert,* our countryman," and this in spite of the fact that he is refuting one of his admired predecessor's theories. In keeping also with the Gilbert tradition is the professor's discussion of the Copernican theory, which he is willing to accept as an hypothesis only: "It is not unknowne to the world, how the greatest Masters of Astronomie, which this age hath afforded, for the more easy salving the apparent anomalar motions of the fixed and erratique cælestiall lights, and avoyding that supervacaneous furniture of the Ancients, do with all alacrity embrace that admirable *Copernican Hypothesis* of the diurnal, Annual, & Secular motions of the earth. . . ." Not the least effect of the great astronomical discovery was the discounting of the "furniture" of the ancients and a distinct lessening of respect for them.[26]

The importance which experiments were beginning to assume is revealed in a volume[27] published two years later. The author, Richard Norwood, who was a teacher of mathematics in London and surveyed the Bermudas for the Bermuda Company, maintained that though the practice and art of navigation had grown to far greater perfection than in any other age, whereby all parts of the world had been discovered, traffic facilitated, and knowledge divulged, yet it still remained imperfect in at least one point, the way of finding distances at sea. To remedy this defect he computed, by means of careful experiments and observations made on land, the length of a degree of the meridian, and so was enabled to reckon with commendable accuracy the diameter and circumference of the earth. Everything is an "experiment" with him. He uses the word again and again, even where we would employ such a word as "calculation." He insists upon the "experimentall resolution" of all problems. In one case he was directly confronted with a choice between authority and the evidence of his senses. Finding that his calculations did not agree with those of the ancients, though somewhat disturbed, he main-

tained his ground: "I resolved not to diminish nor to augment
the number thus arising by observations, measures, and allow-
ances, in respect of the opinions, observations, or measures of
any other man, untill there be made some Experiment more evident and exact then any yet extant." That he held to strict
standards for experimentation, always keeping in mind the possible limitations of an experiment, is quite apparent. Realizing
that in order to compute the circumference of the earth it was
necessary to measure accurately a portion of it, he says, "this
I have endeavoured in the experiment following, which if I have
not handled so exactly in all points as some would desire: (that
requiring more time and charge then I could well bestow) Yet
I doubt not but it wilbe found that I have come very neare the
truth." Again he remarks, "There are onely two things heere
which may seeme doubtfull, namely, the Experiment it selfe, and
the *Hypothesis* of the Sphericity of this Terrestriall Globe consisting of the Earth and Sea." But experimentation had by no
means yet assumed dignity in the eyes of the world. He speaks
of his being censured for the expense incurred by his activity,
since in ten or eleven days he measured the length of one meridian between York and London, and a little later he defends himself against those who think "Mechanicall and bodily exercises
. . . . too meane and unworthy to stoope unto," a prejudice much
more important than appears at first sight, for the mechanical
nature of experimenting was one point which later the traditionalists hammered unmercifully. Imbued with the Baconian spirit
of utility, he praises those "who are so nobly minded for a publique good," and defends applied mathematics against those who
would confine the science to the realm of speculation.

But indeed (as in all other parts of Learning, so in the *Mathematicks*,
especially in their application, or middle *Mathematicks* (as some call
them;) it is necessary with speculation to [join] some actuall and Experimentall practises; the former being emptie and uncertaine without these.
It is true, that the *Mathematicks* afford large fields of delightfull speculations, wherein a man might walk farre with much pleasure: But if from
so many faire flowers he bring home no honey, or from such large fields no
sheaves: I meane if he bring not these speculations to some usefull practises;
neither himselfe nor others are like to receive much fruit by them. . . . But
for my owne part, I acknowledge to have had my living and maintainance
by the *Mathematicks*, and not by *Speculation* onely, but rather by my practice therein; and therefore also I desire (what in me lyes) to make them

fruitfull to my selfe and others; And to that end have spent in some principall parts of the *Mathematicks* neare as much time and means in experimentall practises and conclusions, as in the Speculation.[28]

The last figure to be considered in the Gilbert tradition is one
who later played a great part in propagating the new science, and
who will be mentioned frequently in subsequent chapters. In
1638, John Wilkins published *The Discovery of a New World*,
in which he attempts to show that the moon is inhabited.[29]
Needless to say, this theory possesses no interest for us today,
but the attitude revealed in one or two passages does. Obviously
inspired by Bacon, he proposes as his motive in writing the
treatise the desire to stir up more active spirits in the search for
truth, and he proceeds to point out the chief obstacle standing in
the way:

.... it must needs be a great Impediment unto the Growth of Sciences, for
Men still so to plod on upon beaten Principles, as to be afraid of enter-
taining any thing that may seem to contradict them. An Unwillingness
to take such things into Examination, is one of those Errors of Learning,
in these Times observed by the Judicious *Verulam*. Questionless there
are many secret Truths which the Ancients have passed over, that are
yet left to make some of our Age famous for their Discovery.

The realization of the possibility of future discoveries, the opening of the door on the progress of learning, and the conviction
that the main bar across the door was the feeling that the ancients had encompassed all truths find expression in another passage: "Twere a superstitious, a lazy Opinion, to think *Aristotle's* Works the Bounds and Limits of all Human Invention,
beyond which there could be no possibility of reaching. Certainly
there are yet many left to Discovery, and it cannot be any In-
conveniency for us to maintain a new Truth, or rectifie an ancient
Error."[30] Wilkins' next treatise, *A Discourse Concerning a New
Planet*, 1640, which unequivocally upholds the Copernican theory,
is even more pertinent to our investigation. He shows scant
courtesy for "those barren, empty Speculations about *Materia
prima*, and *Universale*, and such like Cobwebs of Learning; in
the study of which so many do misplace their younger Years,"[31]
for "Whatever the Schoolmen may talk, yet *Aristotles* Works
are not necessarily true, and he himself hath by sufficient Arguments proved himself to be liable unto Error." At this point
there probably came into his mind the idea engendered by the

theory of nature's decay, namely, the superior genius and ability
of the ancients, for he hastens to employ Bacon's dictum concerning the youth and old age of the world:

Now in this Case, if we should speak properly, Antiquity does consist in
the Old Age of the World, not in the Youth of it. In such Learning as
may be increased by fresh Experiments and new Discoveries; 'tis we are
the Fathers, and of more Authority than former Ages; because we have
the advantage of more Time than they had, and Truth (we say) is the
Daughter of Time.[32]

Though Wilkins fully participated in the new spirit of the age
and subscribed to the great discoveries which characterized it, he
is consistently opposed to dogmatic habits of mind, believing
above everything else in an unprejudiced, open-minded, and critical attitude which suspends judgment until the proof is overwhelming. He notes the evils of preconceptions in saying that
astronomical phenomena would be more easily understood, "if
we could with Indifferency attend to what might be said for that
Opinion of *Copernicus,* which is here defended." As regards his
own ideas he is careful to disclaim dogmatic certainty, and to
allow, in the manner of Carpenter, full liberty of belief: "However, there is nothing in this Opinion so magisterially proposed,
but the Reader may use his own liberty; and if all the Reasons
considered together do not seem convincing unto him, he may
freely reject it." He lays it down as an axiom (it was one of
Bacon's) that in all matters "which carry with them any Doubt
or Obscurity, it is the safest way to suspend our Assents; and
though we may dispute *pro* or *con,* yet not to settle our Opinion
on either Side." Furthermore, in evaluating authorities he proposes the only reasonable test: "In weighing the Authority of
others, 'tis not their Multitude that should prevail, or their Skill
in some things that should make them of credit in every thing;
but we should examine what particular insight and experience
they had in those Things for which they are cited." He must have
been influenced to a certain extent by Carpenter, for in a manner
so characteristic of the latter he insists upon a critical undog-
matic attitude of mind which will assent to no old and accepted
idea on authority alone nor reject any new thought from prejudice, but will subject all to examination and investigation, always
suspending judgment until the proof is clear and indisputable.[33]

Bacon was certainly the chief source of Wilkins' ideas. The

latter quotes him twice in *The Discovery of a New World,* and a number of Wilkins' ideas are so close to those of his predecessor as to leave little doubt of their origin. In a work published the year following *A Discourse* he calls Verulam "our *English Aristotle*" and refers with admiration to "that Work truly stiled the *Advancement of Learning.*"[34] Bacon's influence is beginning to emerge in definite outline. But it is hard to believe that Gilbert did not play some part in shaping Wilkins' spirit. As has been noted before, most discussions of the Copernican theory during this period introduce Gilbert, and Wilkins furnishes no exception to the rule. He lists the former with Galileo and Kepler under the title of the "greatest astronomers" who had abandoned the Ptolemaic system. He reprehends Pineda because he does "with so many bitter and empty Reproaches, revile our Learned Countryman Dr. *Gilbert;* in that renewing of this [Copernican] Opinion, he omitted an Answer to the Scripture Expressions."[35]

So we properly close our study of the Gilbertian tradition with consideration of treatises in which the Baconian spirit is predominant. Though Gilbert remains well known to the important scientists of the seventeenth century, Bacon's mighty shadow was to obscure his less eloquent contemporary. But Gilbert's influence during the first forty years of the century was the most significant factor in moulding the liberal scientific spirit which becomes fully apparent shortly after the middle of the century. His earnest advocacy of experiments and the brilliant examples furnished by his work, his stubborn refusal to bow to authority, his scant reverence for the ancients, and his general scientific attitude could not and did not fail to impress his contemporaries and successors. His relationship to the acceptance of the Copernican theory is equally remarkable. His own interpretation of the causes of the earth's motion has passed into the limbo of forsaken scientific theories, but in his day the convincing success of his truly scientific method and experiments in magnetism endowed his name with such prestige that his espousal of the astronomical system inserted an opening wedge in the closed mind of prejudice and made much easier a favorable consideration of the theory. Kepler and Galileo loom much larger in the history and development of the conception, but it is to be seriously doubted if they had more weight in the creation of a favorable attitude to it in England than the Englishman. They furnished truer and more cogent arguments, but "Dr. Gilbert, our countryman," as the proud expression goes, possessed a more intimate entrée into the English heart.

During this time there are in other branches of science and near-science a few faint signs of the stirring of a new spirit, with which in the latter part of the period Bacon may be associated. In general, Galenical medicine continued to reign supreme, though there is some indication that the Paracelsians were making a slight impression upon their adversaries in the war which had been carried over from the preceding century. The anonymous author of *Philiatros*, 1615, seeks to reconcile the two parties with little success, and his reference to Bacon[36] may explain the moderation of his spirit. His attitude toward the future is hopeful: "is Wit and Art in vs dead, that wee can adde nothing? . . . . And haue we not dayly a taste of that trueth in all kinde of knowledge, by how much the more Nauigation increaseth."[37] Others, however, were more severe with the Paracelsians, and their comments manifest some of the stock arguments not only of Galenical medicine but also of conservatives of all ages. Alexander Read, a fellow of the Royal College of Physicians, contrasts Galenical and chemical principles very much to the disadvantage of the latter because they "are meerly factive, commonly called mechanicall, and so unworthy of a Philosopher."[38] This remarkable hostility to any matter requiring practical and manual activity, an activity which later became the boast of the new science, increases in the conservatives as the century advances. Knowledge, they held, must be a matter of the brain only. Perhaps the most representative expression of the conservative attitude toward the chemists is to be found in the writings of John Cotta,[39] which indirectly reveal the views of the chemical doctors as well. To him Galen is "that peerless Physition and Philosopher," and the Paracelsians appear as follows:

The Empericke is he who reiecteth the disquisition of diseases and remedies, their causes, natures and qualities according to iudgment and vnderstanding, and the carefull perpension and ballancing of his action and practise vnto a iust proportion with reason; but onely informeth himselfe by such things as oft appeare euident & manifest vnto sense and experimentall proofe, carrying his heart and vnderstanding (for the most part) in his hands and eyes, taking nothing sure but what he sees or handles; and from the different maners of experience, are numbred seuerall and diuers kinds of experience.[40]

These same Paracelsians, Cotta exclaims, not finding their much vaunted remedies in the books of "(the famous and learned Grecians) with feare and horror endure their very mention," but the Galenists do "from both draw whatsoeuer may in either seeme good or profitable vnto health or physicke vse: from the Grecian deriuing the sound & ancient truth, & from both Greek, Chymicke, or Arabian, borrowing with thankfull diligence any helpfull good to needfull vse." This apparent open-minded and conciliatory spirit finds further expression in his summing up of the question: "Antiquitie hath giuen vs our first lights in all knowledges, succeeding times haue added their seuerall lustres, and our latest posterity hath yeelded also many things not vnworthy their worthy praise. Chiefly to honour the ancient worthies, yet to contemne none, and to view all, is the rarest growth, but truest perfection."[41] Having observed facts contrary to the ideas of the ancients, he maintains that one should reserve "the power and warranted sufficiency within himselfe" to vary from the superstitious imitation of tradition.

And from this worth and vertue hath it come to passe, that many learned and famous men .... have left so many worthy additaments vnto knowlededg and the common good, by their owne speciall proofes & trials of rules, in their peculiar practise oft different from vulgar conceit, vse and custome; vnto whom may not be denied beyond the ordinary bounds, a libertie and dispensation contained within the latitude of safe discretion and art.[42]

In spite of his willingness to recognize some usefulness in observation and experiment, a response to the changing temper of the period, he still remains more on the side of the ancients than of the moderns. "Chiefly to honour the ancient Worthies" well characterizes his position. Precepts, "the rich compiled treasure of the excellent knowledges of many ages and generations," he holds, are essential to the full perfection of every art. He again defines precepts as "the maximes, axiomes, and ancient golden rules of truth, which many ages and aged obseruation from time to time for common good and ease haue commended, compiled and summed methodically into general orders heads and numbers."[43] Here is heard the heart-cry of conservatism—the findings of civilization, the rich heritage of the past, the tried and proved—a cry which unchangingly greets every new movement in the world. Finally, his logic is deductive rather than inductive. Practice, he says, must follow general rules, but from what source they are to be derived he does not state, though the implication is clear—from the ancients.

In the herbals it is perhaps possible to detect signs of defection from antiquity and allegiance to direct observation and examination. In the long title to his huge folio (it contains 1755 pages), *Theatrum Botanicum*, 1640, John Parkinson boasts that his book holds more botanical information than any other by virtue of "the accesse of many hundreds of new, rare, and strange Plants from all the parts of the world." He frankly confesses that he has been free to correct the slips and errors of the ancients, and in answer to all who object that he is "somewhat too tart and quicke," he employs a quotation closely associated with the growing movement toward intellectual freedom: *Amicus mihi Plato, Amicus Seneca sed magis Amica mihi veritas*. Thomas Johnson's preface, dated 1633, to a corrected edition of John Gerard's herbal is interested in showing how conscious some were of the history of learning and of the state of learning in their own day. After mentioning Theophrastus, Aristotle, Dioscorides, Pliny, Galen, and others, he adds,

From these Antients haue sprung all, or the greatest part of the knowledge, that the middle or later times haue had of Plants; and all the controuersies that of late haue so stuffed the bookes of such as haue writ of this subiect, had their beginning by reason that the carelessnesse of the middle times were such, that they knew little but what they transcribed out of these Antients, neuer endeuouring to acquire any perfect knowledge of the things themselues: so that when as learning (after a long Winter) began to spring vp againe, men began to be somwhat more curious, and by notes and descriptions in these antient Authors they haue laboured to restore this lost knowledge; making inquirie, first whether it were knowne by *Theophrastus, Dioscorides*, or any of the Antients.

After discussing the Arabians who followed Galen's method, he says that in the next period

There were diuers obscure and barbarous writers, who by sight knew little whereof they writ, but tooke out of the Greeks, Arabians, and one another, all that they writ, giuing commonly rude figures, seldome setting down any descriptions .... About some 200 yeare ago learning againe beginning to flourish, diuers begunne to leaue and loath the confused and barbarous writings of the middle times, and to haue recourse to the Antients, from whence together with puritie of language, they might acquire a more certaine knowledge of the things treated of, which were wanting in the other.

Here and there in various works of a scientific nature the new attitude appears, somewhat dim and uncertain but not without significance. In *A Discourse of Naturall Bathes and Minerall Waters*, 1632, first published in 1631, Edward Jorden quotes extensively from the ancients, though he says they wrote little about his subject, as well as from modern writers, but he thinks knowledge lies rather in the present and future than in the past.

But in later times the nature and generation of Minerals (from whence the Bathes procede, and from whence the whole doctrine of them both for their qualities, and differences, originals and vse, must bee deriued) being better looked into, and obseruations taken from such as daily labour in the bowels of the earth, for the search of Mines, or such as afterwards prepare them for our necessarie vses; we haue attained to better knowledge in this kinde, then the Ancients could haue, although in all new discoueries there will be defects for succeeding ages to supply, so it falls out in this: *Dies Diem docet: Alpham Beta corrigit.*[44]

Aware of the inadequacy of past knowledge, confident of the value of his own age, and hopeful of the future, he respectfully asks for himself liberty of thought and expression. Although various writers, he says,

haue added much vnto that which was formerly known in this point, and reformed many errors and mistakings in former writers: yet they haue left many things imperfect, doubtfull, obscure, controuerted, and perhaps false, as may appeare in the discourse following. I doe reuerence all their worths, as from whom I haue learned many things, which else I could hardly haue attained vnto; and I acknowledge them to haue beene excellent instruments for the aduancement of learning: yet I hope it may bee as free for mee without imputation of arrogancie to publish my conceits herein, as it hath beene for them, or may be for any other. . . . My end and studie is the common good, and the bettering of this knowledge: and if I shall bring any further light to increase that, I shall be glad.[45]

During the first forty years of the seventeenth century progress, though slow, was made toward the emancipation of science from the domination of the past. Of course, the events of greatest scientific interest in England were the publication of Bacon's various works and of Harvey's discovery of the circulation of the blood, yet during this time the former's influence was rather sporadic, and the latter's discovery seems not to have elicited much response in England until the Puritan régime. Gilbert stands out much more prominently than either, though the slow awakening of the modern spirit was by no means due to him alone. A weakening of confidence in the ancients, a growing appreciation of the importance of experiment and direct observation, and the spirit of independent thought, so characteristic of Gilbert, experience a very perceptible, though gradual, growth during these years. Perhaps if any feature of this movement characterizes the age more than another, it is the revolt from authority and the plea for intellectual freedom, best revealed in Carpenter and Watts, though in most of the treatises mentioned above the note is clearly sounded. This attitude rightly received the greatest emphasis, because liberty of mind was an essential postulate for the advancement of science. Finally, there was more than a casual relationship between the arraignment of religious authority in the Reformation and the revolt from Aristotle's authority in science. The scientific movement, in England at any rate, is definitely connected with religious developments. If Protestantism facilitated the growth of science through its anti-authoritarian bias, the extreme elements among the Protestants, namely, the Puritans, as we shall see in the next chapter, continued to serve the cause by embracing the new science, largely because of its utilitarian value for the "public good."

THE PURITAN ERA

CHAPTER V

THE ADVANCEMENT OF LEARNING AND PIETY

> . . . Universall endevours and the thoughts of a publike Good free
> from partiality; have been lookt upon by many; and those also in
> publike places rather as ridiculous whimsies and projects, then
> matters worth any encouragement; but seeing we are now about a
> Reformation of our Ways, and necessitated to think upon the
> publike Good, even for our preservation from utter ruine and
> confusion; and seeing some such thing as the Advancement of
> Learning hath been oftner, and in a more publike Way, at least
> mentioned in this Nation of late, then in former times, partly by
> the publication of those excellent Works of the Lord *Verulam*
> which at[tempt] such a matter . . . . seeing these things are so,
> and chiefly seeing there is a Reserve of Means purposely kept to be
> employ'd for the Advancement of Religion and Learning, and we
> are inform'd that an Ordinance is drawing up for the effectuall
> disposall thereof towards those ends . . . . we ought not to despair
> of some good issue at last.
>
> John Durie, *A Seasonable Discourse,* 1649.

As we have noticed in the preceding chapter, the influence of
Bacon during the first forty years of the seventeenth century
was much less in evidence than we should suspect from the prestige which he later attained. Toward the end of the period unmistakable signs of his spirit do appear, and earlier his ideas
here and there receive recognition, but it was not until Puritanism seized the reins of government that he rapidly became an
inspiring and dominating force in thought.[1] There were a number of elements in Bacon's philosophy congenial to Puritanism,
which will be pointed out later, but there was one characteristic
common to him and the Puritans which should be mentioned
here, for it is constantly manifested during this period. Bacon
was *par excellence* a reformer. Not only is this fact apparent in
the reformation which he introduced into philosophy and the intellectual outlook of men, but it is also emphasized by the zeal
with which he undeviatingly pursued his purpose, and by the ardent manner in which he sought to drive his ideas home. Needless to say, the most characteristic aspect of the age under consideration was the reforming spirit. The Puritans were out to

reform not only Church and State in their narrow connotations, but almost everything else.[2] While their hands were busy in overturning the old order of things, their brains were equally occupied with new schemes for the present and visions for the future. Bacon may have played some part in the development of this spirit. More likely, however, his influence had to wait until in the general upheaval a more favorable attitude toward his own progressive ideas was established, and a spirit congenial to his own was formed.[3] He certainly did influence the specific direction that some reforms took, and the thesis which I would suggest, even if I cannot convincingly demonstrate it, is that our modern scientific utilitarianism is the offspring of Bacon begot upon Puritanism.

Though the character of the age was predominantly religious, two spirits, both springing largely from Bacon and the new science, and closely related, soon became conspicuous. One may be described as public-spirited and humanitarian; the other as materialistic, utilitarian, scientific. The latter inspired projects of all kinds and especially efforts at educational reform.[4] The first is best revealed in the motives ascribed to, and the goals proposed for, the various schemes advocated. The "public good" becomes practically a slogan in the literature which we shall examine; the relief of man's needs is constantly stressed. These needs are interpreted in a materialistic way; the means of achieving this material benefit are either scientific or born of the scientific spirit. Men were becoming social-minded. Though there are various references to the Commonwealth, the nationalistic spirit is fading, and what we now call society is apparently becoming a motivating force in thought and action. Nowhere is this social and utilitarian spirit better seen than in the many proposals of the day for educational reform.[5]

As early as 1642 John Durie, who spent much of his time in the countries of northern Europe working for the unification of all evangelical churches, published, through the agency of Samuel Hartlib, a work[6] inspired to a certain extent by Comenius, which reveals discontent with traditional education. In discussing the promotion of secular learning, he advocates that plans be laid

for the perpetuall encrease and advancement of Sciences, which should contaíne a full direction concerning the wayes of ordering higher Schools, Colledges, and Universities, and of regulating the exercises, which ought to be

set a foot therein; That men who have attained unto the former degree of
perfection, may improve their talents with advantage one towards another,
and towards the publicke, without vain-glory, and strife, and superfluous
repetitions of matters already discovered; where the manner of writing
books and Treatises, either for the ripping up of the hidden secrets of
nature, or for the examining and rectifying or compleating of the writings
of other men is also to be explained, when first the faults and defects of
Colledges, and Universities are layd open, and the abuse of writing books
is made known, with the wayes of taking a speedy course of reformation in
one and the other.

The defects which he has in mind, as will be seen later, have to do
with the traditional curriculum of the schools, and in the reformation of divine learning proposed by him he excludes particularly disputatious, "bookish" divinity.

The general ideas of Durie find much more definite expression
a few years afterwards in a work by William Petty,[7] later Sir
William Petty, who, besides Wilkins and Boyle, was as much responsible for the promotion of the new science as any of his
contemporaries, and who, after having been appointed by Cromwell professor of anatomy at Oxford, did further service for the
Commonwealth by surveying Ireland. Petty prefaces his proposal with a declaration that it is first necessary to devise some
means whereby scholars can keep in touch with each other, so as
to avoid repetition and gain assistance, and thus by united action
produce tangible results—Bacon's idea of co-operation. "But
this we pass over slightly, though very fundamental to our Business, because the Master-builder [Bacon] . . . . hath done it so
solidly." First, he claims, it is necessary to explode "whatsoever
is nice, contentious, and merely fantastical; all which must in some
Measure be suppressed, and brought into Disgrace and Contempt
with all Men." (This for the next quarter of a century the scientists tried valiantly to do, and partly succeeded.) Second, it is
necessary to sift out of all books containing mechanical inventions "the Real or Experimental Learning." Finally, the men
employed in the various faculties must be given sufficient support
and encouragement, and rewarded according to their achievements.[8]

Of special interest in Petty's scheme are his "Ergastula Liter-
aria" or literary workhouses, in which children "may be taught
as well to do something towards their Living, as to read and
write." The pedagogical method to be employed consists in teach-

ing the pupils to observe and remember all sensible objects and actions, natural or artificial, for children do not need reading before they are acquainted with the things of which they read. The learning of foreign languages is of little importance when there are sufficient books in English for their purpose. Arithmetic and geometry are to be studied because of their usefulness and conduciveness to mental discipline. Even the children of the highest rank should be taught some "genteel Manufacture" such as botany, gardening, chemistry, anatomy, because in this way they will be in a position to make "luciferous experiments" and become patrons of science. Finally, Petty decrees that "no ignoble, unnecessary, or condemned Part of Learning be taught in those Houses of Education; so that, if any Man shall vainly fall upon them, he himself only may be blamed." His next proposal advocates a "Gymnasium Mechanicum," or vocational school, "for the Advancement of all Mechanical Arts and Manufactures." He believed that from contact with all trades and arts new inventions would arise; a perfect and exact history of trades could be written; and "what Experiments and Stuff would all those Shops and Operations afford to active and philosophical Heads, out of which, to extract that Interpretation of Nature, whereof there is so little, and that so bad, as yet extant in the World?" Within the walls of the gymnasium there would be a "Theatrum Botanicum," cages for strange beasts and birds, ponds for exotic fish, repositories for antiquities, a library of select books, agricultural plots, paintings, statues, globes, maps, models of engines, and the like, "So that a Man, conversant within those Walls, would certainly prove a greater Scholar, than the walking Libraries so called, although he could neither write nor read." (Things have achieved ascendancy over words; observation has supplanted the classics.) He also proposes a fully equipped hospital, largely for the purpose of teaching, observing, and experimenting, with a garden, library, chemical laboratory, anatomical theatre, and apothecary shop. He even goes into details regarding personnel and salaries. Among other officers there would be a physician, "skilled at large in the *Phænomena* of Nature," who "shall either dissect, or overlook the Dissection of Bodies dying of Diseases; and, lastly shall take Care that all luciferous Experiments whatsoever may be carefully brought to him, and recorded for the Benefit of Posterity." There would also be a vice-physician to keep the history of the patients' cases, a surgeon to carry on vivisection, and an apothecary to look after the garden. In general, more emphasis is laid upon "luciferous experiments," upon investigation, and upon recording results than upon the immediate interests of the patients. It was the first proposed research hospital.[9]

In nothing which Petty advocated is the practical, materialistic spirit more manifest than in a scheme described under the title *Vellus Aureum sive Facultatum Lucriferarum Descriptio magna,* "wherein all the practised Ways of getting a Subsistence, and whereby Men raise their Fortunes, may be at large declared." Toward this end he recommends a history of arts and manufactures in which "should be described the whole Process of Manual Operations and Applications of one natural Thing (which we call the elements of Artificials) to another," with pictures of instruments and machines. The chief purpose of such a work would be to show "how to apply all Materials that grow in Abundance in this Kingdom, and whereof but inconsiderable Use and Profits are as yet made to more Advantage to the Commonwealth." Finally, there should be a preface "to teach Men how to make the most of Experiments, and to record the Successes of them whatsoever, whether according to Hopes or no, all being equally luciferous, although not equally lucriferous." Petty represents more clearly than anyone else the passion of the age, largely indebted to Bacon, for the concrete, the material, the useful, and the practical. He subscribed to the experimental and observational method of discovering truth, and felt the keenest antipathy to what he considered the linguistic science of the past. That distrust of language and hatred of words, a unique characteristic of early modern science,[10] appears in almost everything he wrote. The history of trades, he claims, would bring it about that

Boys, instead of Reading hard *Hebrew* Words in the Bible (where they either trample on, or play with Mysteries) or Parrot-like repeating heteroclitous Nouns and Verbs, might read and hear the History and Faculties expounded . . . . As it would be more profitable to Boys to spend ten to twelve Years in the Study of Things, and of this Book of Faculties, than in a Rabble of Words; so it would be more easy and pleasant to them, as more suitable to the natural Propensions we observe in them.

Furthermore, such a history would be of assistance to mature men.

Scholars, and such as love to ratiocinate, will have more and better Matter to exercise their Wits upon, whereas they now puzzle and tire themselves, about mere Words and chimerical Notions. They will reason with more Alacrity, when they shall not only get Honour by shewing their Abilities, but profit likewise by the Invention of fructiferous Arts. Sophistry shall not be in such esteem as heretofore, when ever Sense shall be able to unmask its Vanity, and distinguish it from Truth.[11]

In 1649 Hartlib published a second book[12] by Durie, which makes clear the reforming spirit of the times and its relation to Bacon. Profiting by the latter's sharp distinction between religion and science, a distinction, incidentally, which made it possible for the Puritans to embrace his philosophy, Durie maintains that since the application of natural principles to spiritual objects leads to endless confusion, the two worlds should be kept entirely separate, spiritual truths being discerned spiritually, and natural truths naturally. Omitting many of his ideas, including his strong plea for a general educational reformation, we come to the arts and sciences, which, he insists, "must be reformed in the ends and uses, in the rules and precepts, and in the method and order thereof." The ends are two: godliness and "the common relief of mankind from outward miseries," man's bodily and material welfare. Any other end would turn the mind from a profitable object to a vain delight. In regard to rules, methods and the like, he holds that only useful and profitable things are to be taught, and that everything is to be presented practically as well as theoretically. Utilitarianism runs rampant in his mind. He would extend the bounds of learning far beyond the university, and so widen its curriculum that every mechanical art and science would have its professor or lecturer. The chief means he advocates for this extension of learning is an agency for "the Common Good of all Mankind," the purpose of which would be to make good any defect discovered in education by remedies found effective elsewhere or by investigating and searching out the proper means; to disseminate all discoveries made at home or abroad that can in any way result in a "publike Good"; and "to animate the Professors of all Arts and Sciences, by a Correspondency with all that are of any note, to waken them by one another unto all Industry, and to gather up the fruits thereof, to be apply'd unto all the Schools for the Advancement of Learning therein." (The co-operative idea of Bacon is again evidenced here.) In other words, the agent was to act as a clearing house for all new ideas and scientific discoveries, enabling men to avoid repetition, to profit by the discoveries, and to apply the latter "unto the publike uses whereunto they may be most serviceable for the Advancement of Learning." Durie feels certain that if all capable men were influenced "to contribute and suggest what they have gain'd by Meditation, Observation, and Experience to be usefull towards the Advancement thereof," such learning would be established as "will make this Nation not only happy within it self in one Age or two, but a Blessing unto others, and a Praise unto Posterity."[13]

About 1650 Hartlib published a third work by Durie entitled *The Reformed School,* in which exaggerated insistence upon material and actual things and undisguised hatred for the nebulous conceptions and empty notions of the traditional philosophy are again manifested.[14] The author urges against the prevailing mode of education what had now become a widely accepted charge, that children are taught words before they know things,

and when they are taught these things wherin Reason is to be employed, they are lead into a Maze of subtile and unprofitable Notions; wherby their mindes are puft up with a windy conceit of knowledge: their affections taken off from the plainnesse of Usefull Truths; their naturall Corrupt inclinations to pride, vainglory, and contentiousnesse not reformed, but rather strengthned in perversitie; So that they become both unwilling to seek, and incapable to receive any Truth either Divine or Humane in its simplicitie: for their heads are filled with certain termes and empty shewes of learning; which neither containe any substance or solidity of Matter; or give them any addresse by way of Method to make use of that which they know for the benefit of Mankind.[15]

A characteristically puritanic moral element is here injected into the attack on disputatious Peripateticism, and vanity takes its place beside inutility and falsity. Certainly the advancement of both piety and learning necessitated the abandonment of Peripatetic principles. The emphasis here revealed upon the plain, simple, and useful, an attitude which had much to do with the simplification of English prose style, finds frequent expression later in the period. The transforming of the goal of education into a vocational materialistic end, the bread-and-butter ideal of learning conspicuous in our midst today, the disregard of the whole world of man and human personality as a proper object of study, in short, the Philistine spirit, are here definitely taking their rise. "The true End of all Humane Learning is to supply in our selves and others the defects, which proceed from our Ignorance of the nature and use of the Creatures, and the disorderliness of our naturall faculties in using them and reflecting upon them." He repeats the idea that the purpose of knowledge is "to make use of the Creatures for that wherunto God hath made them"; otherwise, it becomes useless and hurtful. This bias leads him directly, as it has led many so-called educational experts of today, into an attack on linguistic study. Languages are of value only as far as they transmit the "Reall Truths in Science," and what might be termed the literary aspects of language-study, "which speak nothing of Reality in Sciences, are to be left to such as delight in vanityes more then in Truths." Stress is laid upon the study of the mother tongue, and the value of foreign languages is strictly confined to utilitarian limits, "as far as their experience in the observations of Things doth go, and no further."[16]

When he becomes more explicit regarding what subjects should be taught children, we are not surprised to find listed the "practical parts" of mathematics, geographical descriptions of the world, astronomical descriptions of the heavens in models, globes, and plain tables, the "experimentall way of Measuring Land," husbandry, gardening, fishing, fowling, and anatomy, as well as history and comparative grammar. We are surprised to find him elsewhere advocating the study of oratory and poetry, but such inconsistency seems only a concession to tradition. The same propensity for the concrete is carried over into his attitude toward logic, in which a pupil should be taught "not *in abstracto* and Notionally, but, by an Example of every kind, and *in concreto*," thus being led to an "experimentall manuduction of his thoughts." Finally he launches a vigorous attack upon syllogistic logic and its evils in contrast with the virtues of experimental induction:

the way of knowing Actually must be experimentally proposed, before the Theoreticall way of apprehending matters be reflexively taught; which is quite opposite to the Practice of those that teach Logick in an ordinary way: who deliver all even to Children Reflexively and Scientifically: before they exercise them in any point of Practice, except in that of a confused, endlesse, unprofitable way of disputing about Notions which is falsly called Logick, by which means those that should be made Scholars of Right Reason, are made habitually wranglers about the Terms of an Art, which they never have been taught: and in stead of making use of their Rationall Facultie, to set it in a way to order their Imaginations aright, they are onely directed and exercised to subtilize their Imaginations, and pride themselves in this mainly to have such conceptions, as are beyond the vulgar Capacitie.

The accurate observation of things, and not the confused ideas gathered from the traditional misapprehension of nature, in his eyes constituted the basis of a sound imagination.[17] In a "Supplement to the Reformed-School" published the same year in *The Reformed Librarie-Keeper,* Durie continues his attack on traditional education, criticizing the universities for their monastic character, their detachment from the world, and their indifference to the public good, together with suggestions for their reformation. He insists that only those should be appointed professors who are public-spirited and who can make a worthwhile contribution to "useful knowledge" or can direct their pupils to such a noble end.[18] He denies the charge that he is seeking to undermine the universities, stating that reformation is his only purpose, and declaring that only those would accuse him of such a design who "can see no further then what wee now ordinarily attein unto; and withal think that there is no *Plus ultra*[19] in nature atteinable above that which they have conceived."

Durie is saturated with the practical and useful idea of education. His utilitarianism is the outgrowth of a passionate love for the public good, which, however, he interprets only in a religious or materialistic way. Realizing that the Peripatetic philosophy contained no utility, and that those who espoused it did so largely for personal reasons rather than from considerations of society, he urges a reformation which would substitute for that wrangling disputatious philosophy useful and real knowledge, knowledge gained by observation and experiment. Imbued with faith in the possibility of progress in learning, he stoutly attacks old attitudes and ideas, and urges the adoption of ends and methods that will remove the obstacles to the future development of knowledge. All this, of course, is reminiscent of Bacon, to the publication of whose works Durie expressly attributes the reforming zeal and the efforts to promote learning in evidence at this time. The dissatisfaction with the old learning, the urging of useful, real knowledge, the emphasis upon things rather than words, the consideration of the public good, and the exaggerated value placed upon utility find a likely explanation in the writings of the great Chancellor.

In the next work[20] to be considered we find an early example of Descartes's influence, though the Baconian spirit is predominant. The author, John Hall, poet and essayist, proposes to show the methods and the ends of the reformation which he advocates. He stresses the importance of the effective advancement of learning in preserving the Commonwealth, and asks,

What better way to your profit, then to command abundance of fruitfull wits, which shall every day bud forth with some invention, serviceable either to the necessities of the poore, or graver magnificence of the rich? when mechanicall knowledge shall be multiplied and abbreviated, and you be able not onely to requite forreigne parts for the curiosities they have lent you; but also invite them hither to be your schollers, when there shall be a confluence of the finest industries among you, and he shall be accounted to want of due civill Accomplishments, that hath not come to perfect them from this place.[21]

Here is a vision of industrialism promising riches to the "godly men," and touched with a humanitarian justification and a materialistic conception of life which almost sanctifies the machine. Mechanical knowledge is made the *summum bonum* of learning. God expects, he says, the Puritans to be the "Foster-fathers of knowledge," who can cause the endowments of the universities to "be converted into uses suitable to the ends of the Donors, and tend rather to a publicke advantage, then to the private fostering of many idle Pedantick Brotherhoods." The learning of the universities was well suited to the times that produced them, but now that the sun has scattered the darkness of those days, and "humane learning hath also been more enlightned . . . . he is not acquainted with the Business of knowledge, that knows not what sensible increases (I had almost said perfections) it hath of late arrived to." So means must be taken to propagate knowledge and for "severe and eager disquisitions of new truths; for knowledge hath no limits nor Land-marks but being ubiquitary, and therefore desirous to diffuse it selfe, she endeavours by all means her promotion and dilatation." A limitless future is opening to the intellectual eyes of men. Hall is keenly aware of the stirring of the new spirit. To him the state "like a wakened Gyant begins

to rowze it selfe up," and God "hath diffused a great and a rest-
lesse Genius in this age, far greater then any hath been of a long
time." It was an age of ferment, when those attitudes and
values which so dominate our world today were assuming form
and outline.[22]

He first attacks the universities because of the linguistic training given the pupils, who are afterwards

delivered over to a jejune barren Peripatetick Philosophy, suited onely (as
Mounsieur Des-Cartes sayes) to wits that are seated below Mediocrity,
which will furnish them with those rare imaginations of *Materia prima*,
Privation, *Universalia*, and such Trumpery, which they understand no more
then their Tutors, and can no more make use of in the affaires of life, then
if 3000 yeares since they had run through all the Hierogliphicall learning of
the *Egyptians*, and had since that time slept in their *Mummy*, and were
now awakened . . . . Then are they turned loose, and with their paper-
barks committed to the Great Ocean of Learning; where if they be not
torne, they returne backe so full of desperation and contempt of their profession, and sad remembrance of their youth so trivially spent, that they
hate all towardly engagements that way, and suffer themselves either to
sinke in a quagmire of idlenesse, or to be snatched away in a whirlpool of
vice. But in case some with much adoe get a shore (for a long or a
far voyage upon these termes they cannot make) and by the foresaid means
stilt themselves into some profession; what deplorable things (unlesse it be
those few which Nature makes for ostentation to be jewells in this earth)
prove they, in filling the world with detestable quacking Empirics, lewd, and
contentious, Gown-men, or ignorant mercenary Divines?[23]

He next proceeds to point out the deficiencies of the universities:
lack of competent men; inability to attract learned men from
abroad; and failure to introduce the experimental method, the
crux of the matter and an inheritance from Bacon.

Where have we any thing to do with Chimistry, which hath snatcht the
keyes of Nature from the other sects of Philosophy, by her multiplied experiences? Where have we constant reading upon either quick or dead
Anatomies, or ocular demonstration of herbes? Where any manuall demonstrations of Mathematicall Theorems or Instruments? Where a promotion of
their experiences [Hall frequently uses this term for experiments], which
if right carried on, would multiply even to astonishment? Where an examination of all the old *Tenets?* Review of the old Experiments and traditions
which gull so many *junior* beliefs, and serve for nothing else but for idle
Priests, to make their Sermons more gaudy?[24]

Among the remedies he proposes are that professorships be increased in number, fellowships be given to competent men, and
the remaining revenues "be changeable and accountable every

yeare: to be disposed of, for examining and pursuing experiments, encouragements of honour, compleating and actuating
some new inventions, supplying the needy ones that really
wanted these wings to take great flights." He further recommends that the licensing of books be abolished, that confiscated
royalist property be devoted to learning, and that learned for-
eigners be honored. He emphasizes, as was customary at this
time, the teaching of pupils by things rather than by words, the
only way to arrive at "solid Ideas and sound representation of
things." Stupid it is, he exclaims, "to rest with the traditionall
wisdome of our Ancestours, and not to looke after further enlargement and advancement." He also takes note, following
Bacon, of those mental limitations which hinder the gaining of
truth, and of the necessity for suspended judgment and a critical
attitude of mind. To find truth, he says, "I could not devise
any better meanes then to make the minde pliant and passible to
any Truth, to free her from all these inquinated prejudices of
education, tradition, or childish observation, and then withall to
plant such a doubtfulnesse in her, as should not easily ascent to
any one thing which was not fortified with strong reason and
right experience."[25]

Somewhat in the manner of Descartes he gives an account of
his efforts to gain knowledge in natural history. In traditional
science, he says, he found nothing but endless controversies and
"vaine speculations, which in all reason of the world I ought to
dis-beleeve." In those who waged sharp warfare against the old
philosophy and claimed to have discovered the real truths of
nature, he found no performance equal to his expectations, for
they proved much better in discovering errors than in finding
truths, and forced nature to their own conceptions instead of
enlarging their conceptions to suit nature. In others who abused
"a Philosophicall liberty" he found nothing but fantastic theories
or probable and neat hypotheses which were foreign to the true
nature of things. From these experiences he drew the conclusion
that, though some headway had been made, much remained to be
done, and that the means to be employed was experimentation.
In a manner reminiscent of Bacon he launches into a discussion
of the improper way to make use of experiments. Since some
men built too much on one small observation and found too many
theories for one and the same experiment, he concluded that "it

was not safe to draw any one principle from any one observation, unlesse there could not possibly be any other reason given for it."[26] Finally, he assures Parliament that if they effect his proposed reformation in education,

you shall see the Taper of a learned Piety burne among us, I hope, like an immortall lampe, fed with refined and sublimest knowledge, whilst all those false lights of ignorance, humane forgery, and superstition shall vanish away. . . . You shall see Nature traced through all her Turnings, to a cleare demonstration of her first cause, and every day bring forth varieties of experiments, either to the reliefe, astonishment, or delight of men; you shall then see us freed from all these fabulous illusions and impostures, which have hitherto beset either Traditions or Cures; and Nature which now disguises her selfe into so many shapes, forced into an open veracity and pure nakednesse. You shall see the number of Arts daily increased, and those we knowe already, wonderfully promoted . . . . You shall have the ways of Education made smooth, and your children with a pleasant successe possessed of all the Treasures of reall knowledge, ere they could have thought they had entred the gates.[27]

Up to this point much of the discussion of educational reform has been rather general in nature. That the chief inspiration for it came from Bacon, either directly or through Comenius, is obvious from the constant and varied use of the expression "the advancement of learning" or its equivalent, and from the all-pervading spirit contained in it. The materialistic, utilitarian, and humanitarian spirit, rising like a genie from the tomb of the great Chancellor, cast its shadow across all these treatises. The learning to be discarded is the inutile study of the classics and the empty disputatious philosophy drawn from them. The proper objects of study are things, not words, nature, not books, and the subjects which deal with nature directly and which possess "useful" value are alone recommended. The observational and experimental method of ascertaining truth is constantly implied and frequently expressed. Yet much of the discussion is abstract, displaying not so much definite schemes and plans of reformation as the desire for reform directed toward certain goals. In the next two works to be considered, however, the Puritan eagerness to transform education is crystallized into a definite proposal to introduce the Baconian experimental science into the curriculum of the universities to supplant the divinity, metaphysics, and Peripatetic philosophy taught there.

The first is a treatise[28] by a certain Noah Biggs, concerning whom I have discovered no more information than the title page affords. The text reveals him as a man of strong Puritan sympathies,[29] a fact which explains his hopeful dedication to the reforming zeal of Parliament. He speaks of the latter's seeking "to wipe off the imputation of intending to discourage the progress of true Learning" (the qualifying adjective "true" really means experimental), and he urges them to reform not only the church but also learning, so that they can refute those who call them the Goths, Vandals, and destroyers of all learning and literature. The nature of the reformation which he advocates is manifested in a series of rapid-fire questions, asked in Hall's manner:

. . . . wherein do they [universities] contribute to the promotion or discovery of Truth? . . . . Where have we any thing to do with Mechanickal *Chymistrie* the handmaid of Nature, that hath outstript the other Sects of Philosophy, by her multiplied real experiences? Where is there an examination and consecution of Experiments? encouragements to a new world of Knowledge, promoting, compleating, and actuating some new Inventions? Where have we constant reading upon either quick or dead *Anatomies*, or an ocular demonstration of *Herbs?* Where a Review of the old Experiments and Traditions, and casting out the rubbish that has pestered the Temple of Knowledge?[30]

To slavish submission to the ancients he attributes the stagnation in the schools; there are no epithets too strong for him to apply to Aristotle, Galen, and other classical authorities. He thinks the remedy lies only in

a thorough and early plowing up the fallow ground of the universities, that she [*sic*] may be laboriously rummig'd in her stupendous bulk of blinde learning, and her rubbish cast out, and no longer be a *Quagmire* of pittiful learned idlenes, to serve for no nobler end then to nurture a few raw striplings, come out of some miserable countrie school, with a few shreds of *Latine*, and to maintain the frothy lectures and mutterings over a few stolne impertinencies & wrackt disputations or industrious scoldings and bawlings of a few yongster *Pedanticks*.[31]

For such educational rubbish he would substitute *experiments,* upon which he places as great an emphasis as Sprat and others did later. Pleading fervently for "freedom of opinion experimently enlarged," he urged Parliament to establish "an Academy of Philosophical freedom." Running through these Puritan attempts to establish the new science on a firm scholastic basis is the constant cry for freedom of opinion, a cry which is expressed in Milton's *Areopagitica.* We should not leave Biggs without commenting on his remarkable style, which is energetic, impassioned, vivid, and concrete, though marred at times by strange and pedantic words, bombast, and involved and incorrect sentences.

The author of the second treatise,[32] John Webster, at one time a Cambridge scholar and later a chaplain in the parliamentary army, proposed, according to Wood, "divers expedients (as he is pleased to stile them) for the reforming of Schools, and for the perfecting and promoting of all kind of Science, yet he was very well known to be one who endeavoured to knock down Learning and the Ministry both together."[33] Webster's book is not very original. He drew on a number of the new thinkers for his material and arguments, but always intelligently and to the point. Conscious that the charge of having secured his ideas from others would be brought against him, he makes no effort to deny it, but glories in being one of a band which, though small, "waged contention with their time's decay."[34] In the "Epistle to the Reader" he says,

And if the things therein contained be hinted at and taught by others, then I only am not *Paradoxical* but they also, and I have produced their testimony, that the world may see how many valiant champions have stood up to maintain truth against the impetuous torrent of antiquity, authority and universality of opinion; and though they be not so numerous yet are they no babes, but strong men, who fight not with the plumbeous weapons of notions, *Syllogism*, and putation, but with the steely instruments of demonstration, observation, and experimental induction, so that I hope I shall not be accused of novelty and singularity, seeing I have so many noble *Heroes* to bear me company.

Undoubting faith in a new belief and ardent allegiance to the great leaders in that belief, which read in their valiant stand against the multitude courage worthy of the heroes of old—and this is not the only example of the epithet "hero" being accorded the leaders in the movement—these qualities, which are perhaps the brightest ornaments of the progressive spirit, are his. To oppose the old and to support unequivocally the new, even at this late date, required courage, and he stoutly defends himself against the charge of audacity and insolence in examining and opposing "that learning, which hath been received and approved for so many years, assented unto and extolled by so many great wits and profound judgements, and defended, and patronized by all the *Academies* in the Universe."[35]

For the most part Webster insists on keeping learning and religion separate. He maintains that divinity cannot and should not be taught in the schools, for religious truths come only through the spirit of God. "Humane knowledge" is excellent in itself, but when it is incorporated with theology, both are corrupted, as in scholastic divinity.[36] Yet the only "humane" learning he advocates has but the narrow purpose of preparing men for a "useful and profitable undertaking in the Commonwealth." Thus Webster is bitterly opposed to the teaching of any subject in its merely speculative aspects. He outdoes his master, Bacon, in this respect. "Is there no further end," he asks,

nor consideration in *Physicks* but onely to search, discuss, understand, and dispute of a natural movable body . . . . Surely natural *Philosophy* hath a more noble, sublime, and ultimate end, than to rest in speculation, abstractive notions, mental operations, and verball disputes: for as it should lead us to know and understand the causes, properties, operations and affections of nature; so not onely to rest there and proceed no further . . . . And secondly, not onely to know natures power in the causes and effects, but further to make use of them for the general good and benefit of mankind, especially for the conservation and restauration of the health of man, and of those creatures that are usefull for him.

He asks again, "Can the *Mathematical* Sciences, the most noble, useful, and of the greatest certitude of all the rest, serve for no more profitable end, than speculatively and abstractively to be considered of?"[37] Does not mathematics exist to enable men to build houses, practice navigation, make war, and in general assist "all mechanick operations"? No pure science for Webster. This crass materialism led Webster, as it led Durie and has led innumerable others since, to depreciate language and its study. To him linguistic knowledge is not real knowledge. Like logic and mathematics it is only an instrument to be used in the acquisition of learning. "The intellect," he says, "is no way inriched" by a mere knowledge of words. At times we think we are reading a treatise by one of our own progressive educationists, as, for instance, when he opposes the teaching of formal grammar on the ground that it can be learned much better from reading and conversation. Of course, Webster has in mind the classical languages, the time spent in learning which he sincerely regretted. The fact that they were the repositories of the learning which he was attacking only deepened that regret. He longed for a universal language which would remove the necessity for Latin.

He proceeds to more specific defects in the curriculum of the universities, the first being logic, upon the difficulty, obscurity, and intricacy of which he enlarges. Aristotle's *Organon* he accuses of being confused and of using terms that should be defined. Since syllogisms depend upon what is already known, they can lead to no new discoveries, but only to "conclusions [which] beget but bare opinations, and putations, no infallible Science. . . . And yet men are puft up with this vaporous, and airy sound of words, growing insolent and confident in the vain glory of *Syllogizing Sophistry*."[38] He elaborates upon the way logic is taught in the universities, piling up in characteristic fashion his opprobrious terms:

As it [logic] is now used in the *Schools* it is meerly *bellum intestinum Logicum*, a civil war of words, a verbal contest, a combat of cunning craftiness, violence and altercation, wherein all verbal force, by impudence, insolence, opposition, contradiction, derision, diversion, trifling, jeering, humming, hissing, brawling, quarreling, scolding, scandalizing, and the like, are equally allowed of, and accounted just, and no regard had to the truth, so that by any means . . . . they may get the Conquest, and worst their adversary, and if they can intangle or catch one another in the Spider Webs of *Sophistical* or fallacious argumentations, then their rejoicing and clamour is as great as if they had obtained some signal Victory. And indeed it is the counsel of the *Arch-Sophister* their Master, to speak ambiguously while they dispute, to obfuscate the light with darkness, lest the truth should shine forth, nay rather to spatter and blurt out anything that comes into the budget, rather than yield to our adversary . . . . O excellent and egregious advice of so profound and much-magnified a *Philosopher!* Is this to be a lover of verity, or indeed to play the immodest Sophister and Caviller.[39]

In the light of this description, we can hardly disagree when he says that such altercations hinder the investigation of truth, the discovery of which requires "more powerful means, than verbal and formal Syllogisms."

And whereas the best part of *Logick* for that purpose is *Induction*, which backt with long experience and sound observation, might be prevalent to discover the working of mother Nature; yet that hath been altogether laid aside, while the glory of *Syllogisms* hath been highly predicated.

After quoting the thirteenth aphorism of the *Novum Organum*, which treats of the inefficacy of syllogisms, he adds, "For whereas we should from particulars proceed to generals, this preposterously laies down universal axiomes without due proof of them, thereby to make good particulars."[40] The two quotations just given contain the substance of all the charges which later were to be preferred against the prevailing mode of reasoning, and when we come to examine the ideas of Sprat, Glanvill, and others, they will seem only like echoes of this Puritan chaplain. Needless to say, this idea of logic was by no means original with Webster. It had appeared both in England and on the Continent, and in no one so conspicuously as in Bacon, from whom Webster drew his chief inspiration and the greater part of his ideas.

The fifth chapter is devoted to mathematics, which he highly extols as excelling all sciences in perspicuity, certitude, and, more important, uses and benefits, but which, he claims, was slighted or superficially handled in the universities. Arithmetic and geometry are neglected, and no progress is made in optics. In astronomy the "rotten, ruinous, absurd and deformed" Ptolemaic system is taught, upon which Webster launches a terrific and rather delightful attack. Cosmography, geography, hydrography, chorography, and topography fare as badly as mathematics, and, save the mark! astrology, which he praises in no insincere manner, to the infinite amusement of Seth Ward, as we shall soon see. Finally, those technical subjects which depend upon mathematics, such as architecture, and which in his eyes are of more use, benefit, and profit than anything studied in the universities, are totally ignored. Utility remains the supreme standard for judging the excellence of educational subjects.

The next chapter, an important one entitled "Of Scholastick Philosophy," will be discussed later, but we should remark here that he treats the matter by constantly contrasting modern discoveries with ancient errors, and the fruitfulness of the inductive experimental method with the barrenness of futile, empty, verbal Peripateticism. The seventh chapter, dealing with metaphysics, ethics, politics, economics, poetry, and oratory, is short and evidently lacking in interest to Webster. The metaphysics of the schools is full of "fruitless, and vain Chimæras" useless and unprofitable to man, mere vain opinion "void of scientifical demonstration," in contrast to which he praises highly the *mens cogitans* of Descartes. Aristotelian ethics contains nothing practical. Much of Aristotle's and Plato's economic thought needs purging. In political science Bodin and Machiavelli,

and divers other modern authors may duly challenge as much praise in this point, as that of *Aristotle*, which the *Schools* do so much adhere to and magnifie, yea even our own Countreyman master *Hobbs* hath pieces of more exquisiteness, and profundity in that subject, than ever the *Græcian* wit was able to reach unto, or attain; so that there is no reason why he should be so applauded, and universally received, while more able pieces are rejected, and past by.[41]

Throughout the whole volume runs this tendency to match ancients with moderns, to depreciate ancient learning and to praise the achievements of his own age. In Webster is revealed the same spirit which had animated Hakewill and was to animate Glanvill and Wotton in upholding modernity in its controversy with antiquity.

In the chapter dealing with "Customes and Methode" Webster criticizes certain definite faults: the slothfulness and idleness of students; the requiring of a definite length of residence for a degree; the quarrelsome disputations, and the fact that students "never go out by industrious searches and observant experiments, to find out the mysteries contained in nature"; the use of Latin instead of English in scholastic exercises; subserviency to Aristotle and slavish adherence to antiquity in general; and the lack of an independent spirit in thought and of freedom in pursuing truth. His view of the English language is worth while, and suggests the part that science, and Puritanism, too, played in depreciating Latin and in recognizing the importance of the mother tongue. Besides the great loss of time incurred by the study of Latin, which, after all, is only a key to the erroneous science of the past, and besides the fact that the study of nature requires no mastery of languages, another evil in the study of the classics is that while pupils study foreign languages, "our own lies altogether uncultivated, which doubtless would yield as plentiful an harvest as others, if we did as much labour to advance it." The Greeks and Romans used their own tongues, "yet we neglecting our own, do foolishly admire and entertain that of strangers, which is no less a ridiculous than prejudicial custome." For this reason he definitely advocates

That care may be had of improving, and advancing our own language, and that arts and sciences may be taught in it, that thereby a more easie and short way may be had to the attaining of all sorts of knowledge: and that thereby after the example of the *Romans* we may labour to propagate it amongst other nations, that they may rather be induced to learn ours, than we theirs, which would be of vast advantage to the Commonwealth, in forrein Negotiations, Trading, Conquest and Acquisitions, and also of much domestick advantage within our own territories. For if we should arrive at any extraordinary height of learning, and knowledge, though we should but speak and write in our own mother tongue, then would other nations be as earnest in learning it, and translating our books, as former ages have been in labouring to attain the language, and translate the books of the *Græcians*, and *Romans*, and we at this day of the *French*, and *Germans*.[42]

It was not only nationalistic pride and natural affection for the mother tongue that inspired this proposal, but also the utilitarian spirit of practical considerations and the desire to prevent the loss of time necessitated by studying the classical languages. As the old science sank in importance and prestige, the languages in which it was contained fast lost their utility and desirability. Had there not been felt a need for some medium of international communication, the use of Latin would have disappeared much more rapidly than it did.

The last three chapters are devoted to pointing out some remedies for the defects exposed. The author's puritanic nature is revealed in his recommendation that the Scriptures be not made the subject of man's reason, but be laid aside in scholastic exercises. He would make the concession, however, that what can be discovered of God and supernatural things by the power of reason and the light of nature may be handled as a part of natural philosophy "(unto which it doth belong) because it is found out by the same means and instruments that other Sciences are." He proposes that the English language be cultivated and that Comenius' grammarless method of teaching foreign languages be introduced. The principles of syllogistic logic should be clearly demonstrated and induction improved, especially as regards some way of rectifying the delusions and fallacies of the senses.[43] He would have mathematics introduced more generally into the schools, "And I could also wish that the sound, and *Apodictical* learning of *Copernicus, Kepler, Ticho Brahe, Galilæus, Ballialdus*, and such like, might be introduced, and the rotten and ruinous Fabrick of *Aristotle* and *Ptolomy* rejected and laid aside."[44] He advocates the following changes in the teaching of natural philosophy: that Aristotle's natural philosophy and astronomy be eradicated, for his errors have only been multiplied by his commentators; that physical learning be introduced; that the philosophy of Plato, as represented by Ficino, of Democritus as represented by Descartes, and of Epicurus as represented by

Gassendi be taught; that Gilbert's "excellent" magnetical philosophy be examined and the proved truths accepted; that pupils be trained in laboratories; and that Galenical medicine be abolished and that of Helmont taught. Metaphysics he would reform largely by introducing the philosophy of Descartes, who had "shaken off the loose and superfluous questions, notions, and frivolous *Chimeras* thereof," so that it may become useful and beneficial. And finally he urges that ethics be made more practical and less philosophical.

More important, however, than any of the recommendations mentioned above is what he has to say of Bacon, for he perceived that the secret of all the new discoveries in nature and the key to all truths in the varied branches of science were the method. To him in common with other Baconians observation and experiment are the open sesame to the closed mountain of truth. Bacon's method, he says, must be introduced, for

It cannot be expected that *Physical* Science will arrive at any wished perfection, unlesse the ways and means, so judiciously laid down by our learned Countrey-man the Lord *Bacon*, be observed, and introduced into exact practice; And therefore I shall humbly desire, and earnestly presse, that his way and method may be imbraced, and set up for a rule and pattern: that no *Axioms* may be received but what are evidently proved and made good by diligent observation, and luciferous experiments; that such may be recorded in a general history of natural things, that so every age and generation, proceeding in the same way, and upon the same principles, may dayly go on with the work, to the building up of a well-grounded and lasting Fabrick, which indeed is the only true way for the instauration and advancement of learning and knowledge.[45]

To achieve this goal, which is clearly Bacon's universal natural history, he saw it was necessary to introduce radical changes into the educational policy of the universities; nor did he hesitate to urge these changes, the most important of which was the introduction of laboratories,

That youth may not be idly trained up in notions, speculations, and verbal disputes, but may learn to inure their hands to labour, and put their fingers to the furnaces, that the mysteries discovered by *Pyrotechny*, and the wonders brought to light by *Chymistry*, may be rendered familiar unto them: that so they may not grow proud with the brood of their own brains, but truly to be taught by manual operation, and ocular experiment, that so they may not be sayers, but doers, not idle speculators, but painful operators; that so they may not be *Sophisters*, and *Philosophers*, but *Sophists* indeed, true Natural *Magicians*, that walk not in the external circumference, but in the center of natures hidden secrets, which can never come to pass, unless they have Laboratories as well as Libraries, and work in the fire, better than build Castles in the air.[46]

The importance of dealing sensibly with concrete and material things, of actually expending physical energy in pursuit of truth, which the Puritans extolled as industry in contrast to the non-physical activity, which they did not hesitate to brand as laziness, of those who sought truth in books and the mind, finds frequent and emphatic expression during the third quarter of the seventeenth century. Here seems to be the origin of that powerful though half-concealed belief that activity of the mind, no matter how strenuous, is futile and akin to laziness unless associated with physical energy or material objects. The traditionalists, on the other hand, proud of an intellectual aristocracy, were unstinted in their scorn of "manual operations," and clung tenaciously to the idea that truth was for the most part to be discovered by the operations of the intellect. Had they been content to relegate the world of physical nature to manual study, and to lay emphasis upon the importance of sheer mental activity in discovering the truth of man, or human truth, their position would have been rendered more nearly impregnable. As it was, beaten from outworn and fallacious theories of nature, they witnessed the rise of a materialism in Hobbes, which, wedded to the mechanical philosophy, spread from the sphere of nature to the world of man, and which today in conscious theory and unconscious feeling would make man and nature one and the same. One more result of this one-sided conception of the way to discover truth is revealed in the praise of the simple artisan and in the tendency to extol him as one more learned than the bookish man, because he is in constant physical contact with material things. The spirit of a work democracy has entered the world; it is still with us in twentieth-century America. Doers not thinkers; democrats not aristocrats.

Webster's treatise is the most important expression of the new scientific outlook between Bacon and the Restoration. It is clear, practical in most of its suggestions, and comprehensive. Many of the great scientific discoveries in his own and the preceding age find enthusiastic acceptance and frequently energetic defence in his volume. Copernicus, Brahe, Gilbert, Galileo, Kepler, Harvey are heroes in his eyes. He read Descartes with interest and approval. But to Bacon above all he paid whole-souled allegiance. He accepted the Baconian philosophy with all its connotations, and his words reveal understanding of the inductive method and complete consciousness of its importance and efficacy. He is scientifically minded; his gross materialism is only evidence, from the seventeenth-century point of view, of how completely scientific he was. The strong utilitarian element, which he shared with his age, though still prevalent does not so completely dominate the world of science today. His attitude toward the ancients, his characterization of, and attack on, the old science and philosophy left little for the later defenders of the Royal Society to say. Finally, the spirit which turns from the past to face the future, which insists on the freedom of investigation and thought, and which burns with reforming zeal, in short, the spirit of seventeenth-century science is his in good measure.

The significance of Webster's book is indicated by the answers it provoked, of which I have discovered three, and there probably were more. The most interesting of these is the *Vindiciae Academiarum* of Seth Ward and John Wilkins.[47] Webster seems to have been ignorant of, or to have overlooked, the fact that only a few years before the publication of his treatise Puritan influence had installed in various positions at Oxford men who were as much interested in the new science as Webster himself, and who were more truly scientific, such as Ward, Wallis, Petty, Goddard, and Wilkins. Furthermore, at the very time when Webster was writing, these men, together with some others, were faithfully putting into practice the Baconian philosophy by meeting weekly to carry on experiments. Naturally they were inclined to defend the university of which they were a part, but being Baconians themselves, their scientific outlook and their attitude toward the traditional philosophy were pretty much Webster's. Thus they were placed in a quandary, in which they found it necessary either to combat the new science, to which they also paid allegiance, or to deny the truth of Webster's description of the universities, too much of which they knew was correct. They solved the problem by answering the somewhat irrelevant and less important charges, by conceding where concession was absolutely necessary, and, when no other recourse was possible, by resorting to laughter. Wilkins denies that the universities are so subservient to Aristotle that they will admit discussion of noth-

ing against his philosophy, and insists that general liberty of
discussion is granted and all hypotheses are entertained—the
atomical, the magnetic, the Copernican—for the universities are
"ready to follow the Banner of truth by whomsoever it shall be
lifted up."[48] Ward seeks to discredit Webster by claiming that
he stole all his thunder from Helmont, Bacon, and Gassendi,
though he himself had only the highest admiration for the last
two. He laughs at Webster's exaggerated emphasis upon the
useful and practical in natural philosophy and mathematics, and
he defends as far as he dares syllogistic logic, though claiming
at the same time that induction is taught in the schools, by pointing to mathematics, the most deductive of sciences. (Webster
never dreamed of making his estimate of syllogisms applicable
to mathematics. He had in mind natural history only.)

Induction is ridiculously applyed to Mathematicall truths, and Syllogisme is
to be applied to Physicks; it was a misfortune to the world, that my Lord
Bacon was not skilled in Mathematicks, which made him jealous of their
Assistance in naturall Enquiries; when the operations of nature shall be
followed up to their Staticall (and Mechanicall) causes, the use of Induction will cease, and Syllogisme succeed in the place of it, in the interim
we are to desire that men have patience not to lay aside Induction before
they have reason.

This last statement brings him round to Webster's position, for
in the eyes of the early scientists the "interim" must have seemed
long. Ward flatly denies that arithmetic and geometry are neglected and optics not advanced, but as regards astronomy he does
not claim that the Copernican theory is preferred, but rather he
describes the method followed, in which all astronomical theories
appear to be placed on the same level.

The Method here observed in our Schooles is first, to exhibit the *Phenomena,* and shew the way of their observation, then to give an account of the
various Hypotheses, how these Phenomena have been salved, or may be
(where the Æquipollency or defects of the severall Hyppotheses are shewn.)
And lastly to shew how the Geometricall Hypotheses are resolvible into
tables, serving for calculations of Ephemerides, which are of quotidian use,
and if *Mr. Webster* have any thing to amend in this method, and will afford
it our Professor, I will undertake he will be thankful for it.

The only way he can find to answer Webster's attack on scholastic philosophy is to emphasize the freedom of discussion allowed
and practiced in English universities, and to show how many of

once accomplished. There is one thing which this sort of Pamphleteers insist on, which as it is pursued by my *L. Verulam*, so it carries weight with it, but is very impertinently applied, either as an exception against us, or as a generall rule to be imposed upon us in our Academicall institutions. It is, that instead of verball Exercises, we should set upon experiments and observations, that we should lay aside our Disputations, Declamations, and Publick Lectures, and betake our selves, to Agriculture, Mechanicks, Chymistry, and the like.

It cannot be denied but this is the way, and the only way to perfect Natural Philosophy and Medicine: so that whosoever intend to professe the one or the other, are to take that course, and I have not neglected occasionally to tell the World, that this way is pursued amongst us. But our Academies are of a more generall and comprehensive institution, and as there is a provision here made, that whoever will be excellent in any kind, in any Art, Science, or Language, may here receive assistance, and be led by the hand, till he come to be excellent; so is there a provision likewise, that men be not forced into particular waies, but may receive an institution, variously answerable to their genius and designe.[51]

An analysis of this passage reveals that Ward recognized the desirability of a reformation in education, and that though he felt antipathy to the somewhat frenzied and uncouth nature of the Puritan treatises, he perceived the justice of their criticism and its relation to Bacon. But he maintained that they were wrong in bringing their charges against "us" (does he mean the small group of experimenters to which he belonged?) and in trying to turn the whole university into scientific and vocational schools. He readily admits that in scientific studies Webster is right. He differs, however, with Webster's insistence that divinity be abolished and the classical languages and literature slighted. Ward is concerned, in behalf of the university, to make his position appear as far removed from his opponent's as possible, yet actually their ideas are not in most respects very far apart, and one feels that he is not so much displeased with the proposals as he is critical of the proposer.

From the ranks of the conservatives also there were answers to Webster which show little or no appreciation of the new science. In *Histrio-Mastix*, 1654, Thomas Hall finds much merriment in the suggestion that students work in chemical laboratories, and he speaks derisively of the scientific and vocational studies proposed in the treatise: "These, these, if you will believe *Mr. Webster*, are the only excellent studies of Academians."[52] Such subjects he does not hesitate to brand as low, inferior,

empty things compared with logic, ethics, and the like, an evaluation which reveals the inevitable clash between the naturalistic
and humanistic spirits of the age, but which does not find as
frequent expression as we would expect. He elevates syllogistic
logic far above induction, and ascribes to the followers of Aristotle the evils which Webster attributed to the philosopher himself. Edward Leigh's *A Treatise of Religion and Learning*, 1656,
in a way foreign to our interest, advocates against Webster the
importance of linguistic knowledge, logic, and rhetoric for a
preacher.[53]

The intensity of the opposition to the universities as then constituted can hardly be imagined now. It really began with the
Reformation in opposition to scholastic divinity (the Puritans
frequently draw upon the early reformers for arguments and
authority), and appears early in the period under discussion.[54]
Wood says that in 1653 Parliament "considered among themselves of the suppressing Universities and all Schools of Learning, as heathenish and unnecessary. And the matter being discussed some days, they would have effected it, but that some of
them of better judgments gave a stop to their frenzy."[55] This
was the time of the strongest antipathy to traditional learning,
and various attacks on, and defences of, the universities appeared
in this and the following year.[56] Wood again states that in 1657
there was another outbreak against universities and human
learning, and in 1658 the insolence of the Anabaptists in Oxford
became so great that Cromwell found it necessary to send soldiers to save the university from destruction. And still again this
hostility broke out in 1659, when Parliament passed a resolution
"that the Universities and Schools of Learning shall be so
countenanced and reformed, as that they may become the Nurseries of Piety and Learning."[57] It was the teaching of divinity
and the belief that secular learning was essential to the ministry
that aroused the resentment of the Puritans, since they held that
the Bible was sufficient in itself to teach all religious truths.
The mixing of "humane" with divine learning, the blending of
Christian doctrine with heathenish philosophy,[58] and the lack of
usefulness in traditional science constituted the charge laid at
the door of the universities and the *causa irae* of the Puritans.
To the more fanatical factions destruction of the colleges seemed
the only remedy, but to the saner and more influential elements,

as represented by Durie, Hall, Webster, Petty, Biggs, and others, the solution seemed to lie in the elimination of theology from the curriculum, and in the reformation of secular subjects through the abolition or modification of traditional knowledge and methods, and through the introduction of experimental science and technical and more useful studies at the expense of what might be called the humanistic subjects.[59]

In this way the scientific program of Bacon found a natural place in the educational reform which the Puritans sought to establish, even if it did nothing to inspire the desire for change, by no means a certain assumption in the light of the quotation which heads this chapter. "The Advancement of Learning" becomes the "Advancement of Piety and Learning." Divinity is to be banished. Experiments are to take the place of classical science. Humanistic studies in general, with the possible exception of history, are to yield in importance to such practical and useful subjects as mathematics, geography, chemistry, and the like. Vocational courses are to be introduced because of their value to the individual and to mankind.[60] There is evidenced throughout the whole period a spirit only too familiar to our academic world today, which insists upon the practical and useful in education, which emphasizes scientific subjects at the expense of humanistic, and which would corrupt the curricula of our colleges of liberal arts with professional courses and vocationalism.

There was much to recommend Bacon and the new science to the Puritans. Bacon's definite separation of science and divinity made it possible for the Puritans to embrace his philosophy without any compromise with religious convictions, and the germs of the deadening materialism imbedded in his attitude had not yet developed their full virus in the metallic philosophy of Hobbes, nor had the potentialities in Descartes's dualism, dangerous for an age the spirit of which yearned for the mechanical and material, become apparent. The element of reform so pronounced in Bacon's writings was in many ways congenial to their desire to reform; his break with antiquity harmonized with their efforts to cut loose from the moorings of the past. They possessed a common enemy in Aristotle, who in Puritan eyes had corrupted divinity, and who in Bacon's mind had effectually put a stop to the advancement of knowledge by the intellectual servility which his great name commanded. They were both anti-authoritarians.

The general ferment which accompanied Puritan efforts to break with tradition rendered the time ripe for new ideas; Bacon's philosophy presented a definite program for one significant reform. Most of all, the utilitarian seeds of Bacon's philosophy fell upon most fertile ground in a movement sponsored in great part by artisans, merchants, mechanics, and townsmen. "Serve God and grow rich" might have been the sign by which they hoped to conquer. Serve God by studying the Bible untainted by secular philosophy; grow rich by embracing the experimental philosophy which held out a rich promise of material benefits. The injunction to measure learning by the standard of tangible and practical usefulness, implied in the writings of the great Chancellor, became operative with the Puritans. Finally, the opinion which saw in the artisan and mechanic, no matter how mean the spirit or impoverished the mind, and merely because of contact and dealings with material things, a more laudable type of industry and a nearer approach to truth than in men whose minds were stored only with the lore of the past, could hardly fail to appeal to the people whom it exalted. (There is similarity, and it was noticed, between the worker in the shop and the worker in the laboratory.) Well might Sprat say that the experimental scientists preferred "the language of Artizans, Countreymen, and Merchants, before that of Wits, or Scholars."[61]

A few more general attitudes Bacon and the Puritans held in common. The germ of the idea of progress, implicit in the very title of Lord Verulam's great work, thoroughly infected the Puritans, as may be seen not only in the many repetitions or variations of the expression "Advancement of Learning" but also in the spirit which animates many of their treatises. The belief that there is no closed door to human progress, that learning can grow and man's condition improve, if not infinitely, at least to some far-off felicity, together with a feeling of obligation to assist the process, was responsible for much of the enthusiasm of the reformers. Like Bacon they turned from the past and faced the future. The duty of promoting Bacon's utilitarian science found sanction in another value which, if not new, became more potent than ever before, namely, the "public good" or the welfare of society. Here was laid the basis for a motive which has grown and grown until it threatens to supplant all others. But in order to put a more sanctified face on the extreme emphasis placed upon material benefits, they seized upon a humanitarian argument. The welfare of the poor became an important justification of a materialistic program. There is the title of a treatise of this period, which, though the text is little to our purpose, expresses succinctly the relationship between Baconianism and humanitarianism conspicuously revealed in the Puritans: *The Relief of the Poore: and Advancement of Learning Proposed.*[62]

The importance of the Puritans in making Bacon's works popular and in disseminating and advocating his ideas, values, and attitudes is perhaps greater than we realize. Before 1640 he seems to have made no remarkable impression upon men's minds; at the Restoration he had become by far the greatest influence of the time. His name and works were the subject of endless adulation and unvarying tribute. Furthermore, Puritan support of experimental science is seen not only in the educational treatises described above, but also in the small group of experimenters at Oxford, composed in part of men who in 1645 had helped to establish a loosely organized society in London for discussion of the great scientific discoveries of the Renaissance and for trying experiments themselves. The most important members of the Oxford group owed their positions in the university to Puritan patronage; in fact, they secured their appointments, we may be sure, because of their interest in Baconian science: Wilkins, Wallis, Petty, and Goddard.[63] Since all but Petty had been members of the London group, their scientific activity must have been known to Cromwell. These men were in no way fanatical Puritans, but their allegiance was definitely to the Puritan side. There is other evidence of the government's interest in experimental science. On May 18, 1649, Hartlib wrote Boyle, "Fauxhall is to be sett apart for publick uses, by which is meant making it a place of resort for artists, mechanics, etc., and a depot for models and philosophical apparatus." It was also proposed that "experiments and trials of profitable inventions should be carried on" which will "be of great use to the Commonwealth."[64] Furthermore, on July 20, 1653, Parliament established a committee "for the advancement of learning."[65] In view of the title of the committee and the educational treatises discussed in this chapter, its purpose must have included some kind of Baconian reform. The next year Hartlib speaks of an

"Honourable Committee for the Advancement of Learning," as if it were actually functioning.[66] The frequency with which the names of Hartlib and Durie appear in the minutes of the Council of State and other state papers indicates more than a casual interest in their designs.[67] Durie was appointed librarian of the library at St. James's, was commissioned to translate Milton's *Eikonoclastes* into French, and was praised for his good services "and present likelihood of usefulness." Hartlib was given frequent grants of money and a pension, employed by the Committee for Trade, appointed on the Committee for Durham College, and commended for his good services in public undertakings. The Puritans by no means monopolized science (witness Scarborough, to mention no others), but before the return of Charles II they were more responsible for its growth than anyone else. At any rate, every attitude of importance which later characterized the Royal Society and its defenders had been clearly expressed and defended before the Restoration.[68]

CHAPTER VI

THE REVOLT FROM ARISTOTLE AND THE ANCIENTS

Let it not be told in *Gath*, nor published in *Askelon* that the Genius of the *English Nation* now made a *Common-wealth*, should be so low, so base, and so beggarly, [as] to daunce after the pipe of meer whistlers, to be the Hinch-boys of *Aristotle*, and confine themselves to the principles of those, who are as a dark lanthorne in a thick night.

Noah Biggs, *Matæotechnia Medicinæ Praxeos*, 1651.

Much of the material discussed in the preceding chapter is definitely related to the present subject. Since the traditional learning of the universities was largely based upon antiquity, any criticism of it was really an attack on the ancients. Almost all efforts at educational reform advocated the abandonment of ideas derived from the classics; conversely, any attack on the classical writers had some bearing upon educational matters, for if the science and philosophy of the past were erroneous, it followed as a natural consequence that they should not be taught in the schools. Yet since the movement toward educational reform was essentially Puritan, and the arraignment of antiquity only partially so, and since in some treatises educational implications are faint and opposition to the ancients much stronger and more definite than in most educational writings, the division into separate chapters seems justifiable, even though there is a close affinity between this and the preceding chapter, and some overlapping is inevitable.

The attitudes which generally accompanied criticism of the traditional philosophy have already been noticed, but at this point it may not be amiss to sum them up briefly. First was the spirit of adventure, of finding out what lies beyond the closed boundaries of knowledge, of widening the limits of acquired truth, together with the faith that such expansion was possible. Another attitude stressed the need of an unbiased and critical mind and of freedom of thought and discussion. If servility to the authority of the ancients precluded examination of traditional beliefs, no hope could be held out for increased knowledge, and if assent were too easily granted, without clear proofs and demonstrations, no certainty would reside in learning. Still another value is revealed in the utilitarian standard, conspicious in the preceding chapter, which was applied with telling effect to classical philosophy. Utility was rapidly rising to that ascendancy over other values which is evident today, and which at first found justification in the conception of the public good, sometimes narrowed, though not so frequently as in the educational treatises, to a humanitarian aspect.

As we have already seen, John Wilkins early in his career had seen fit to oppose scholastic philosophy, daring to question the truth and reliability of Aristotle's works, and to assert that mere multitude of authorities had no weight with him.[1] In espousing the Copernican theory he struck at the very roots of ancient authority, and in doing so he challenged the defenders of antiquity, one of whom was not slow in answering. Alexander Ross was one of the most redoubtable upholders of the intellectual faith of his fathers, and his defence of traditional science reveals the nature of the attacks on it as clearly as they themselves. Educated at Aberdeen, he was made, through the influence of Laud, one of the chaplains of Charles I, and later became master of the Southampton Free School and finally vicar of Carisbrooke on the Isle of Wight. The history of the controversy between the ancients and moderns reveals no more consistent loyalty to the lore of the past than is discovered in his writings.[2] Though wedded to Peripateticism, he introduces into his arguments many apposite observations of his own, and generally proves to be a rather clear though hard-headed reasoner, to whom Aristotle offered the best explanations of phenomena. To him Aristotelianism was a live philosophy, which he applied to what he observed. Thus he was inevitably and hopelessly on the wrong side as regards most of the new ideas of the day. He characterized the Copernican theory as "false, absurd, and dangerous . . . . having neither truth, reason, sense, consent, antiquity, or universality to countenance it," the last two representing the conventional tests of conservatism. He saw in Wilkins' idea of philosophical liberty "the loosing of the reines to exorbitant wits, to run headlong into every kinde of absurdity." He opposed Wilkins' use of Bacon's dictum, *juventus mundi, antiquitas saeculi*:

You *are the fathers (you say) in such learning as may be increased by experiments and discoveries, and of more authority then former ages*. Why doe you not tell us plainly, that you are fathers of learning, as well as in learning? but indeed you are not the fathers of learning, you are onely fathers of your new discoveries and fresh experiments; that is, of new, fond, and savourlesse phansies: and why you must be of more authority then former ages, I see no reason.[3]

Continuing to denounce the new astronomical theory, he remarks, "how fruitful this later age of the world is of new and frivolous opinions."

Ross was quite aware of the stirring spirit of his times, in which

the Dictates and Opinions of the ancient Champions of Learning, are sleighted and misconstrued by some modern Innovators; whereas we are but children in understanding, and ought to be directed by those Fathers of Knowledge: We are but Dwarfs and Pigmies compared to those Giants of Wisdom, on whose shoulders we stand, yet we cannot see so far as they without them: I deny not but we may and ought to strive for further knowledge, which we shall hardly reach without their supportation. I disswade no man from inventing new; but I would not have him therefore to forget the old, nor lose the substance whilst he catches the shadow.[4]

So saying, he falls upon Sir Thomas Browne's *Vulgar Errors*, in which many false ideas having their roots in antiquity are contradicted, and he declares that Browne's "assertions are refuted, and the ancient Tenents maintained." Yet the spirit of the age weighs heavily upon him, as he labors

to vindicate the ancient Sages from wrong and misconstruction, thinking it a part of my duty to honor and defend their reputation, whence originally I have my knowledge, and not with too many in this loose and wanton age, slight all ancient Doctrines and Principles, hunting after new conceits and whimsies . . . . I pitie to see so many young heads still gaping like Camelions for knowledge, and are never filled, because they feed upon airy and empty phansies, loathing the sound, solid and wholesome viands of Peripatetick wisdome, they reject *Aristotles* pure fountains, and digge to themselves cisternes that will hold no water . . . . [and] go awhoring as the Scripture speaketh after their own inventions . . . . let us return to our Masters house, there we shal find pure fountains of ancient University learning.[5]

In the appendix he falls foul of Bacon, Harvey, and Fernelius, toward all of whom, however, he maintains a dignified and respectful attitude. Bacon's method he does not attack, but rather isolated ideas and opinions expressed in *Sylva Sylvarum*, in answering which Ross is as wrong as his illustrious predecessor. Yet in spite of his effort to preserve a tolerant and liberal attitude

he cannot conceal his strong dislike for "this sordid and phan-
tasticall Age," in which "brainsick Sciolists" are busy shaking
the foundations of true learning and piety. The "new philosophy" he accuses of being only the discarded errors of antiquity
tricked out in new terms, and he laments the fact that "These
new Philosophers, as if they were wiser then all the world besides, have like fantastick travellers, left the old beaten and
known path, to find out wayes unknown, crooked and unpassable,
and have reduced his [Aristotle's] comely order into the old
chaos," such, for instance, as considering heat to be motion
rather than a quality. After claiming that he honors the men
themselves, he feels constrained to point the right way out to
those who go astray, and if he himself is out of the right path, he
is consoled with the thought that "I follow the conduct of the
most and wisest Philosophers, so that . . . . I am not alone; and
better it is to go astray with the best then with the worst, with
company then alone."⁶

In astronomy above everything else the inadequacy of ancient
theories was most glaringly apparent. So it is not surprising
to find one astronomer, Jeremy Shakerly, proclaiming that "our
new experiments" inform "us of the insufficiency of ancient suppositions," and give "an example to our followers to handle our
conceits, as we have done theirs who went before us," a succinct
statement of one essential of scientific progress. The labors of
the best astronomers, he asserts, "have taught us how great a
thing experience is, and what danger there is in conformity to
ancient rules. Little could be expected from them who were
guided by so few; yet their performances were beyond expectation." Then the spirit of the age stirs within him: "More may
be required from our times, whose subsidiary knowledge is
greater. This last age (by the blessing of God) doth enjoy the
benefit of more admirable and usefull Inventions, then any, or
almost all before it, and still new ones are added to the old, and
the welcome tidings of Mathematical novelties daily delight us."
The future holds out its promise: "And indeed what shall we
mortals now despair of? within what bounds shall our wits be
contained?" The achievements of his age, which elsewhere he
ascribes to the experimental and observational method, assume
large proportions in his eyes:

Difficult, remote and envious things are now grown pervious to humane industry. We have seen the spots of the Sun, and its conversion about its own axis; we have seen the laterall Guardians of *Saturn* and *Jupiter,* the various Phases of *Mars,* the horns of *Venus* and *Mercury,* the mountains and Seas of the *Moon;* we have seen the generation of Comets, the apparition and disparition of new Stars amongst the Planetary Orbs; we have seen innumerable Cohorts of fixed Stars; yea, even the small constellation of the Pleiades, beautified with divers lights impossible to be numbred; we have seen the Rainbowlike colours of divers lucid Globes, and the motley outsides of the Dog-star and *Orion.* O Heaven and Stars! how much hath our age triumphed over you! Neither doth our victory end here, still new miracles adde to the number of the old, and no day passeth without a triumph.

He proceeds to ask a very pertinent question, one that was demanding its answer of the age with growing insistence:

Why then shall we subject ourselves to the authority of the Ancients, when our own experience can inform us better? why do we not break their bands asunder, and cast their cords far from us? certainly the Ancients (were they now alive) would not condemn our choice. *Aristotle* that referred the studious in his time to *Eudoxus* and *Calippus* for satisfaction, could not dislike our Arguments against the corruption of the visible Heavens. *Ptolomy* that founded his Hypothesis upon Observations, would not be angry if our Observations perswade us to another hypothesis then he had constituted.

He concludes with an echo of Bacon's view of the need of time and co-operation for the advancement of science: "Nor is it indeed possible that one age, or one man should perform all things requisite in this Science. Many things are impossible ever to be found out, and many things reserved to the discovery of after ages."[7] The new thinkers of the age were buoyed up with the hopeful belief that "It is not one Age, much lesse one man, that is able to restore *Astronomy."* Shakerly does not hesitate to assert his own intellectual independence, refusing "to be overruled with the authority of any Writer, unlesse his Reason have a greater authority then his Name, and Judgement tread on the heels of Invention."[8]

There were no more clamorous declaimers against the ancients than were found in the rabble of alchemists, astrologers, magicians, and the like, who formed a tatterdemalion army under the aegis of Bacon, and went forth to do battle with the war-cry of nature, experiment, and observation. They maintained a constant barrage and sulphureous smoke-screen on the entrenched positions of the ancients. There is no clearer evidence of the way in

which the Baconian spirit was penetrating to the darkest and
most distant recesses of the period than is manifested by the way
these outcasts argued violently for a purely physical explanation
of their ideas and emphasized the importance of experimentation
and observation of nature instead of conformity to the ancients.[9]
Their theories are absurd and their language generally mere
jargon, but they sought justification for the former in those
values and attitudes which were establishing themselves in the
world. It is possible they did the cause of science more harm
than good; certainly authentic scientists drew up their skirts
from them (witness Ward's mocking assault on Webster's advocacy of astrology, previously discussed). Yet they swelled the
chorus of the hymn of hate directed against antiquity and voiced
a loud if ignorant allegiance to the experimental method. I can
lay no claim to having read most or the greater part of these
impossible treatises, but shall discuss a few representative ones.

The magicians were careful to dissociate their activity from
all diabolical or occult suggestions and to base it squarely upon
physical laws and forces. One declares that, as his book "is really
Magica, so it is truly φυσικά," and another defines magic as
"the Connexion of natural Agents and Patients, answerable each
to other, wrought by a wise Man to the bringing forth of such
effects as are wonderfull to those that know not their causes."[10]
They constantly distinguish between natural magic, to which they
subscribed, and black magic. Nature being the professed object
of their study, and the experimental method beginning to have
more weight with the world, it was inevitable that they should
subscribe to the method and cast their lots against the ancients.
Aristotle's principles, says Thomas Vaughan, are but fancies
which "stand more on our *Concessions,* then his *Bottom,*" and
from which his followers have derived nothing but notions and "a
meer *Tympanie* of *Termes.*" He rejoices at what he considers
Telesius' and Campanella's victory over the Stagirite. After
answering the objection that the new philosophy is *new* and Aristotle's *old,* and disclaiming "any *Clamorous opposition* to their
Patrone [Aristotle]," he repudiates any desire to have men swear
allegiance to his own dictates, but wishes them only to "compare
my *Conclusions* with *Nature,* and *examine* their *Correspondency.*" He gives us good Baconian doctrine when he says, "If
thou wilt not quit thy *Aristotle,* let not any *prejudice* hinder thy

further search; Great is their *Number* who perhaps had *attain'd* to *perfection,* had they not *already thought themselves perfect.*"[11] Traditional philosophy he holds imperfect and false, being "built on meer Imagination without the least Light of Experience," and he urges the sons of Oxford "to looke beyond *Aristotle,* and not confine their Intellect to the narrow, and cloudy *Horizon* of his Text."[12] In like manner, Samuel Boulton, though fair to the ancient philosopher, expresses the wish that "Physicians would follow nature, and leave off one to swear himselfe a slave to *Galen,* another to *Avicen,* and a third to *Paracelsus:* These men were great, but when they strove to defend their owne Opinions, they often erred."[13] Boulton frequently refers to experience and experiments, his final injunction being, "Follow Nature, and diligently mark her operations, and be acquainted with her footsteps."[14]

There is little doubt that the authentic but more mysterious natural phenomena attracted the attention of the upholders of magic. Hardick Warren gives as examples of the secrets of Nature in which magicians were interested the action of the loadstone and the relationship of the moon and the tides.[15] Furthermore, in emphasizing the importance of observation and experiments they were contributing something to the promotion of science. Thomas Vaughan advises the Aristotelians "to use their Hands, not their Fansies, and to change their Abstractions into Extractions; for verily as long as they lick the shell in this fashion, and *pierce* not *experimentally* into the *Center* of *things* . . . . they cannot know things substantially, but onely describe them by their outward effects and Motions, which are subject, and obvious to every Common Eye."[16] When he urges them not to let their "Thoughts feed . . . on the *Phlegmatic,* indigested Vomits of *Aristotle,* [but to] look on the green, youthfull, and flowerie *Bosome* of the *Earth,*" he is voicing the central idea of science from Bacon to Sprat.[17] In another work Vaughan flays Aristotle and the Peripatetics, calling them fumbling, confused "Scriblers,"[18] whose philosophy possesses "no firm *Fundamentals in Nature.*"[19] The magicians felt, or pretended to feel, as keenly as anyone else the necessity of abjuring the authority of the ancients, though they show poor taste in the abuse which they heaped upon the latter. In general they reveal at least a superficial affinity with the new science in hostility to the ancients, attention to nature, insistence upon concrete results, and emphasis upon experiments. They tried to stay in the best company.[20]

One more aspect of their revolt from Aristotle should be noticed. Peripatetic theories of nature receive from them no more condemnation than the Aristotelian element in scholastic divinity, an attitude characteristic of the Puritan era. Vaughan calls the Peripatetic schools "A *spleen* first bred, and afterwards promoted by *Disputes,* whose damnable *Divisions* and *Distinctions* have minc'd *one Trueth* into a Thousand hæreticall *whimsies,*" and he holds that "some *Divines* believe onely for *Aristotle's* sake, if *Logic* renders the tenet probable, then it is *Creed,* if not tis *Alcoran.*"[21] Galen was honest enough to admit his ignorance of the soul, he says, but Aristotle "who had not so much honestie is voic'd *Prince of Philosophers,* and the positions of more *glorious Authors* are examin'd by his *Dictats,* as it were by a Touchstone; Nay the Scripture it self is oftentimes wrested, and forc'd by his Disciples to vote a *Placet* to his Conclusions."[22] As is usually the case, St. Paul's condemnation of "vain philosophy" is conspicuously employed. Perhaps no one factor played a greater part in discrediting Aristotle's authority than the outcry against medieval divinity. In the Puritan efforts to reform education, as we have seen, one of the chief charges preferred against the Grecian was his corruption of theology. Hotly opposed to the introduction of any alien element into the study of the Bible, the Puritans could not but be sympathetic with science's hostility to the Stagirite.

An interesting attack on ancient authority is discovered in the work of an obscure man whose real or assumed name was Agricola Carpenter. Judging from some of the eulogistic prefatory poems, which stress the author's break with antiquity, and which indicate how widespread the opposition to the past was becoming, Carpenter appears to have been a young man, and judging from the treatise itself, he seems to abuse Aristotle and tradition because he himself wished to advance something new. It may possibly be intended as a burlesque of the atomic writings of the day. The book deals with the soul, and is written in the style of Sir Thomas Browne. After complaining that Aristotle's *dixit* had so prevailed in the past that the schools prostituted their reason to slavish adherence to his philosophy and introduced his αὐτὸς ἔφη to confute experience itself, he proudly declares, "I am a Free-born Subject; and the Law of Nature which is indispensible knows no slaverie!" Reason rather than authority is his guide, "for none (I think) since the Magick of the Apple, could challenge to themselves the Popes Chair .... I could never resolve my Reason into the Opinion of Antiquity, or surrender my Experience upon a different determination; the bare assertions of the most Authentick have never been embraced by me as Oracles."[23] Neither tied to the wheel of authority nor wedded to any opinion, he boasts a critical attitude of mind which incites him to "disturb the tranquillity of those unlimitted Oracles of Antiquity." He echoes Bacon in saying that intellectual slavery and the belief that truth has already been discovered have debarred many from making discoveries. In a verbose manner he praises experiments and depicts a hopeful vision of progress when men become intellectually free:

.... a soul that is not tyed to intellectuall slaveries, in the Geography of knowledge will confute tht conceit of *Hercules* Pillars, and a *Ne plus ultra* is but [a] fantastical Dream to his more nice and subtle scrutinies .... Truth is not ingrossed by aged Parents; there is an *America* of knowledge yet unfound out, discoverable by the endeavours of some wiser *Columbus,* and the promised fertility of succeeding Ages. Those unsuccedable attempts of our incultivated Intellectuals, must needs stand condemned by the more sublimed Reason and acute Philosophy of Posterity: Whereby our Primitive imparement of additional Degrees of a betterment may obtain a parole from ignorance, and procure a Reverse to the first perfection. For if this be true, that the order and course of Nature is Circular, and that souls in Revolution of time receive a sensible improvement in their Cognition. Then surely succession in Nature must needs carry Degrees of Perfection, and the last man must be an *Adam.*[24]

The parallel here drawn between expansion in geographical knowledge and progress in general is not merely a figure of speech. It occurs frequently during this period and indicates the subtle influence which the discovery of new lands had upon men's vision of the future and upon their desire to break with the past. Carpenter sees in the perfection of man that "far-off divine event to which the whole creation moves."

The most noted opponent to ancient authority at this time in England was, of course, Thomas Hobbes, who had as much influence as any of his contemporaries in loosening the shackles that bound the human mind. Writing in a plain, perspicuous

style, and expressing ideas revolutionary, startling, and easily
understood, he came to be the most widely read, frequently
quoted, and violently discussed man in the country. Though he
was completely discomfited by both Boyle and Wallis in scientific and mathematical controversies, and though he could never,
much to his disappointment and chagrin, edge into the Royal
Society, his words carried ever-increasing weight. His affilia-
tion with some of the leading scientific men in France and his
admiration of the great discoverers of the first half of the century, such as Kepler, Galileo, and Harvey, whose views he accepted, to say nothing of his revolt against authority and his
belief in direct observation of nature, give him a place in the development of the new science. Yet it must be confessed that he
was more intent on meditating upon the experiments of others
than on experimenting himself; from the purely scientific point
of view, he suffered from too much speculation and too little
experimentation.[25] Though he based the truth of natural philosophy ultimately upon experiment and observation, he was too
fond of trusting his mind beyond the reaches of his senses. His
relegation of experiments to a subordinate position, his proneness to rely too much upon reason, and his contempt for un-
philosophical mechanics and mere experimenters, for the
mechanical as opposed to the philosophical, which led him to
attack the Royal Society, were partly responsible for his es-
trangement from English scientists.

We are chiefly interested, however, in his position in the controversy between the moderns and the supporters of antiquity.
To begin with, though he does not violently attack Aristotle and
the ancients, he expresses no sympathy with their views and
denounces the authority which they commanded. After stating
that astronomy began with Copernicus and natural philosophy
with Galileo, he answers the question whether there were no
natural philosophers in Greece as Bacon and others had answered it:

There walked in old *Greece* a certain Phantasme, for superficial gravity,
(though full within of fraud and filth) a little like Philosophy; which
unwary men thinking to be it, adhered to the Professors of it, some to one,
some to another (though they disagreed among themselves) and with great
Salary put their children to them to be taught, instead of Wisdome, nothing
but to dispute; and, neglecting the Laws, to determine every Question
according to their own fancies.[26]

He objects to many of Aristotle's theories, civil and scientific, such as his democratic ideas, his conception of "visible species," his idea of gravity as an appetite to rest, and the like. To him the logic of the Grecians consists of nothing but words, and Aristotle's philosophy is vain. But he is more intensely concerned with the slavish adherence to Aristotle's philosophy characteristic of the universities, which, he claimed, had reached such a pitch that the philosophy taught there was only "Aristotelity." The evil of this servitude he perceived to lie in the obstacle placed in the way of the advancement of knowledge. Hard it is, he says, to root out inveterate opinions when confirmed by the authority of the most eloquent writers. He joined with the Puritans in the movement for educational reform; in fact, his criticism of the universities attracted more attention than theirs, though he employs much the same arguments. Especially did he emphasize and condemn the corruption of theology by Aristotle's philosophy, citing with much gusto St. Paul's "vain philosophy." No Puritan ever inveighed with more vehemence against the evils of this corruption than Hobbes, the atheist. One feels, however, that he is not so much moved by solicitude for Christian doctrine as by indignation at so many meaningless terms discovered in school divinity. Catholicism, he says, was banished from England partly because the schoolmen had brought Aristotle into religion.[27]

The most consistent and important contribution Hobbes made to the cause of modernity was his driving assault on the schoolmen, whose "empty words," such as species, substantial forms, incorporeal substances, instinct, antipathy, sympathy, occult qualities, were anathema to him. In none of his works does he ever pass by an opportunity to rap them, but it is in his controversy with Bramhall, Bishop of Derry, that his attitude is most clearly disclosed. In excoriating scholastic terminology he was in truth attacking Aristotelianism, on which this terminology was in part based, and for this reason was arguing the case of the moderns against the ancients. He rendered distinct service to the new thought of his age by insisting upon mental clarification. Even if the old conceptions possessed some life for former philosophers, at this time they were dwindling to mere extinct terms which, in the eyes of the new thinkers anyway, gave an empty sensation of knowledge without the reality thereof. This fact lies at the base of the scientist's distrust of language and their craving for a plain manner of expression.[28] "I would have him," Hobbes says of Bramhall, "quit both his terms and his profession, as being in truth . . . . not at all profitable to learning, but made only for an essay to the learner; and the divine to use no word in preaching but such as his auditors, nor in writing but such as a common reader, may understand. And all this, not for the pleasing of my palate, but for the promotion of truth."[29] The habit of building up ideas from words rather than fashioning ideas from the things which the words represent, with the consequence that all true relationship between language and reality is lost, receives his severest condemnation. "In this place," he says, "the Bishop hath discovered the ground of all his errors in philosophy, which is this; that he thinketh, when he repeateth the words of a proposition in his mind, that is, when he fancieth the words without speaking them, that then he conceiveth the things which the words signify: and this is the most general cause of false opinions."[30] And again, "But to him that thinketh not himself upon the things whereof, but upon the words wherewith he speaketh, and taketh those words on trust from puzzled Schoolmen, it is not only hard, but impossible to be known."[31] Hobbes attacked Aristotelianism because neither its ideas nor its language seemed to possess any vitality, and he reveals more clearly than anyone else at the time the craving for conceptions based upon actual observation of things, and for a language more adequate to the ideas so derived.[32]

The attack on the ancients, as well as other characteristics of the scientific movement in England, was by no means independent of foreign thought.[33] Gassendi, Descartes, Helmont, and others certainly left their mark on England. Although the present study ignores for the most part foreign influences, two or three translations made about this time warrant consideration. The first is a translation of Descartes's famous treatise, *A Discourse of a Method,* 1649, in the preface to which Descartes is styled the first philosopher "of this age." It is needless to do more than call attention to this revolt from the authority of the ancients and advocacy of a critical attitude. The second is a translation of another of Descartes's works, *The Passions of the Soule,* 1650, in the "Advertisement" to which the translator assails the Aristotelians, emphasizes the necessity of discoveries, and praises the new philosophers, especially the man whose work he is translating, to the skies. Some think that knowledge cannot be advanced beyond the ancients, but, he declares, there are many things yet to be found out in physics. He follows Bacon in maintaining that "the too great reverence born to antiquity, is an errour extreamly prejudicial to the advancement of Sciences." Until this is removed, no new learning can be acquired. He mentions inventions such as printing and the compass, "which were not found out till these latter ages," as an argument that knowledge may advance far beyond the "Pillars of Hercules." Gilbert, Kepler, Galileo, and Harvey he praises at the expense of Aristotle, declaring that "a man may, without much pains, imagine a body of Philosophy, lesse monstrous, and grounded on conjectures more conformable to truth, than that which is extracted from the writings of Aristotle." Experiments are the chief requisites in his eyes:

most men conceive not how necessary experiments are, nor what expence they require; those who, not stirring out of their study, nor casting their eyes on any thing but their books, undertake to discourse of nature, may well tell how they would have created the world, had God given them authority, and power to do it; that is, they might describe Chimera's that have as much Analogy with the imbecilitie of their wit, as the admirable beauty of the Universe, with the infinite puissance of its Maker.

This plea for experimentation is followed by a reference to its probable source: "I have also seen the Advancement to Learning and the *New Atlantis,* of my Lord Chancellour *Bacon,* who, of all them that have written before you [Descartes], seems to me the man who had the best notions, concerning the method to be held to bring the Physicks to their perfection."

The last translation to be considered is that of a treatise of Comenius entitled *Naturall Philosophie Reformed by Divine Light,* 1651, which declares that Aristotle's authority should not be tolerated in Christian schools as an absolute authority, but that men should be free to philosophize. "Why do we not, I say, turn over the living book of the world instead of dead papers. . . . If a Monitour or Suggestour [is needed], we have more and better then *Aristotle,* experience (of the various and occult Mæanders of nature) being multiplyed in the processe of so many ages." Owing to the remarkable growth of science, Aristotle's tenets have been shown to be uncertain and false.

Now hee that will may see *Campanella* and *Verulamius*, (for it may suffice, to have shewed these Hercules, who have happily put to their hands to the subduing of Monsters, and cleansing *Augias* stables; and to have opposed them to those, whom the authority of *Aristotles* vainly swelling Philosophy holds bewitched:) and feel how farre *Aristotles* assertions are often from the truth, and this is the cause why it seems convenient, that *Aristotle* with all his heathenish train should be excluded from the sacred Philosophy of Christians.

He continues at some length his attack on the mingling of Aristotle's philosophy with divinity. Turning again to science, he insists that it is demonstration rather than speculation, persuasion, or authority which should determine truth. "Let no man be compell'd to swear to his Masters words, but let the things themselves constrain the intellect: Nor let a Master have any more credit given him, then hee can demonstrate in very deed, that hee is to have."[34] Comenius' significance lies in his appreciation of the progressive thinkers of his age, especially Bacon, and in his attack on authority, Aristotle, and school divinity. His treatise, which in other respects is insignificant, voices the current criticisms of Aristotle and Peripateticism.

In earlier chapters we have seen how the Paracelsians kept up a running fire on the ancients, and during the Puritan era their descendants, the chemical doctors, largely inspired by Van Helmont, continued the battle with increased fury. The times, perhaps, afford no more picturesque arraignment of Aristotle than is found in a treatise by Noah Biggs, which has already been in part considered. In answer to what he says is the opinion of most men, that all knowledge of any value in science is to be found in the ancients, he draws upon Bacon in declaring that "the too great Reverence born to *Antiquity* is an error extreamly prejudicial to the advancement of *Science* . . . . so prejudiciall, that till it be rejected, it is impossible any *new* learning can be acquired; which may be one reason to prove, that we are far from knowing all we are capable of."[35] He makes use of a figure popular in the past and destined to be popular in the future with the new thinkers, when he accuses the English of acting

as if our *Ancestours* resting places, were to be like *Hercules* pillars, inscrib'd with a *Ne plus ultra;* as if they had attain'd to the *Meridian* of *all* knowledg, by the fix'd *North-pole* of all *perfection;* and on that *Axletree* must turn the whole *Globe* of posterities knowledge, with the whole *Hierarchy* of lesser and greater increasings; as if the wits and spirits of the *present* times would serve for nothing, but to go a wool-gathering in the wildernesse or wild fields of the foregoing *sheep-shearers;* or were fit for nothing, but to sit with their hands in their pockets; in a lazy, and implicite conformity to this medical statue of Æsculapius; and rather then to correct, adde, encrease, perfect, purg and reform the present *mode* and *practise,* had rather subscribe to it, and be an obedient son of the former *Physick-fathers* . . . .[36]

Strange it is, he cries, that Europe has taken no thought whatsoever (a palpable exaggeration) to escape from the bondage of a heathenish philosophy and the tyranny of Aristotle.[37] Longing for the time when it will be considered a disgrace to be well read in Galen,[38] he delivers stout blows upon those who think that by reading Galen and Aristotle they have encompassed the whole of nature, whereas they neither have discovered any truths themselves nor do they permit others to do so.[39] The universities he hammers unmercifully because they possess the authority to impose upon men the worst of servitudes, ancient medicine and philosophy,[40] and because their scholars, resting at "Hercules Pillars" in a lazy lethargy and with comfortable revenues, prevent industrious souls from advancing knowledge.[41]

Biggs with sweeping blows disperses medical superstitions and legends, bearing down as hard on some of the traditional chemical hallucinations, such as potable gold, as he does on Galenical physic. With extraordinary virulence he assails antiquity, authority, custom, avarice, ignorance, intolerance, slothfulness, cowardice, servility, and the like, advocating in their place experiments, to which he has constant recourse in proving his points. His faith resides only in the experimental method. As has already been remarked, his style is highly individualistic. Figures of speech are numerous and picturesque; of particular interest are his frequent geographical metaphors, generally employed with reference to the widening of knowledge beyond circumscribed bounds, which, as has been noticed, argue a definite relationship between the spirit of the new science and that of previous explorations.[42]

Another Helmontian who directed some rude blows against the ancients was George Starkey, a close friend of Robert Boyle, according to his own statement, and an empiric who made himself famous by vending quack medicines, according to the *Dictionary of National Biography.* Though he indignantly denied the charge of empiricism, he certainly shows more wisdom in what he condemns than in what he recommends. Nothing but applause need greet his denunciation of the dangerous nature of such remedies as bleeding, vomiting, violent purges, cooling drafts, the withholding of water from fever-stricken patients, and the terrible omnibus receipts of the pharmacopoeia of the day. He emphasizes simplicity rather than bewildering and meaningless complexity. Yet he is almost as absurd in the extravagant, though seemingly sincere, claims he makes for chemical medicines; he shrouds himself in too much mystery, his ideas regarding the Archeus verging on the occult; and his experiments seem to have been carried on in the dark. Yet he was one with the moderns in his revolt from the ancients and the schoolmen, in his emphasis upon the experimental study of nature, and in his forward-looking spirit which had caught the vision of progress. He might almost be considered a martyr to science, for he died in 1666 from the plague, contracted while he was dissecting a corpse. The chemical doctors later took great pride in pointing out that during the great scourge they had remained at their posts while the Galenical physicians had fled the city. It must be admitted that the former evince a zealous spirit which manifests a wholehearted loyalty to their own brand of experimental science.

Starkey begins his treatise[43] with a history of the rise and development of medicine, in which he emphasizes its corruption by Galen and his followers, the Arabians, until, show being separated from substance, it became firmly established in the universities, protected by law, and favored with emoluments. The science became mere speculation far removed from reality: "Medicine therefore being made too gross for them by reason of the necessary dependency it hath on practise, they suckt out the universalities of it with the universalities of all other natural practical Arts, as Geometry, Astronomy, Uranomancy, Geography, Arithmetick and the like, which they moulded up into one aery lump of Natural Philosophy." The leader of the "Rabble was *Aristotle,* who though he in many places severely carped at *Galen,* yet forasmuch as they agreed in the main, namely in point of much apparency with little or no truth, they were both accepted and successively read, *Aristotle* as the father of Philosophy, and *Galen* of Medicine"; and they established their sway over the whole of Christendom. "Thus posterity being drawn along with the two Coach-horses, with the authority of *Galen* and *Aristotle,* admiration of these was a badge of their Academical loyalty; and whosoever should dare to swarve from these, was branded with a note of Infamy; and so being looked upon as Heterodox, was the object of scorn and derision."[44]

Starkey himself had gone through the conventional university mill, his vivid description of which has been quoted in a note on a previous page. He found the common philosophy "totally rotten," his skill in logic and philosophy later proving beneath contempt. He furnishes us a Cartesian account of his intellectual development:

I therefore rejected *Aristotle* and all his fictions, against whose fallacious shew I wrote with a pen dipt in salt and vinegar, yet without gall, a Treatise called *Organum novum Philosophiæ* [the title reveals the source of his inspiration]: but before I could pitch on what way to turn (for knowledge I desired too immoderately) I wandred through many pensive hours, and waking nights, till at length I got som Chymical Authors.

Those then I perused and noted with much diligence, not so much out of desire to rifle their Hesperian Garden, as to suck out of their principles some solid truth; for truth I knew was uniform. Wherefore as many experiments as I could try I tryed, and took nothing upon any mans trust, so as to build any thing on it, or to draw any conclusion from it: I invented many sorts of Furnaces, procured what Glasses were possible, with all manner of Simples, Mineral and Metalline especially (which I most esteemed) in these I spent my time for several years, and I may say without boasting, that if ever any in the world were an insatiable prosecutor of experiments, I was one.[45]

Under the influences which were beginning to exert their force upon the age, his experience must have been by no means unique. Certainly others were revolting in the same manner, a fact which the lamenting testimony of the conservatives discloses. Yet little impression seems to have been made on the universities, and it is to "the mighty force of education" that Starkey attributes the blindness of men to the weakness and absurdity of received opinions.[46]

Starkey eagerly points out to his readers the difference between the Helmontian and the Galenist: "one is the immediate follower of Nature, and of his Author, so far only as he doth bear faithful witness to her operations: the other is the immediate follower of him whom he makes the Captain of his Notions, to whose placits if Nature will not comply, he will either force her or adieu."⁴⁷ He elaborates upon his description of the latter by asking: "Is not subscription the top stone of the *Galenical* Art? Is not *Galens* method to this day retained and defended, according to which all created Doctors are ingaged to go."⁴⁸ He castigates the laziness of those who are satisfied to turn over the leaves of a book, preferring "rather to subscribe, then to undergoe the trouble and pains of search and inquiry."⁴⁹ In no uncertain tones he assures us that it "is not reading of *Aristotle* that will make a Philosopher, or of *Galen, Hippocrates, Avicen, Mesue* . . . . or the like, that will make a Physician."⁵⁰ No, proficiency in science requires a quite different kind of mortal, "a mental man, patient, laborious, and one who is not niggardly in expences, such a man must toyl without wearisomness; and although after several years searching, with the expence of many pounds, he hit not what he aims at, yet must he still patiently proceed, which task is more *Herculean*" than that of a Galenist.⁵¹ The industry and self-sacrifice required of an experimental scientist find frequent expression in the third quarter of the century, but no treatise discloses ardent loyalty to the ideal of the experimental and observational method more clearly than the following passage, which glorifies even Starkey's confused conception of true science:

Yet I know that although I consume and spend whatever moneys I can borrow from my bare necessity, or at utmost my most absolute conveniency, in Furnaces, Coals, and Glasses, with the Bee making Honey, but not for my self, yet the experience, which through Gods blessing this industry hath brought, doth and will bring me, will make my name live, when the names of hundreds, that bark and snarl at me, and load me with unworthy reproaches, shall lie buried in perpetual oblivion . . . . what ever I do get I lay out in future discoveries, and all to do good to an ungrateful generation: oft times running in debt for conveniencies, and necessaries, and sparing out of my belly to finde out new experiments in Medicine.⁵²

Posterity in labeling him a quack has proved as ungrateful as his own generation, but I, at any rate, am glad to call attention to a spirit which was one of the factors making possible the rise of modern science, and which was possessed in varying degrees by the followers of Bacon.

The *Academiarum Examen*, 1654, of John Webster, which has already been discussed in part, contains one chapter devoted exclusively to an attack on Aristotle and the Peripatetics.[53] Speaking with the utmost scorn of them, Webster echoes Hobbes in declaring that the philosophy subscribed to by the schools is "meerly Aristotelical." He remarks that the Stagirite, being only a man, might have erred; that his teachings are opposed to Christian doctrine; and that they have in the past by no means been secure against the criticism of those who supported other philosophies. There is no reason, he asserts, that Peripateticism should have pre-eminence in the philosophical world, for it is not certainly known what Aristotle wrote, and the works ascribed to him are so obscure that their meaning is dubious, as may be proved by the many factions among the Peripatetics. Attacking the inadequacy, errors, impiety, and contradictions of the prevailing philosophy, he is especially severe with its verbal and speculative nature, which makes it only a labyrinth of intricate terms. Moreover, since Aristotle's philosophy, as he views it, is entirely lacking in observations and experiments, his followers themselves shirk the arduous task of experimenting. Chemistry is worth more to the schools than all the Aristotelian philosophy, but because of the domination of Galenical physic, based on Aristotle's idea of the four elements, no progress in medicine has been achieved. An advance in anatomy has been made, owing to Harvey's discovery, yet there has been little improvement in practice and application. His parting shot is that the universities have made nothing of Gilbert's magnetic philosophy, although in fifteen hundred years nothing equal to it has been found out.

Some of Webster's efforts to expose the faults of scholastic philosophy are beside the point, but what he was trying to do was to break down the blind veneration of Aristotle, subserviency to a name, slavery to authority, and to establish in men's minds the need of truth based upon experimental investigation. He is really opposing his version of the modern method of experimentation to reliance upon the outworn philosophy of the ancients. "*Aristotle's ipse dixit*," he says, "ought no longer to pass for oracles, nor his tenents for truths before others,"[54] for his

Philosophy is meerly verbal, speculative, abstractive, formal and notional, fit to fill the brains with monstrous and airy *Chymæras*, speculative, and fruitless conceits, but not to replenish the intellect with sound knowledge, and demonstrative verity, nor to lead man practically to dive into the internal center of natures abstruse, and occult operations: But is only conversant about the shell, and husk, handling the accidental, external and recollacious qualities of things, confusedly, and continually tumbling over obscure, ambiguous, general and equivocal terms, which are onely fit to captivate young *Sciolists*, and raw wits, but not to satisfy a discreet and wary understanding, that expects *Apodictical*, and experimental manuduction into the more interiour clossets of nature. Here in the *Schools* is found no such thing, but objurgations, and clamours, fighting and contending . . . . like blind, or madmen, not knowing where they wound others, nor where they are hurt or offended themselves.[55]

Finally, he notes what Bacon had called attention to, namely, that too much doting on antiquity breeds the belief that nothing is left for the moderns to discover. Quoting his predecessor to the effect that there will always be something for later ages to find out, he makes use of Bacon's paradox, *antiquitas saeculi, juventus mundi,* to show that the moderns are in a position to know more than the ancients.

In one intellectual movement, to a certain degree associated with science, we find a modified hostility to the ancients, or better, to dependence upon their authority. In general, Descartes's influence was more powerful with the Cambridge Platonists than was Bacon's. Yet one, Nathaniel Culverwell, shows more evidence of being influenced by the English philosopher. Indeed, one section of his work[56] is devoted to a discussion of one of the chief obstacles to the advancement of learning stressed by Bacon. "The great and noble *Verulam* much complains (and not without too much cause) of those sad obstructions in learning, which arose upon the extreme doting upon some Authors, which were indeed men of rare accomplishments, of singular worth and excellency, and yet but men, though by a strange kinde of *Ἀποθέωσις,* a great part of the world would have worshipt as Gods." The canonizing of some profane authors, he continues, has prevented much knowledge from coming to light. Though professing great respect for Aristotle he condemns severely the desire

to set him up as a Pope in Philosophy, as a visible head of the truth militant, to give him a negative voice, to give him an arbitrary power, to quote his texts as Scripture, to look upon his works as the irreversible decrees of Learning, as if he had seal'd up the Canon, so that whoe're addes to him, or takes one word from him, must be struck with a present *Anathema* to condemn all for Hereticks that oppose him, for Schismaticks that depart from him, for Apostates that deny him.[57]

On another occasion he refers to what "that great advancer of Learning" says of the progress made by the mechanical arts, such as architecture, manufactures, printing, and navigation, in contrast to the stagnation of the liberal sciences due to the domination of antiquity. "For [with the former]" there's "no limiting and restraining men to Antiquity, no chaining them to old Authors, no regulating them to I know not what prescribed formes and canons," since in them men do not labor under the injunction not to build better than their ancestors.[58] Bitterly condemning the imprisonment of Galileo because of his mathematical calculations and use of the telescope, he launches an incensed attack on the Catholic Church for its persecution of heretical beliefs, its corruption of Christian doctrine with Aristotle's philosophy, and its general subjection of reason to tradition and authority. Largely inspired by Bacon and imbued with the spirit of the new science, Culverwell makes an eloquent plea for intellectual independence. His words also furnish evidence of the affinity between the revolt against ecclesiastical and theological authority and that against Aristotle's domination of science. As we shall see later, Sprat definitely attributes the latter movement to the former, and certainly at the roots of both revolts lay the desire to search out the truth for one's self. Modern science is peculiarly associated with the rise of Protestantism.

In the essays of the period we find evidences of the Bacon-inspired attack on antiquity. One who signs himself "T. C." (probably Thomas Culpeper) condemns the time spent on studying the classical languages, and advocates clearer and more mathematical demonstrations of scientific theories, so that those who are "beaten from *Aristotle*" may have something to hold to. He inveighs against the traditional disputatious philosophy which was concerned only with notions and shadows while true knowledge lay encumbered with a throng of distinctions,

like a true diamond, in a heap of counterfeit Jewels? why should we acquiesce in Authority, which, though most necessary to Law and Gospel, is the bane of Arts, and a *ne plus ultra* to knowledge? Some I confess, have very auspiciously begun a Reformation in this behalf; but it is in the embrio, and will require a Colledge of the most learned to finish it; indeed it is almost like the translation of the Bible, work fitter for national Synods, then private men.[59]

The Baconian flavor of this passage inheres not only in the idea
that the authority of antiquity is a bar to scientific progress,
but also in the belief that co-operation is essential to the reformation of science, while his suggestion of a college looks forward
to the Royal Society. A more extended attack on the blind worship of the ancients is afforded by Francis Osborne's *A Miscellany,* 1659, which in condemning the disputatious philosophy
and magisterial authority of the universities laments the fact
that traditional errors are propagated and truths new and out
of the beaten path are suppressed. From long commerce with
books the native invention and ability of the moderns are stifled;
they grow diffident of expressing their own opinions out of
reverence to an over-awing antiquity. Thus an effectual bar to
the advancement of knowledge is set up, and scholars are

tyed to *Cultivate* and confine their *Travels* within the Circuit of a *Wilder-
ness,* stocked with *insignificant Termes* and such *crabbed Allusions* and
improbable Conclusions, that have no better Plea for their *Authority* then
the exorbitant veneration which *Tradition* rather than *Merit,* hath awarded
Antiquity above the *equal,* if not. *better Judgements* found in *our Age.* Nor
is there hope of any further or more happy *Progress* till these *rotten Car-
casses* be *removed* out of the way, or *covered* from their Eyes that want not
Hearts and abilities to *jogge* on better without them.[60]

Making use of a figure quite familiar to us by now, he asserts
that the ancients erected "their *Pillars* upon the borders of
Philosophy under such an *Imperious Injunction,* as none, till of
late have ventured to discover any thing beyond them." Yet he
is fully convinced of the law of progress, that one age makes an
advance over the preceding, for "how many *New* and more
usefull Arts are now as it were daily *invented.*"[61]

The movement which we are tracing frequently involved not
so much a criticism of classical writers as a condemnation of
the authority vested in them. Yet the surest way to undermine
this authority was by pointing out the insecure basis upon which
it rested. Again, distinction is frequently drawn between the
ancient philosophy and its later corruption in what had become
the traditional learning. In the latter, however, antiquity participated so largely that the distinction, even when expressed,
is often lost sight of. The one element which is found in every
phase of the break with the past was the conviction that freedom
of thought was essential to any possible intellectual progress.

Generally this progress was interpreted in terms of the new science, but not always. An excellent case in hand is revealed in two orations delivered by Edward Bagshaw, second master of Westminster School under the redoubtable Mr. Busby, in Christ Church, Oxford, respectively upon taking up and laying down the office of censor.[62] There is no suggestion of the new science in his words, for he believed that the study of nature was vain. Furthermore, he praises Aristotle highly, as well as the ethical writings of the ancients in general, and advocates a relearning of his philosophy after all the intermediate purveyors of it have been swept aside. It is servility to his authority which enrages Bagshaw, and which he attempts to undermine by gleefully pointing out errors in the ancient philosophy such as the conception of *materia prima*, form, and the like. Against the scholastic philosophy he is very severe: it consists of nothing but idle and vain disputations; it is sterile and full of quarrels and unprofitable questions. The position of dictator which Aristotle assumed was unusually obnoxious to him. He has driven out, says Bagshaw, all other philosophers, and rules supreme, especially in Christ Church. Here a temple and altar have been erected to the Stagirite as to a philosophical deity, to whom also festive days are consecrated and solemn pomp and annual panegyric dedicated. Here he walks and enjoys his praise as if he were the *loci genius*.

Bagshaw is keenly sensitive to the evils of authority and hidebound tradition. To him the matter of supreme importance is the liberty of philosophizing, which he loudly declares he will not permit to be taken from him, for "nec toleranda esse ab Ingeniis ea Imperia, ubi Mens subjecta est." He explains the lack of such freedom on the ground that men are loath to give up errors of long standing rendered dear from long use, and will not suffer themselves to be separated from that in which they have so long and so comfortably slept. Furthermore, the mind with great difficulty surrenders the impressions of early education, and freely acquiesces, as if from satiety, in that which it first imbibed. Finally, he indignantly declares that if any are so lazy and of such a debased mind as to think it glorious to err with many, they resemble cattle rather than men. This ringing cry for intellectual liberty is characteristic of the age. In England its first elaborate expression was Carpenter's *Philosophia libera*, 1621, and it grew in magnitude during the period of Puritan supremacy, as witness Milton's *Areopagitica* and other treatises of similiar nature.[63] The importance of this freedom was, perhaps, more keenly appreciated by the Baconians than by anyone else, for the introduction and acceptance of the new scientific method could be achieved only through the liberty of criticizing the old philosophy and of promoting the new. But the scientists also perceived the wider value of intellectual emancipation for every kind of intellectual progress.

A corollary of this cry for freedom is the insistence upon a critical as opposed to a dogmatic attitude of mind. Bacon had proclaimed the necessity of suspending judgment upon scientific theories and explanations until they were thoroughly demonstrated, and later in the century Descartes re-enforced this caution to an even greater degree. In England this view received its most enthusiastic and elaborate expression in a treatise[64] by a young man of Oxford, Joseph Glanvill, who later became a rector in Bath, and who after joining the Royal Society near the end of 1664 became its most ardent defender. Though his work is largely devoted to a plea for scepticism and to an elaboration of the reasons for uncertainty and lack of progress in knowledge, which involves an attack on Peripateticism and the worship of Aristotle, there is hardly a single important trend in the new thought of the age which does not traverse it. Though chiefly inspired by Descartes, it refers frequently and with praise to Bacon,[65] while Harvey, Gassendi, Hobbes, More, Digby, and others are duly noticed. Glanvill at this time was associated with the Cambridge Platonists and reflects French more than English scientific thought in that he is more concerned with philosophical conclusions drawn from scientific investigations, with philosophical systems than with concrete experiments or the experimental method, though he obviously subscribed to the latter. Moreover, his interest lies mainly in pure science, for the element of utility occupies a subordinate place in his work, in contrast to the whole English trend of science. After he joined the Royal Society, however, his point of view underwent a great change. He speaks in much less enthusiastic terms of Descartes; his praise of Bacon becomes extravagant; and the experimental method together with the usefulness of the new science receives all the stress.[66] His evolution discloses in a nut-shell the essential difference between the scientific spirit of England and that of France in the seventeenth century.

Glanvill prefaces his attack on the ancients by saying that although Bacon had "handsomly" exposed the defects and errors of ancient learning, he will venture some considerations of Peripateticism, which has survived all other philosophies of the past, in order to show "how little our adherence to *Antiquity* befriends Truth," and especially "how groundless are the *Dogmatists* high pretensions to Science."[67] In contrast to the sceptical spirit of the new science, that of traditional learning was dogmatic, and Glanvill seeks to undermine this dogmatism by showing the erroneous nature of the philosophy which inspired it. He adds nothing new to what had by now become the stereotyped charges against the authority of the past. Reverence for antiquity stops all intellectual progress; had ancient authority prevailed universally, "*Hercules* his Pillars had still been the worlds *Non ultra*," and such discoveries as Galileo's would never have been made.[68] He marshals the well-worn evidence against Peripateticism: it is only a verbal, notional philosophy, a huddle of words and non-significant terms, which have no counterpart in reality, and upon which the intellect feeds as a chameleon on air. From the charge of empty verbosities he passes to a criticism of the litigious nature of the philosophy, which retards instead of advancing science, since the subjects for debate are chosen for their controversial possibilities, and other subjects capable of some demonstration are neglected. He attacks Aristotle's verbal explanations, which do not explain, and exposes his astronomical errors revealed by the telescope. He echoes Bacon in declaring that Aristotelianism has proved itself inept for discoveries, since it has not led to a single invention. In contrast, the new science inspires him, as it inspired others in this period, with appreciation of, and confidence in, his own age and in future scientific progress. All arts, he says, are capable of improvement, and "there is an *America* of secrets, and unknown *Peru* of Nature."

Methinks this Age seems resolved to bequeath *posterity* somewhat to remember it: And the glorious Undertakers, wherewith Heaven hath blest our Days, will leave the world better provided then they found it. And whereas in former times such generous free-spirited Worthies were, as the Rare newly observed *Stars*, a single one the wonder of an Age: In ours they are like the lights of the greater size that twinkle in the *Starry Firma-*

ment: And this last Century can glory in numerous *constellations*. Should those Heroes go on, as they have happily begun; they'll fill the world with *wonders*. And I doubt not but posterity will find many things, that are now but *Rumors*, verified into *practical Realities*. It may be some Ages hence, a voyage to the *Southern* unknown *Tracts*, yea possibly the *Moon*, will not be more strange then one to *America*. To them, that come after us, it may be as ordinary to buy a *pair* of *wings* to fly into remotest *Regions;* as now a pair of *Boots* to ride a *Journey*. And to conferr at the distance of the *Indies* by *Sympathetick* conveyances, may be as usual to the now comparatively *desert* world into a *Paradise*, may not improbably future times, as to us in a *litterary* correspondence. . . . And the turning of the now comparatively *desert* world into a *Paradise*, may not improbably be expected from late *Agriculture*. Now those, that judge by the narrowness of former *Principles*, will smile at these *Paradoxical expectations:* But questionless those great Inventions, that have in these later Ages altered the face of all things; in their naked proposals, and meer suppositions, were to former times as *ridiculous*. To have talk'd of a *new Earth* to have been discovered, had been a *Romance* to *Antiquity:* And to sayl without sight of *Stars* or shoars by the guidance of a *Mineral*, a *story* more absurd, then the flight of *Daedalus*. . . . *Antiquity* would not have believed the almost incredible force of our *Canons;* and would as coldly have entertain'd the wonders of the Telescope. In these we all condemn *antique incredulity;* and 'tis likely Posterity will have as much cause to pity *ours*. But yet notwithstanding this straightness of shallow observers, there are a set of enlarged souls that are more *judiciously credulous* . . . .[69]

This essentially was to be the spirit of the nineteenth century, of *Locksley Hall* and, indeed, of all science-inspired belief in progress. In the chapter following this glowing "vision of the world and all the wonder that would be" is found the story of the Scholar Gypsy, which some two centuries later was to inspire Matthew Arnold's poem. The ironic fulfilment of the prophecy is sufficiently apparent.

> O born in days when wits were fresh and clear,
>   And life ran gaily as the sparkling Thames;
>   Before this strange disease of modern life,
> With its sick hurry, its divided aims,
>   Its heads o'ertaxed, its palsied hearts was rife—
>   Fly hence, our contact fear!
> Still fly, plunge deeper in the bowering woods!
> Averse, as Dido did with gesture stern
>   From her false friend's approach in Hades turn
>   Wave us away, and keep thy solitude.

Such was the triumph of science in Arnold's eyes!

As is true of most of the assaults on Aristotle, Glanvill makes much of the incorporation of Aristotelian philosophy in school divinity. "A *Schoolman*," he says, "is the Ghost of the *Stagirite*, in a Body of condensed Air: and *Thomas* but *Aristotle sainted*."[70] He quotes Luther's condemnation of that theology, and ascribes to the latter's litigious and disputatious nature the lack of unity in the Christian church. The "notional theologues" are analogous to the notional scientists.

Several chapters treat, together with the evils of dogmatism, of the difficulty of attaining certain knowledge.[71] The concluding chapter is devoted to a justification of philosophy, in which he relies heavily on the great scientific thinkers of his age. It is their example which inspires him to oppose the theory of nature's decay, an idea which had by no means lost its enervating influence, and which in a subtle manner subbased the opinion of the superiority of the ancients.

And me thinks, those generous Vertuoso's, who dwell in an higher Region then other Mortals; should make a middle species between the *Platonical* θεοί and *common Humanity*. Even our Age in variety of glorious examples, can confute the conceit, that souls are equal: And the sole Instances of those illustrious Heroes, *Cartes, Gassendus, Galileo, Tycho, Harvey, More, Digby;* will strike dead the opinion of the worlds decay, and conclude it, in its *Prime*.[72]

This is not the first example of the "heroizing" of scientists, an attitude which clearly anticipates modern hero-worship of the masters of science. The shift of values shown by it is an important indication of the change in values which has revolutionized human thought.[73]

The revolt from the ancients arose from many causes. The old explanations had failed to satisfy the spirit of inquiry born of the Renaissance. The traditional philosophy seemed so far removed from reality that it evaporated into mere verbal mists, and also the discoveries of this and the preceding century had shown many traditional ideas to be erroneous. Copernicus, together with Kepler and Galileo, Gilbert and Harvey, had so impressed upon the age the increase of modern knowledge over ancient that the prestige of the latter was sorely shaken. Nor should the part which Paracelsus and his followers played in discrediting antiquity be overlooked. Inspired by recent discoveries and sustained by an unwavering faith in the possibility of progress, new thinkers were eager to widen the bounds of knowledge.

The realization that superstitious reverence for antiquity, embodying the disheartening belief that nothing new was to be discovered, was an effectual bar to the advancement of learning, incited them to vociferous iteration that there were no Pillars of Hercules with their *ne plus ultra* for modern thought, and to vigorous assaults upon the ancients. This eager desire to increase knowledge and this active impatience with all its hindrances, coupled with the anti-authoritarian disposition characteristic of the Reformation, led them to insist upon the importance of a critical mind unbiased by ancient prepossessions and upon the right to philosophize with utmost freedom. They realized that only under such conditions could the inadequacy of ancient science be made manifest and new truths discovered. Finally, their firm faith in, and enthusiastic allegiance to, the Baconian method, with its insistence upon experimentation and utilitarianism, sharpened their antipathy to an inutile science gathered from confused or fallacious books, created the antithesis of works and words, and moved them to contrast the strenuous activity of direct investigation with the lazy ease of mere reading. Other reasons there undoubtedly were for the opposition to the past, such as the mingling of Aristotle's philosophy with divinity and also the hostility of his ideas to Christian doctrine, but they did not operate so universally as the others. In almost every aspect of the revolt Bacon's influence is apparent; to him more than to any other single man were due its spirit and success.[74]

In a larger way, the clash between the two forces was really, as Rigault pointed out long ago, between conservatism and liberalism, between progress and tradition.[75] The stock arguments of conservatism since the world began appear in the defence of antiquity. We should honor the ancients because they are the fathers of our knowledge and have received the sanction of many ages and of all the universities. We should be slow to discard the findings of civilization. There were an irrational hatred of novelty and an equally irrational love for the old and tried. Individual wits who would pit their own ideas against the tested truth of the past should be curbed in their pride. The conservatives even steal an argument, in which they were not far wrong, from their opponents in claiming that the religious factions which were disrupting England were due to new-fangled ideas and the revolt from traditional thought. The stock liberal attitudes also appear: the evil of the old, the need of change, emphasis upon freedom of thought and discussion, enthusiastic loyalty to new ideas which inspire exorbitant hopes and extravagant visions of the future. The perennial clash between these two forces reveals itself in the rise of the new science, but it is also to be discovered in all the intellectual and social upheavals of modern times.

CHAPTER VII

PROJECTS, INVENTIONS, AND THE PROGRESS OF SCIENCE

> . . . . there are two very distinct Ends that Men may propound to themselves in studying Natural Philosophy. For some Men care only to Know Nature, others desire to Command Her: or to expresse it otherwise, some there are who desire but to Please themselves by the Discovery of the Causes of the known Phænomena, and others would be able to produce new ones, and bring Nature to be serviceable to their particular Ends.
>
> Robert Boyle, *Certain Physiological Essays*, 1661.

The two aspects of science, which Bacon clearly pointed out and which we today recognize under the titles "pure" and "applied," have to do with knowledge as an end and knowledge as a means. Bacon himself laid emphasis upon the latter, and the seventeenth century followed him in that respect. The utilitarian spirit, born of the humanism of the Renaissance but narrowed and emphasized by the Baconian philosophy, though generally going hand in hand with scientific thought, at times diverges somewhat and manifests itself in phenomena not obviously scientific. So the utilitarianism of the day is by no means restricted to the search for useful knowledge through observation and experiment, but with equal clearness is revealed in the craze for inventions and in the enthusiastic promulgation of projects. Inventions, of course, are really part and parcel of science, and were so regarded in Bacon's scientific program, but projects may possess a less obvious relationship to science. Yet the desire to organize all possible forces for the practical good of mankind, to introduce what might today be called a scientific method in meeting the everyday needs of man, and to lay a new emphasis upon the active and practical as opposed to the meditative and speculative can hardly be dissociated from the scientific movement of the period. Those who were seeking man's salvation not in books borrowed largely from antiquity but in more efficient marshaling of the factors making for his practical good are not far removed from the scientists themselves. Furthermore, such projectors as Comenius, Durie, and Hartlib, all men of Puritan sympathies, were in close touch with the Baconian movement.

It was the last of these men who, as it were, epitomized the period in this respect, a man upon whom the *Zeitgeist* cast such a spell that at the expense of all his energies and resources he remained its thrall to the very end. If we may believe the first editor of Boyle's works, Samuel Hartlib's mother was an English woman, and his father a Polish merchant who had moved to Prussia.[1] According to Birch, Hartlib came to England in 1630,[2] where he "soon distinguished himself by his great zeal for the improvement of natural knowledge, and making it useful to human life."[3] His works on agriculture secured for him Cromwell's favor and in 1646 a pension of £100 from Parliament, which was trebled the next year.[4] From that time on he was a tireless worker in the promotion of knowledge and the public good. He became, as it were, a clearing house for all new ideas, discoveries, and inventions. No sooner were these made known to him than he passed them on with enthusiasm to someone else. Full of schemes for promoting learning and the public good, he must have harried the government with his ideas of organizations and instrumentalities toward that end. He maintained a constant foreign correspondence in order to bring the new ideas of the continent to England.[5] His house was called by a contemporary "the center of all usefull and solid Learning,"[6] and his friend Beale eulogized him as "the sedulous advancer of ingenious arts and profitable sciences."[7] Milton's characterization of him in *On Education* has done much to keep his name alive.[8] Boyle in praising Hartlib's enthusiastic interest in "the invisible college" assured him that the members of the latter felt a sincere concern in all his vicissitudes. He could claim among his numerous friends Milton, Boyle, Gassendi, Beale, Ward, and Petty. He was especially interested in Petty, encouraging him in his scientific endeavors and trying in every way to assist him. Agriculture, anatomy, chemistry, and experimental science in general were the object of much of his activity.[9] He turned his kitchen into a chemical laboratory and was interested in another "universal laboratory" "to prosecute really philosophical studies" for the "health and wealth of all mankind," in connection with which, he says, "a general chemical council, not far from *Charing-cross*, sits often."[10] To the revolution which was taking place in science he was keenly alive. He expressed pleasure that the scholars in Oxford were sensible of any reformation in learning and he went into detail regarding the development of "experimental learning" at Cambridge.[11] At the beginning of the fifth decade of the century he was interested in "an institution for the advancement of learning" and in the late fifties he expressed the hope that Richard Cromwell would do more to promote it than his father had done.[12]

The fact that Bacon was Hartlib's chief inspiration is apparent in all the latter's activities. Not only in his strong utilitarian and social-minded spirit, in his constant use of the expression "advancement of learning," and in his emphasis upon experiment, but also in definite acknowledgments Bacon's in fluence reveals itself. One of the purposes of his "Office of Publicke Addresse" is "To put in Practice the Lord Verulams Designations, *De Augmentis Scientiarum,* amongst the Learned," and he repeats that the agent is to act "for the advancement of Divine and Humane Learning, according to the Counsell and Designe of Lord *Verulam,* to whose structure, by their joint advice, every yeare some stones should bee added."[13] Surely this is building on the foundation laid by Bacon, to whom, indeed, Hartlib attributed many of his ideas.[14] In his ambitious schemes Hartlib comforts himself with the hope which Bacon holds out relative to his natural history, that all "those things are to be held possible and performable, which may be accomplisht by some Persons, though not by every one; and which may be done by the united labours of many, though not by one apart; and which may be firnisht by the Publique Care and Charge, though not by the Ability and Industry of Particular Persons."[15] Furthermore, the writers of treatises sponsored by Hartlib often recognize in Bacon the source of their inspiration.[16] Perhaps there was no one at this time, with the exception of Boyle, of whom the spirit of the great Chancellor took more complete possession.

Though Hartlib's interests were as wide as science itself, there were three subjects to which he was especially devoted. One was agriculture; another was some kind of organization for the advancement of useful learning; and the third was what he called "An Office of Publicke Addresse in Spiritual and Temporall Matters."[17] In his exposition of the last, he gives the office a twofold function: to serve as "an Addresse of Communication for Spiritual matters" and as "an Addresse of Accommodations for Bodily matters." The latter, in the description of which utilitarian zeal runs wild, was to be an employment bureau, advertising agency, real estate exchange, and the like, in short, a means of bringing sellers and buyers together. Its purpose is further elaborated upon in a treatise[18] which appeared a year later, in which it is stated that the office should contain four "Registers." The first would be for the poor, either in the way of finding employment or, in the case of incapacity for work, of listing possible benefactors, names of physicians willing to do charity work, legal advisers, and scholarships for poor students. The association of humanitarianism with early science is here emphasized by the needs of the poor being the first to be provided for.[19] The second register, for commerce and bargains, was to furnish information about money, rate of exchange, food, drink, hotels, property for rent, household goods, libraries, booksellers, etc. The third was a register of persons and actions in all offices and relations, in which are included an employment bureau for women, ministers, lecturers, professors, teachers, physicians, clerks, nurses, and others; a matrimonial agency; a newspaper giving foreign and domestic news, notices of marriages, deaths, births, and the like, and hours of departure of carriers and messengers; a room for checking parcels; and a tourist agency where strangers might receive their mail, be informed of the best inns, and buy travelers' cheques and hotel coupons. The last was a register of ingenuities and matters commendable for wit, worth, and rarity. "Here then, if any hath a feat in any Science which is extraordinary. Either a new discovery of a Truth, or an Experiment in Physick, Mathematicks, or Mechanicks; or a method of delivering Sciences or Languages, not ordinarily known, and very profitable; or some intricate Question and difficulty which he would have resolved by the most experienced in any or all Arts," he may come. Mathematical, astronomical, and other scientific instruments were included under this head.[20] The insight into social needs revealed in the proposals given above is clearly borne out by the many instrumentalities devised by our world to meet these very needs.

The "Office of Addresse of Communications" Hartlib considered much more important than the other. Its object was information in matters of religion, learning, and ingenuities, "which are Objects of Contemplation and delight unto the Mind,

for their strangenesse and usefulnesse unto the life of Man."[21]
Its purpose in religion, inspired, perhaps, by Durie, was to prevent schisms by a unification of all protestant churches. In
learning its function was threefold: to carry out the scientific
program outlined by Bacon; to further Comenius' educational
reforms; and "to offer the most profitable Inventions which he
[the agent] should gaine, unto the benefit of the State, that they
might be Publikely made use of, as the State should think most
expedient."[22] It is especially emphasized that the agent in charge
of this work, by domestic and foreign correspondence and by
other means, must keep in touch with the latest profitable undertakings, books, discoveries, and inventions in the interest of
those to whom they may prove useful.

. . . hee [the agent] should yearly once at a certaine time bee obliged to
give up the account of his Annuall Negotiation, to the Professors of all
Sciences in both Universities, and to the Heads and Masters of Colledges
and Halls, who should bee made a Speciall Committee and appointed,
according to their severall Faculties, or all jointly to meet, and to take into
Consideration the things which he shall produce: that such peeces as shal
deserve to bee put into the Publike Libraries, to bee made Common unto
Scholars, or otherwise published in Print for the benefit of every one, may
by their advice bee applyed unto their proper Uses; for the advancement
of Divine and Humane Learning, according to the Counsell and Designe of
the Lord *Verulam,* to whose structure, by their joint advice, every yeare
some stones should bee added. And to this effect a more speciall Way
of Concurrence and Correspondency amongst the Professors and Heads
of Colledges themselves should in due time bee contrived and proposed.[23]

Hartlib evidently wished to enlist the services of the universities
in behalf of the Baconian philosophy, a fact which is also apparent in his suggestion that the office of address of communications should be located at Oxford. The other office, of course,
was to be placed in London. It was also proposed that Parliament
appoint Hartlib superintendent general over all offices of address in England and Wales, that £200 be allowed him, and that
a "great house" be given him for an office.[24]

There are two expressions which with slight variations occur
with such frequency throughout these propositions that they
stamp themselves indelibly upon the mind of the reader: the
public good and the advancement of learning. One may venture
to assert that Hartlib was animated by no other motive than
this philanthropic and scientific passion. The relationship be-

tween the two is quite apparent, and equally obvious is their relation to Bacon. Hartlib's whole outlook upon life was conditioned by the scientific program of Lord Verulam with its utilitarian and social implications. Fired with the zeal of a fanatic, he left no stone unturned in pushing his public-spirited projects to completion. He seems to have been entirely selfless. He must have found comfort in the sentiment expressed by his friend Petty:

> For most men will not intend any Publick Ayme till they can secure their own Interests, and see a way to get advantage by that which they call the Publick: but wee [Hartlib and Petty] shall never ayme at this; our delight shall be, that all may be advantaged, and the Publick Interest of the Commonwealth setled, although it should be to our cost and disadvantage: For we know the promise, that if we faint not, and become not wearie in welldoing, we shall reap in due time the fruit of righteousnesse.[25]

So Hartlib was improvident as regards his own interests, spent all his money on enterprises calculated for the public good, and when his pension was stopped by the royal government at the Restoration, he was reduced to dire straits, had to beg assistance from his friends, and ended wretchedly. He was interested almost entirely in the utilitarian aspect of science, but no one ever justified his creed by his life more conscientiously than Hartlib.[26] Hartlib can hardly be called a scientist in the usual sense of the word. Rather he was, as has been said, a disseminator and encourager of scientific information, as well as an enthusiastic propagandist for the new science and the great good it could confer upon the world.[27] He was full of schemes for the advancement of science and its application to the needs of life, but the greater part of his activity lay in passing on to the world, either in print or privately, the scientific findings of others. This fact is clearly revealed in his concern with agriculture. What is perhaps his most significant agricultural publication is only a collection of letters from various anonymous writers, some from abroad, telling of agricultural experiments and observations somewhat in the manner later adopted by the *Philosophical Transactions*.[28] In the same way he was responsible for only the publication of the anonymous *A Designe For Plentie, By an Universall Planting of Fruit-Trees: Tendred by some Wel-wishers to the Publick,* a work permeated with the utilitarian spirit, the author of which, Hartlib tells us in the preface, was an aged clergyman who had experimented with trees for many years.[29] But the man in whom Hartlib had the greatest confidence, and whose works he vigorously promoted was Cressy Dymock. The first of these, *The Reformed Husband-Man*, 1651,[30] while pointing out many defects in the usual method of cultivating corn and expatiating at length on the greater returns secured from land by the author's method, does not reveal that method, but insists upon payment for doing so. Nor is this the only treatise of the day to suggest a pecuniary motive, though Hartlib was singularly free from the latter. Neither does he seem to suspect it in Dymock, for in the preface to the work under discussion he says,

the Excellency of these *Proposals*, is, that they seek to impart to all, even to the whole *Nation*, the *Meanes* of all *Plenty* and *Riches*: and they offer a course which may improve by *Experimental Industriousness*, some rare known *Advantages* of *Husbandry*, unto an infinit increase (whereof the possibility is in nature) beyond that which at present is attainable in the ordinary way. For the *Authour* presumes not to have exhausted all that may be found out in nature: But by that which *God* hath imparted unto him beyond others, he perceives that a greater perfection of knowledge may be gained; if only the *Observations* which he hath made, be followed.

Here is faith in the experimental method as practically applied to human needs; here also the essence of social utilitarianism which insists upon the public good as the highest of human motives. The same is true of Hartlib's preface to Dymock's *A Discoverie for Division or Setting out of Land*,[31] which is full of a utilitarian humanitarianism and complains of the lack of appreciation of those who love humanity. The treatise itself is insignificant, containing communications from various men, mostly anonymous, regarding the method of caring for cattle, raising corn, and the like.

In view of all this interest in agriculture it is not strange to find Hartlib sponsoring a proposal put forward by Dymock to establish an agriculture college.[32] In the preface, the former asserts that it is only the narrowness of men's spirits that is responsible for their misery, for if they would look beyond themselves to co-operation with their fellow-men, "we could not be without *Luciferous Employments* for our selves; nor *Unfruitfull* to our neighbours, as now for the most part we are, only because we mind not the *Objects* of that *Industriousness*, which without a mutuall Concurrence cannot be advanced." The need of co-operation, ultimately traceable to Bacon,[33] is in large part responsible for the scientific and near-scientific activity of the Puritan era. The Puritan bias of this attempt to establish a school of agriculture is apparent in Dymock's words. Especially do his ideas reveal the rank materialism characteristic of the Puritans, to whom the element of utility in Bacon's philosophy made an irresistible appeal.

In humane affaires, and which relate not immediately unto God; nothing doth more tend to the good & wel-being of a *Nation* (God giving his blessing thereunto in an humble and right use of it) then plenty of *Food* and *Raiment*, and of all other *Merchantable Commodities* to send abroad; which will not faile to returne the prosperity and hapinesse of other *Nations* again in exchange. And surely a *Nation* thus blessed can want no *Earthly Comfort.*[34]

It is not necessary to go into the details of his proposed college, except to point out that the experimental method is emphasized above traditional methods; instead of relying upon the past, progress in agriculture is to be sought through principles founded upon experiments and true scientific induction. In his school, he says,

some may teach, some learne, and all practise the whole and every part of this so *honourable* an *Art,* so *deep* a *Mystery,* and that not onely in the more customary and *Common way;* but according to the most *Excellent Rules,* that *Ingenuity* and *Experience* gained by *rationall trials* & real *Experiments* have or can attaine to; and so the honour, wealth, and happiness of this *State* may be multiplied, even before it self is aware, and the duller members thereof wonne by emulation or example to such practises for their *own private* & *Publique* Good.[35]

In all the various projects and proposals, the list of which given above is representative rather than exhaustive, occur again and again the catchwords of the period: public good, public-spirited, relief of the poor, advancement of both learning and piety, usefulness, and experiment. The emotional intensity with which the Puritans embraced the utilitarian and social correlatives of Bacon's philosophy is matched by the strength of their allegiance to experimentation as the test of the validity of truth. The complete absence of references to the classics, or any calling upon authority, is striking. Truth has now come to rest upon an experimental basis. Nature and not books is the object of their study. They had learned their lesson well of Bacon. In them also the scientific spirit entered practical fields and sought through

intelligent organization and rational instrumentalities to meet in
an efficient and scientific manner the everyday needs of man.
They had not yet learned the word "efficiency," but its spirit is
quite apparent.

The same attitudes are revealed in the craze for inventions,
which bulked as large in their eyes as in ours. And Hartlib was
as zealous in promoting them as other things. His friend Dymock
relinquished his agricultural interests long enough to advertise
a creation of his as *An Invention of Engines of Motion Lately
Brought to perfection*,[36] an advertisement which reveals Baconian inspiration in a two-page translation from the *Novum
Organum* relative to the importance and nobility of inventions.
The author makes due acknowledgments of assistance from Hartlib, and lays claim to a philanthropic motive in announcing
his invention. The treatise, which possesses no significance,
speaks with enthusiasm regarding the virtues of the engine, but
gives no idea of the engine itself,[37] the reason for which neglect
may lie in Dymock's unwillingness, so he says, to make it known
until the laborers supplanted by his machine could find other
employment.[38]

Always keeping an eye open for anything new in the way of
inventions, no sooner did Hartlib learn of one than he broadcast
it among his friends or sought to make it effective. Evelyn
speaks of going "to visit honest and learned Mr. *Hartlib,* Public
Spirited and ingenious person who had propagated many Usefull
things and Arts," and who "was master of innumerable curiosities, and very communicative."[39] Hartlib's letters contain frequent references to inventions. From Beale he received an
account of a newly invented water-clock, and he writes Boyle
enthusiastically of some recently invented carriages, of which
he had heard from a Dr. Horne. He tells his same correspondent
of hearing Dr. Chamberlain boast of having devised an instrument to enable vessels to move against wind and tide, and remarks that Mr. Morian had promised him a large account of a
remarkable contraption for catching and condensing sunbeams.[40]
His only venture into the field, if it can be called an invention,
was a rather feeble attempt at a universal language which he
described in a treatise entitled *A Common Writing,* 1647, an
early step in a movement which gathered strength until it

reached its climax in Wilkins' *Essay towards a Real Character, and a Philosophical Language,* 1668.[41]

The urge to invent was abroad in the land. The few instances cited in this chapter are hardly an adequate indication of the vogue which appears everywhere we turn. Petty was especially stung by the inventive bee. During the Restoration his double-bottom boat created a furore, as Pepys makes clear in his *Diary.* In 1648 he put forth a description of some kind of multigraphing machine in *A Declaration Concerning the newly invented Art of Double Writing,* for which he received a patent from Parliament. Hartlib, who thought that the lack of success of this machine had impaired Petty's reputation, expresses the wish that he "would shew himself to the world by some rare piece or other."[42] When Durie journeyed to Germany for the ambitious purpose of establishing the unity of all protestant churches, he also declared his intention of gathering "All inventions, and Feats of Practise in all Sciences," "such as may bee profitable to the health of the body, to the Preservation and Encrease of wealth by trades and mechanicall Industries, either by sea or Land, either in Peace or Warre."[43] Inventions rank next to religion in importance. In the light of this great interest in inventions, we are not surprised to discover a history of them, perhaps inspired by Bacon's proposal of a history of trades, published near the end of the period under discussion, in which dials, clocks, watches, spheres, globes, air and water machines, printing, drawing, painting, spinning, weaving, the compass, telescope, microscope, and a host of other things figure.[44]

The man who best expressed the inventing temper of the age was Edward Somerset, second marquis of Worcester. Only a few years after Bacon's death he set up a laboratory, especially devoted to mechanical inventions, in charge of which he placed Caspar Kaltoff, who served him faithfully to the end of his life. Between 1638 and 1641 Worcester exhibited a supposed perpetual-motion machine before Charles I and other illustrious men. His interest in the invention continued unabated up to his sojourn with the exiled court in France, 1648-52. Upon his return to England in the latter year he was committed to the Tower, from which he was released three years later. Yet imprisonment did not interrupt his activities. Hartlib writes to Boyle, May 8, 1654, "The earl of *Worcester* is buying *Vaux-hall* from Mr. *Trenchard,* to bestow the use of that house upon *Gaspar Calehof* and his son, as long as they shall live; for he intends to make it a college of artisans."[45] After his release from the Tower the marquis did everything possible to secure aid for his inventions, especially his "water-commanding engine," which may have been the first steam engine. Through Secretary Thurloe he even solicited the aid of Cromwell.[46] He seems to have been as fanatically devoted to his inventions as Hartlib was to his projects. Certainly he spared no expense. In his petition to Charles II he claims to have spent over £9,000 on his laboratories and more than £50,000 in carrying on experiments, a princely sum in those days. In 1663 he received a patent for his water-commanding engine, from which he expected to derive large profits, but though it was exhibited to the public for some time, it was not taken seriously by important people.[47] Though he does not mention Bacon, Worcester is an outstanding example of the hold which the inventive spirit, fostered by Verulam, was fixing on men.[48]

Turning from the practical aspects of the scientific movement, we find that distinct progress is being made in the development of the thought and spirit of the new science. Three great discoveries were especially influential in opening the eyes of mankind to the new science and in undermining the traditional reliance upon the authority of antiquity. Two, those of Copernicus and Gilbert, have already been touched upon. The other, Harvey's discovery of the circulation of the blood, first announced in 1616 but not published until 1628, was somewhat slow in making headway in England, but at the beginning of the period under discussion it comes into prominence. Harvey himself was no enemy to the ancients. Blind to the discredit cast upon the latter by the discoveries of the Renaissance, and unresponsive to the attacks made upon them by Gilbert and Bacon, to say nothing of continental thinkers, he remained a Peripatetic to his death.[49] Yet he reveals the new attitude in the emphasis placed upon experimentation, so nobly vindicated in his own work. His injunction was "to study Nature by way of experiment," and in his published work of 1628 he asserts that for nine years or more he had confirmed his views by multiplied demonstrations, thus putting into practice his own doctrine that "Without the due admonition of the senses, without frequent observation and reiterated experiment, our mind goes astray after phantoms and appearances." He was by no means indifferent to the unprofitable habit of relying on traditional learning. "The method," he says, "of investigating truth commonly pursued at this time, therefore, is to be held erroneous and almost foolish, in which so many inquire what others have said, and omit to ask whether the things themselves be actually so or not." And he expresses himself in stronger terms to Dr. George Ent: "I have often-times wondered and even laughed at those who have fancied that everything had been so consummately and absolutely investigated by an Aristotle or a Galen, or some other mighty name, that nothing could by any possibility be added to their knowledge."[50] It is the attitude revealed in these quotations, as well as his monumental discovery, which creates our interest in him. His emphasis upon experiment and the exposure of the errors of antiquity effected by his discovery did much to encourage the new science to break with the ancients.

An illustration of this last statement appears in a published defence of his discovery, Dr. George Ent's *Apologia pro Circuitione Sanguinis. Qua respondetur Æmulio Parisano Medico Veneto,* 1641. In his dedication to Harvey the author inveighs against those who believe that all knowledge is to be found in the ancients "as if antiquity were the only mark of truth," and he states that "it is considered impious to wish to deprive them [the ancients] of this absolute sway. Thus modernity is shackled to antiquity and those who would revolt are vilified. If a new discovery is put forth, buttressed with demonstrations, then envy declares that it had been discovered by the ancients or poorly demonstrated." This, he says was Harvey's fate. A better example of the latter's influence upon the medical thought of the time is discovered in Thomas Willis, known for his work on diabetes. In a treatise,[51] published in 1660, he bestows unstinted praise upon the discovery of the circulation of the blood, attributing to it the collapse of ancient medicine on the ground that it discredited the idea of the four humours and thus changed the method of curing diseases. Now, he claims, it does not behoove wise men to close their eyes to the light, preferring to err with the ancients rather than be right with the moderns, for Harvey has remade medicine from the ground up, in which process he has rendered false and obsolete what has been derived from the classics. To the conservatives in medicine who feared to change from tradition in matters affecting human lives, he retorts that it is only by following the example of Hippocrates in cultivating observations and experiments that medicine can progress.

The discovery did more than any one thing to break the bonds placed on medicine by the past, though Bacon's influence in this respect was also very strong. Furthermore, it vindicated the observational and experimental method and powerfully promoted its development and utilization.[52] The attitude characteristic of Harvey begins to be manifested in other medical treatises, such, for instance, as Francis Glisson's *A Treatise of the Rickets,* 1651.[53] Disclaiming any selfish motive in publishing the work, the author asserts that he was incited

by this Consideration only, That becaus we are not born for our selves, we might make these [experiments] (such as they are) common, which in som measure may advance the health of Infancy . . . . and likewise propagate an Encreas unto Learning: with this hope also, That by this Example we may invite the Wits of other most learned men to make inquisition into the Essences of Diseases and their Causes, and to examin these our Labors that posterity may enjoy them yet more perfect.

The treatise, which truly purports "to tread in unbeaten paths," contains a most careful and accurate description of the disease, based upon direct observation. This emphasis upon experiment and observation, together with a declaration of faith in them, is found in other passages: "We will in this Chapter briefly and faithfully declare those things which we have hitherto Observed by long experience and frequent Dissections, namely, Those things which we have seen with our eyes, and handled with our hands." In the following quotation we catch an echo of Bacon regarding hasty conclusions gathered from insufficient experiments: "Not to make a rash judgment from the inspection of one or two bodies, but first by a reiterated and sedulous experiment, to be able to distinguish what things perpetually occur, what for the most part, and what but seldom, in the dissected bodies that have perished of the same Disease."[54] Certainly we are moving rapidly under the influence of Bacon and Harvey to Sydenham and the establishment of clinical medicine.

Not only medicine but science in general is advancing from its dependence on books to intellectual independence and first-hand observation. And with this advance Bacon's spirit marches *pari passu.* The influence of his philosophy, as well as the great prestige of his name, is everywhere to be observed, but perhaps nowhere more clearly than in the next two works to be considered. The author of the first, Ralph Austen, a writer of various works on gardening, reveals an almost comical fear of disagreeing with Bacon, the "Learned, and incomparable Author," who "hath left unto men such *Rules,* and helps in all kinds of Learning, that they will be much wanting to themselves, if *Arts,* and *Sciences* improve not, very much above what they have been in former ages."[55] Though some may think him too bold in differing from "the Judgment of so *Eminent,* and *worthy an Author,*" he holds that Bacon himself would approve of his action, since the latter urges in the *Advancement of Learning* that speculative men write the results of their experiences, and since Verulam was not moved by self-interest but sought only truth "for the *good of future generations.*"[56] His own experiments and observations and "the publique profit" constitute his apology for disagreeing with Bacon, whose name was now one to conjure with.[57] The method is ranked higher than, in Austen's eyes, its author.[58]

The writer of the second treatise is probably the most modest and humble follower of Bacon to be discovered in the annals of science. Joshua Childrey, who became archdeacon of Sarum in 1664, tells us that he "first fell in love with Lord Bacon's philosophy in the year 1646, and tried several experiments (tho' such as I now reckon be not of any moment) in 1647, 48, 49, 50." But two years before the Restoration he set about in earnest and "bought me as many paper books of about sixteen pages apiece as my Lord Verulam had histories at the end of his *Novum Organum.*"[59] Thus equipped he proceeded to record the "natural rarities" of all England, Scotland, and Wales in a book, the title of which indicates the source of his inspiration.[60] Paying high tribute to Bacon, his "Master . . . . from whom I received my first light," and whose precepts he religiously followed, he proposes his work as a supplement to the *Novum Organum.* He does not forbear pointing out the injury done science by the authority of Aristotle and the scholastic philosophy. Not content with the limits observed by the treatise, he expressed an ambitious intention, worthy of Bacon's comprehensiveness, of publishing *Philosophical Rarities of the World,* an undertaking in which he had made considerable progress, and which he believed would contribute much to the advancement of learning. Childrey's volume describes, shire by shire, the inhabitants, soil, crops, mineral resources, climate, topography, plants, animals, antiquities, and strange phenomena of all kinds. His nature reveals a mixture of conflicting elements. Though at times manifesting a superstitious and credulous spirit (he believed in astrology, though sceptical of many of its principles and advocating a scientific method of testing it), he frequently maintains a critical and scientific attitude. In his opinion one advantage to be derived from science was the destruction of superstition. In fact, Childrey presents an excellent example of the struggle between the budding scientific spirit of the age and the inherited chaos of superstition and fear. Nature is still mysterious, but there is a disposition on his part to assign all phenomena to natural causes. His work furnishes a clear picture of the impact of Bacon's philosophy on a representative mind of the age.

It is a long step from Childrey to Robert Boyle, though they did have in common admiration of Bacon and faith in his philosophy. Boyle dominated science in England during the third quarter of the century. The "father of chemistry," the discoverer of an important law in physics, and the promoter of medical knowledge, he occupies a significant place in the development of science. His scientific curiosity seems to have been first aroused, while he was still in his teens, by attendance at the meetings of the London group of experimenters, to which he referred in 1646 as the philosophical or invisible college. Certainly by the beginning of the next year he was conducting chemical experiments in Ireland.[61] From that time on, his life was given over to experimenting and writing, and though his choice of subjects was wide, little that he wrote does not reveal his scientific attitude. But he was more than an experimenter and discoverer; above everything else he was a missionary in behalf of the new science, a proselyter of those who held to the scholastic philosophy, and a propagandist and popularizer of the experimental method. By his own fruitful investigations he demonstrated the value of the experimental philosophy, and he gave it impetus by the sincerity and indefatigability of his propaganda. To his first published scientific work, *New Experiments Physico-Mechanicall, Touching the Spring of the Air, and its Effects,* 1660, in which he laid the foundation for his important law of gases, he definitely assigned the purpose "of promoting Experimental and Useful Philosophy."[62] "Experimental learning," or "experimental philosophy," is the key-word to his work, and the results of his own research are put forward in the hope of inciting others either to repeat his own or to make new experiments. The importance of experimentation as the criterion of truth finds varied expression. Those, he says, who deny the possibility of a vacuum do not base their opinion on experiments, which have proved to him the opposite.[63] He refuses to accept any idea until it has been verified by an experiment. Nor does the fact that his theory, buttressed by his frequent trials, is opposed to the scholastic idea of nature's abhorrence of a vacuum disturb him in the least.[64] But though at almost every turn he finds himself opposed to the traditional philosophy,[65] and though he was a resolute champion of the Baconian method, he mentions the schools, Peripatetics, Aristotle, Galen, and others with little animosity, for dogmatic condemnation was repugnant to his calm, critical spirit.

Another aspect of his work, significant in view of his avowed purpose of promoting the experimental philosophy, was the publicity with which his experiments were conducted. He expresses pleasure at having "such judicious and illustrious Witnesses of our Experiment" as Wallis, Ward, and Wren, who freely offered explanations of it. On one occasion he speaks of Wilkins' bringing some of his inquisitive friends to his lodgings, and on another he sends for Wallis to witness a demonstration. He takes the trouble of noting that some of his work was performed "to the no small wonder of some famous Mathematicians, who were Spectators of some of these Experiments." He furnishes proper entertainment to a visiting "Virtuoso of quality" by permitting him to participate in an experiment. He is distinctly gratified by the wonder of bystanders (of whom, indeed, he frequently speaks) at some strange results of his experiments. The bystanders, however, who included women as well as men, were not always passive. In one case a bird was placed in a container from which the air was being pumped,

Which sort of Experiments seem so strange, that we were oblig'd to make it several times, which gain'd it the Advantage of having Persons of differing Qualities, Professions and Sexes, (as not onely Ladies and Lords, but Doctors and Mathematicians) to witness it . . . . [The bird] had probably [not] liv'd much longer, had not a great Person, that was Spectator of some of these Experiments, rescu'd him from the prosecution of the Tryal.

On another occasion the bird would have died but for "the pitty of some Fair Lady's (related to Your Lordship) who made me hastily let in some Air at the Stop-cock."[66] Surely Boyle had adopted an effective means for securing publicity for the experimental method and belief in its efficacy, for it is not hard to imagine the extent to which the bystanders talked about the "show."

The purpose of Boyle's next publication, written in the time of the Commonwealth, was even more obviously propaganda for the new method. Inspired by Bacon's caution against erecting systems on insufficient experiments and by his assertion that science is not the work of one man nor of one age, Boyle advocates the patient pursuit of experimentation in small fields as opposed to the hasty erection of general theories, or even principles and axioms, on too slight an experimental basis. His purpose is not so much to make public the demonstrations described in the treatise as to emphasize the importance of what he calls the "experimental essay," that is, collections of experiments for the explanation of which no theory need be evolved, and which are still valuable, even though wrong explanations may be suggested for them.[67] Boyle was influenced by the comprehensiveness characteristic of Bacon's philosophy in thinking that all the evidence must be in before a generalization should be drawn. To him the compilation of experiments and observations was the great necessity. Experimental truths, he says, are of most concern to mankind, and not the least impediment to the *"real* advancement of *true* Natural History" is the tendency to compile whole scientific systems which are suggested by only a few experiments, and the imposing titles of which deter modest men from publishing their findings because of the implication that there is nothing left to be discovered. Wise men, he claims, realizing that no one is in possession of even a small part of the data necessary to the formulation of a comprehensive theory in chemistry, anatomy, and the like, are content only to publish their experiments and let systems wait. He makes an exception to his general caveat against theories in the case of textbooks or books written to record what has been discovered so far, but he stipulates that such works be written by a knowing and very intelligent person, and after there has been considerable progress in knowledge since the last book of the kind. He also permits the temporary adoption of a working hypothesis, since he recognized its importance and necessity in investigation, as well as the fact that an absolute suspension of reason was impossible. In upholding experiments, rather than elaborate reasoning, as a sure means of refuting erroneous opinions, he cites the efficacy with which a simple observation or experimental discovery has completely overthrown some errors of the longest standing, such as the uninhabitability of the torrid zone, the solidity of the heavens, and the conduction of the blood to the heart by the veins. He pays proper tribute to the part played by the great discoveries of his and the preceding century in vindicating the scientific method and in destroying the authority of antiquity, when he says that the erroneous ideas mentioned above are no longer accepted because of "those new Discoveries with which they are inconsistent, and would have been abandon'd by the Generality of Judicious Persons, though no man had made it his businesse purposely to write Confutations of them."[68] The refutation of opinions by reason alone he looked upon as a great impediment to the advancement of true science. Though he realized that the learned framing of elaborate systems secured one a much greater reputation,[69] yet for himself he was willing to sink private interest for the good of science:

I know also that the way to get Reputation, is, to venture to explicate things, and promote Opinions . . . . whereas by my way of Writing to which I have condemn'd my self, I can hope for little better than to passe for a Drudge of much greater Industry than Reason, and fit only to collect Experiments for more rational and Philosophical heads to explicate and make use of. But I am content, provided Experimental Learning be really promoted, to contribute ev'n in the least plausible Way to the Advancement of it, and had rather not only be an Underbuilder, but ev'n dig in the Quarries for Material towards so useful a Structure, than not do something towards the Erection of it.[70]

In a way Boyle opposes experimentation to rationalism. Like Bacon, he felt that the first duty of a scientist was to contribute data for a universal natural history, and to be very hesitant to reason upon this data, since a sufficient amount had not been gathered for general theories. Like Bacon, too, he believed that many errors of the past and the present were due to the desire of men to explain before they knew what was to be explained.[71]

For this reason he stressed the supreme importance of experimentation, the chief purpose of which was to furnish data, and depreciated to a certain extent scientific theories. He was the missionary *par excellence* of the experimental philosophy.

The desire clearly to distinguish experiments from explanatory theories determined his attitude toward the mechanical, or, as he frequently calls it, the corpuscularian philosophy, and in this distinction is clearly seen the difference between the relations of Bacon and Descartes to the scientific movement in England. Though Boyle in general subscribed to the atomic theory, even writing a number of treatises in support of it, he by no means considered it absolutely certain. He disclaimed that he was wedded to atomic principles, adducing as proof that he had at times explained phenomena by other theories. There were other speculations in natural philosophy, he declared, besides the atomic, though he admitted that for the most part he had recourse in his explanations to "the Magnitude, Figure, and motions of Atoms." He conceded that explanations based upon matter and motion gave greater satisfaction to the intellect, but he was critical enough to hold that other explanations were not to be despised, wherein effects were deduced from obvious and familiar qualities or states of body such as heat, cold, fluidity, etc., which seemed to be atomic phenomena. It is interesting to see to whom he thinks the mechanical philosophy is especially adapted:

I think it therefore very fit and highly useful, that some speculative Wits well vers'd in Mathematical Principles and Mechanical Contrivances, should employ themselves in deducing the chiefest Modes or Qualities of Matter, such as are Heat, Cold, etc. and the states or conditions of it, as fluid, firm, brittle, flexible, and the like, from the above-mention'd most primitive and simple Affections thereof. . . . But I think too, we are not to despise all those Accounts of particular Effects which are not immediately deduced from those primitive Affections of Atoms.

The opinion that the mechanical philosophy is a matter of speculation, which seeks to satisfy the mind rather than the needs of life, finds fuller development in his treatise. He insists that no one should refrain from writing what he thinks, even though it may not conform to the atomic philosophy, for

there are two very distinct Ends that Men may propound to themselves in studying Natural Philosophy. For some Men care only to Know Nature, others desire to Command Her: or to express it otherwise, some there are who desire but to Please themselves by the Discovery of the Causes of the known Phænomena, and others would be able to produce new ones, and bring Nature to be serviceable to their particular Ends, whether of Health, or Riches, or sensual Delight. Now as I shall not deny but that the Atomical, or some such Principles, are likely to afford the most of satisfaction to those speculative Wits that aim but at the knowledge of Causes; so I think that the other sort of men may very delightfully and successfully prosecute their ends, by collecting and making Variety of Experiments and Observations, since thereby learning the Qualities and Properties of those particular Bodies they desire to make use of, and observing the Power that divers Chymical Operations, and other wayes of handling Matter, have of altering such Bodies, and varying their Effects upon one another, they may by the help of Attention and Industry be able to do many Things, some of them Strange, and more of them very useful in Humane life.

It is the utilitarian standard which raises the experimental philosophy in importance above the mechanical. Boyle points out that many "artificers," ignorant of the latter, have given the world many more useful inventions, such as the compass, than any "speculative Devisers of new Hypotheses, whose Contemplations aiming but for the most part at the solving, not the encreasing or applying, of the Phaenomena of Nature, it is no wonder they have been more ingenious than fruitful, and have hitherto more delighted than otherwise benefitted Mankind." The danger of reasoning out theories on insufficient evidence and the lack of usefulness in purely theoretical science depress scientific theories below the worth of mere sense-observation, though he believes with Bacon that when theories are established upon a much wider experimental basis, they may disprove "all Imputation of Barrennesse."[72]

The stress laid upon the importance of observation and experiment in opposition to hypotheses devised to explain them appears again in Boyle's attitude toward the chemists of his day. The latter, as we have seen, had been the most implacable foes of the ancients and the most vociferous advocates of experiments, and so they continued throughout the third quarter of the century. Though belief in the transmutation of metals had not vanished (Boyle himself was touched with the illusion), and though universal remedies were still talked about, the chief activity of the "pyrotechnists" was concerned with the discovery and improvement of chemical medicines. Yet the chemists had inherited theories, which they dogmatically insisted upon as explanations of their experiments, and the chief of which declared that all

matter was composed of salt, sulphur, and mercury. Against
these theories Boyle speaks with little reserve. They are obscure,
ambiguous, narrow, unreliable, and inconsistent. Even at that,
however, they are superior to the four elements of the Aristotelians and their unintelligible substantial forms. But he is careful to distinguish between the theories of the chemists and their
experiments. For the latter he has nothing but praise,[73] and
well he might, since his first scientific activity seems to have been
in chemistry, and he calls himself a great lover of chemical experiments.[74] It is their experiments, Boyle explains, that have
secured them a more favorable hearing than the atomists, a clear
indication of the growing faith in experimentation.

.... since the Mechanical Philosophers have brought so few Experiments
to verifie their Assertions, and the Chymists are thought to have brought
so many on the behalf of theirs, that of those that have quitted the un-
satisfactory Philosophy of the Schools, the greater Number dazl'd as it
were by the Experiments of Spagyrists, have imbrac'd their Doctrines instead of those they deserted; For these Reasons, I say, I hop'd I might at
least do no unreasonable piece of service to the Corpuscular Philosophers,
by illustrating some of their Notions with sensible Experiments, and manifesting that the things by me treated of, may be at least plausibly explicated without having recourse to inexplicable forms, reall Qualities, the
four Peripatetick elements or as much as the three Chymical Principles.[75]

The supreme importance of experiment, revealed in his finding
fault with the mechanical philosophers for not bringing more
experiments to support their theory, receives frequent expression from Boyle. Just as he condemned the Aristotelians for
their failure to bring the evidence of their senses to test the
authority of the Stagirite, he likewise refuses to accept certain
theories of the chemists until their experiments "exact" his approval.[76] Had the "spagyrists," he says, proposed their ideas
only as notions, he would offer no objection, but he refuses to
accept them as authoritative principles.[77] Boyle perceived the
potentialities of chemical experiments, and in emphasizing their
importance to science in general beyond their usually narrow
application to medicine, in rejecting the chemical principles for
mechanical explanations, and in widening the field of chemical
investigation beyond analysis by fire, he was laying the basis of
modern chemistry. In doing so he placed all the stress upon
experiments.

Though the mechanical philosophy offered Boyle the most satisfactory explanation of natural phenomena, he considered it an hypothesis only, which could by no means be considered absolute.[78] The fact that more and more varied observations could be explained by it, and, more particularly, the fact that it possessed greater clearness, which he calls the prime requisite of any hypothesis, won his approval of it as an hypothesis only.[79] Clarity of conception, which Descartes emphasized, and upon which Hobbes insisted as much as did the members of the Royal Society, was one of the key values of the period. But as clear as the mechanical theory was, it was still a product of the reason, and reason divorced from the constant evidence of the senses was an erratic guide, subject to all kinds of mistakes, as scholastic philosophy only too well proved. As has been said before, a persistent distrust of reason appears in many of the treatises being considered.[80]

In seventeenth-century opinion the one factor more responsible than anything else for fallacious reasoning was the lack of sufficient data. Thus the need of heaping up experiments and observations, stressed by Bacon years before, was declared again and again. Experimental philosophy remains a thing distinct from the mechanical, and Bacon, who was the chief sponsor of the former, far outweighs in importance Descartes, who lent his great influence to the latter. Boyle associated the utilitarian element in science with the experimental learning only, looking upon the Cartesian theory as speculation which might satisfy the mind but which could contribute nothing to man's material welfare. Needless to say, the scientific movement in England in the third quarter of the seventeenth century, with its strong utilitarian bent and reiterated stress upon experiments, was largely inspired by the great Chancellor. So when Sprat came to write the *History of the Royal Society*, he almost ignored Descartes and exalted Bacon.[81] Undoubtedly the Baconians had in mind some far-off day when generalizations could safely be drawn from an abundant supply of observations and experiments embodied in a universal natural history, but the high goal with them was to increase the data, though tentative theories might be proposed. In the next chapter the distinction between Cartesianism and Baconianism will again be touched upon.

Bacon exerted a pronounced influence upon Boyle. In fact, no other scientist of the period was so imbued with the Verulamian spirit. Every mention of Sir Francis is instinct with praise, and he is mentioned again and again in Boyle's writings. He is most frequently "excellent Verulam" or "illustrious Verulam," but the unvarying tribute finds varied expression: "so great and so candid a Philosopher," "that great Ornament and guide of Philosophical Historians of Nature," "one of the most judicious Naturalists that our Age can boast," "That great Restorer of Physicks," "our famous experimenter," "the first and greatest Experimental Philosopher of our Age," and "the great architect of experimental history."[82] The experiments in *Certain Physiological Essays*, Boyle says, were first collected as a continuation of the *Sylva Sylvarum*. He frequently refers to Bacon for evidence or justifies himself by Bacon's example.[83] He discusses Bacon's ideas with approval, and when he dissents, it is with the utmost reverence.[84] The above are only a fraction of the citations which might be produced. Finally, Boyle's insistence upon experimentation, his aversion to theories based upon insufficient experiments, and his conception of the need of collecting comprehensive data, as well as the pronounced utilitarian element in his view of science, go straight back to Verulam. He is the greatest scientist who can be placed without reservation to the credit of the Lord Chancellor.[85]

Boyle's earnest ambition to uphold and publish the experimental method is prophetic of the enthusiastic defence of, and propaganda for, the experimental philosophy, which followed hard upon the establishment of the Royal Society. The founding of this Society was the most important scientific event of the period under discussion. One of the distinctive features of Bacon's scientific program was the provision which urged the necessity of co-operation for the advancement of learning. His insistence upon comprehensive data and his stern opposition to the hasty establishment of theories could not but make imperative the joint labors of many men. Nor was this idea without great influence in the seventeenth century, for one of the most remarkable aspects of early science in England was the tendency of experimenters to congregate. Though here and there we find isolated men trying to follow the new philosophy by their own unaided efforts, certainly the most significant English scientists of the middle of the century joined forces early in their careers. Bacon's dictum, that the advancement of science was not the work of one man nor of one age, which we find frequently cited at this time, inspired the movement as well as hope for the future of science. In addition, the model of scientific co-operation under governmental supervision, furnished by Solomon's House in the *New Atlantis,* was the parent of numerous schemes which were finally realized in the Royal Society. Though in the first half of the century there is apparently a tendency to class Bacon's plan with Plato's *Republic* and More's *Utopia* as a dream hardly worthy of serious consideration, during the Puritan régime the *New Atlantis* came to be viewed in a more serious light and stimulated others to devise schemes more or less like it.[86]

The earliest of these schemes, and also the last, for its author never relinquished the idea, was by our friend Samuel Hartlib, with whom this chapter began. It appeared in 1641 under the title *A Description of the Famous Kingdome of Macaria,* and Macaria figures in the closing years of Hartlib's life. The dedication to Parliament indicates that Bacon furnished the inspiration of the conception and also designates the public good as the motive of the work, a motive which likewise appears in the author's desire to assist Parliament in the "good reformation" under way. Among the blessings of Macaria was a great council with five subcouncils devoted to husbandry, fisheries, domestic and maritime trade, and new plantations, the functions of each being described. For medicine there was a "Colledge of experience, where they deliver out yeerly such medicines as they find out by experience; and all such as shall be able to demonstrate any experiment for the health or wealth of men, are honourably rewarded at the publike charge, by which their skill in Husbandry, Physick, and Surgerie, is most excellent." This is later referred to as "the Societie of Experimenters." As we would expect from Hartlib, this treatise places much stress upon experiments, the public good, and reformation, fully revealing the philanthropic and utilitarian spirit so conspicuous later in his life.

Hartlib never abandoned his conception. It is possible that part of it was embodied in his Office of Public Address, which has already been discussed. Toward the end of the Commonwealth it takes the form of some kind of institution of learning or philosophical college. He writes to Boyle, June 1, 1658, "My son *Hartlib* has had a very large conference with Mr. Secretary about the business of learning; and finding him full of frivolous

suspicions and allegations, he hath notably pleaded for the justice
of the affair, and doth not mean to give it over, till he be convinced into better resolutions."[87] A manuscript entitled *A
Memorial for Advancement of universal Learning*, undoubtedly
drawn up by Hartlib, gives the plan of this philosophical college,
named Atlantis, which discloses the nature of the project, Durie's
and Hartlib's part in raising money for it, its relationship to
Bacon, and its connection with the Office of Public Address.[88]
The project possessed enough worth to secure enthusiastic praise
and offers of assistance from Boyle.[89] In a letter dated August
10, 1658, Hartlib complains to the latter of losing two hundred
and fifty pounds, which were diverted to a design of Petty's, and
at the same time speaks of his having "been made to hope these
two years" for the success of his plan.[90] In January of the next
year, reverting to the name he had first used for the proposal, he
expresses the hope "that Macaria will have a more visible being"
soon.[91] And again on February 13, 1660, he writes Worthington
that he has been again and again secretly assured that Macaria
will be made an actuality.[92] The letter concluding his correspondence with Boyle contains another reference to Macaria, after
which follows a long silence, due probably to the Restoration.[93]
In the latter part of 1660, however, we hear of Macaria in connection with another society, probably similar in nature but with
greater emphasis upon the religious purpose, which went under
the name of Antilia.[94] In this society Hartlib was much interested
for a time, and he experienced keen disappointment when it came
to naught from lack of financial assistance.[95] His friend Worthington, much disgusted with the pretentious designs of Antilia,
mentions another society "more desirable (and less pretending to
wonders) which in good time may find a place in the earth."[96]

Schemes seem to have been forming thick and fast,[97] for while
Hartlib was recovering from his disappointment over the An-
tilians he was put in touch with another proposal of a more scientific and less religious nature. "I have received," he writes
Worthington on December 17, 1660, "some other papers, that
have been confided to me, holding forth almost the same things
as the other Antilia (for be not offended if I continue to use this
mystical word) but, as I hope, to better purpose." This last
project was sponsored by a Swedish nobleman, Lord Skytte, who
"hath devoted the spending of his life and estate after this man-

ner," and who had approached Charles II about a royal grant in support of it.

Thus much is certain, that there is a meeting every week of the prime virtuosi, not only at Gresham College in term time, but also out of it, at Mr. Ball's chambers in the Temple. They desired his Maj.[esty's] leave that they might thus meet or assemble themselves at all times, which is certainly granted. Mr. Boyle, Dr. Wilkins, Sir Paul Neale, Viscount Brouncker, are some of the members. Mr. Wren is chosen Register. I look upon this society as a previous introduction of the grand design here represented. His Maj.[esty] is sayd to profess himself one of those virtuosi."[98]

Although Lord Skytte was one of the group, his design evidently did not meet with the approval of his associates, for nothing came of it.[99] The "prime virtuosi" were already embarked upon a project which, ignoring Hartlib and the dream he had cherished for a score of years, was soon to be adumbrated in Cowley's *Proposition,* evidently prompted by the new organization, and a little later realized in a definite and royally patented society, which became the center of the new philosophy in England and a cynosure of scientific eyes on the continent. There is something pathetic in the fact that after so many years of hopes, disappointments, and tireless activity Hartlib should not have been associated with the activities which led to the founding of the Royal Society, and that he was not taken into the confidence of the group. "That which you mention about Dr. Cowley's Book of Proposals will oblige me to take more notice of him,"[100] he writes to Worthington, June 26, 1661, unconscious of the stream of events which were moving directly to an end dear to his heart. During the nine remaining months or less of his life there is little or no evidence of any participation on his part in the rapid evolution of the scientific movement.[101]

The reason, I think, for the sad anti-climax of Hartlib's career is to be discovered in the easily comprehended desire of science to conceal the evidence of the circumstances which attended its early growth. It had burst its Puritan cocoon and was eager to hide the discarded shell. Wallis, Wilkins, Petty, and others, though later twitted by the enemies of the Royal Society for their previous Puritan affiliations, were able to make their way safely to the other side, probably because of their less fanatical and higher intellectual qualities. Too much religion was involved in Hartlib's and Durie's schemes. Science wished to strike piety out of the slogan "the advancement of learning and piety. " Further-more, Hartlib's association with the Puritan government had been too close and his commerce with it too active to enable him to find a friend at court. The patronage of science had passed from the "godly men" to a king who was decidedly of this world. Yet those fundamental elements in the scientific movement which found expression with the Puritans did not disappear, but have survived even to our own day.

Schemes for a philosophical college, published and unpublished, were by no means confined to the man who has figured so much in this chapter. In a letter to Boyle, September 3, 1659, Evelyn outlines one, the principal purpose of which was "the promoting of experimental knowledge." He goes into detail regarding grounds, buildings, servants, members, laboratories, a museum, experimental gardens, and the like.[102] The next year one who has been plausibly identified as Robert Hooke published a con-tinuation of Bacon's *New Atlantis*,[103] the first part of which, after a eulogistic dedication to Charles II, gives a summary of Bacon's fiction. The second part is largely devoted to laws but touches also on manners and customs, in which the public good and the materialistic elements in life are emphasized. "R. H." would establish colleges for almost every subject,[104] but he is chiefly interested in inventions and discoveries. He gives an ex-tended account of the pompous ceremony with which an inventor of asbestos writing paper is honored by the Father of Solomon's House.[105] A college of agriculture, rendered familiar to us by Hartlib, is also described.

In a way, the climax of these proposals for a scientific institu-tion was reached in Abraham Cowley's *A Proposition For the Advancement of Experimental Philosophy*, 1661, which embodied a "philosophical college."[106] In the conclusion the purpose of the institution is described as follows: to examine and test what past ages have handed down and to explode the false; to recover the lost inventions of the ancients; to improve arts; and to discover new arts. In the preface Cowley condemns the attitude which holds that the ancients have left nothing to be discovered, and he stresses the necessity of subordinating reason to the evidence of sensible objects. His outline of the organization of the college calls for twenty professors, four of whom would travel constantly in Europe, Asia, Africa, and America to collect the multitudinous and varied data called for by Bacon's plan; a president, "a Lover of solid Learning, and no Stranger in it"; and eight trustees of similar qualifications. Provision is also made for a chirurgeon, apothecary, druggist, keeper of instruments, and various and sundry servants. His plan of the buildings and grounds specifies a schoolhouse, a library, a gallery of pictures or statues of discoverers and inventors of useful things, a natural-history museum, an anatomy chamber with skeletons, an apothecary's chamber, a mathematical chamber with instruments, great laboratories, experimental gardens, a zoo for animals for purposes of experimentation, an observatory, and underground vaults for carrying on experiments. All successful experiments were to be listed in a register kept under lock and key, and signed by those making them.

The chief characteristic which distinguishes Cowley's proposal from those already discussed is the educational provision. The other conceptions were largely concerned with the co-operation of scientists for the discovery of scientific truths and the promotion of useful inventions. The poet, however, seems to consider the scientific instruction of youths of equal if not greater importance.[107] In this conception he was only following the movement in which educational reformers like Durie, Petty, Hall, Webster, and others mentioned in a previous chapter had combined Bacon's philosophy with the pedagogical ideas of the Puritans. To the two hundred and sixteen boys for whom the school made provision, Cowley would have the professors teach "all sorts of Natural, Experimental Philosophy": mathematics, mechanics, medicine, anatomy, chemistry, history of animals, plants, minerals, agriculture, architecture, art military, navigation, the mysteries of all trades and the improvement of them, manufactures, natural magic, "and briefly all things contained in the Catalogue of Natural Histories annexed to My Lord *Bacon's Organon*." In keeping with the Puritan tradition no theology but God's commandments and works was to be taught. Likewise in keeping with this tradition are the author's complaint that students waste six or seven years in the study of words only, and his advocacy of a method whereby words and things could be learned at the same time.[108] Towards this latter end, he proposes as the student's apprenticeship in natural philosophy that he study those classical works, such as Virgil's *Georgics*, which contribute to a knowledge of nature. In short, the demands of science would for the most part determine the curriculum.[109]

It was not, however, only in proposed organizations that Bacon's idea of co-operation for the advancement of science proved fertile. Simultaneously with the desire to experiment arose the tendency to gather together for that purpose. The history of these groups of experimenters, later merging into the Royal Society, is well known but may bear another recounting. Our chief authority for their early history is John Wallis, Savilian professor of geometry at Oxford from 1649 to 1703, who has left us two accounts.[110] According to him the Royal Society had its *"first Ground and Foundation"* in the year 1645, "if not sooner," when "divers worthy Persons,[111] inquisitive into Natural Philosophy, and other parts of Humane Learning; and particularly of what hath been called the *New Philosophy* or *Experimental Philosophy*," held weekly meetings "at a certain day and hour, under a certain Penalty and a weekly Contribution for the Charge of Experiments, with certain Rules agreed upon amongst us." At first they met at Goddard's lodgings in Wood Street, because the doctor kept a man for grinding glasses for microscopes and telescopes, or at the Mitre in the same street. A little later they met at the Bull-head in Cheapside, and in term time at Gresham College at Samuel Foster's lecture on astronomy, after which they retired to Foster's lodgings or some place near by. In order "to avoid diversion to other discourses, and for some other reasons," they barred all discussion of theology and state affairs, restricting their discourse to "Philosophical Inquiries": physic, anatomy, geometry, astronomy, navigation, statics, magnetics, chemistry, mechanics, and natural experiments.[112] The great discoveries which ushered in modern science, some of which had been but slowly making their way in the world for over a century, as well as those more recently brought to light, were the topic of much of their conversation.[113] These discoveries, as Wallis found necessary to explain, were not so well known or accepted then as they were when he wrote his account. They also discussed "other things appertaining to what hath been called *The New Philosophy;* which from the times of *Galileo* at *Florence,* and *Sir Francis Bacon* (*Lord Verulam*) in England, hath been much cultivated in *Italy, France, Germany,* and other Parts abroad, as well as with us in *England.*" Certainly these early experimenters were in touch with the spirit of Lord Verulam.

This group was not only ultimately responsible for the Royal Society, but it also inspired the greatest English scientist of the period to engage in experiments. A young man, only a year or two returned from the continent, was invited to their meetings and became an ardent convert to the new philosophy. In a letter from London, dated October 22, 1646, Robert Boyle gives an account of his activities: "The other humane studies I apply myself to, are natural philosophy, the mechanics, and husbandry, according to the principles of our new philosophical college, that values no knowledge, but as it hath a tendency to use," and he asks his correspondent to assist in collecting information on agriculture, "which will make you extremely welcome to our *invisible college*."[114] Evidently their community of spirit was sufficiently strong and their organization compact enough to justify the title "college." The utilitarian standard, which they applied to the evaluation of truth, and the democratic spirit, revealed in the passage to be quoted next, are identical with the Puritan values discussed in a previous chapter. With this attitude Boyle was in complete agreement, being both proud to be considered one of them and eager to pursue their method. "The best on't is," he writes from London to Francis Tallents, a fellow of Magdalene College, Cambridge, early in 1647,

that the corner-stones of the *invisible*, or (as they term themselves) the *philosophical college*, do now and then honour me with their company, which makes me as sorry for those pressing occasions, that urge my departure, as I am at other times angry with that solicitous idleness, that I am necessitated to during my stay; men of so capacious and searching spirits, that school-philosophy is but the lowest region of their knowledge; and yet, though ambitious to lead the way to any generous design, of so humble and teachable a genius, as they disdain not to be directed to the meanest, so he can but plead reason for his opinion; persons, that endeavour to put narrow-mindedness out of countenance, by the practice of so extensive a charity, that it reaches unto everything called man, and nothing less than an universall good-will can content it. And indeed they are so apprehensive of the want of good employment, that they take the whole body of mankind for their care.[115]

This was the light in which Boyle viewed the men who first inspired him with scientific zeal. It is not strange that Hartlib should have been interested in their activities, and they in turn should have taken notice of him.[116]

In 1648 and the following year the "philosophical college," as
its members termed it, or the "invisible college," to use Boyle's
title, lost three of its members, Goddard, Wilkins, and Wallis, to
Oxford University. The others continued to meet as before at
Gresham College, at Mr. Rooke's lecture after the death of Mr.
Foster. Those going to Oxford organized another group of experimenters there, consisting, besides themselves, of Seth Ward,
Savilian professor of astronomy, William Petty, professor of
anatomy, Ralph Bathurst, Thomas Willis, "an eminent physician
in *Oxford*," and "many others of the most inquisitive persons in
Oxford."[117] This group, known as the "philosophical society,"
met weekly at Petty's lodgings in an apothecary's house, because
it afforded an opportunity of inspecting drugs, until Petty left
for Ireland in 1652, their meetings being "numerous, and very
considerable. For, beside the diligence of Persons, studiously
Inquisitive, the Novelty of the Design made many to resort
thither; who, when it ceased to be new, began to grow more
remiss, or did pursue such Inquiries at Home." After Petty's
departure, the society lost some of its members and met less
frequently at Wilkins' lodgings in Wadham College.[118] In fact,
Wallis says that the set meetings ceased entirely at Oxford and
were held in London, but on February 17, 1657, Petty writes
Boyle, "I have not, amongst all my intelligence, heard better news,
than that the club is restored at *Oxford*."[119] When in 1659
Wilkins removed to Cambridge to become master of Trinity
College, the club met at the lodgings of Boyle, who had come to
Oxford in 1654. The group at London continued to meet as before at Mr. Laurence Rooke's lecture, after which they adjourned
to some place near by. In 1658, the year in which a number of
the members of the Oxonian society removed to London and
joined the society there, Gresham College was taken over by
soldiers, and the group was scattered, not to reconvene until
after the Restoration, when its members were augmented by the
"accession of divers Eminent and Noble Persons."[120]

It is to this group that the immediate rise of the Royal Society
is to be traced. On November 28, 1660, an organization was
effected, temporary officers elected, rules drawn up, and a tentative list of members determined upon.[121] With this organization
the society made sure and rapid progress, and soon attracted
the notice of the most eminent men in England, even the King

himself, who had acquired, perhaps in France, an interest in science, which proved a boon to the scientific movement in England. Only a few months after the Restoration, Charles found it convenient to drop in at Gresham College to look through a telescope, and he threatened to pay Boyle a visit. In December of the same year he freely grants the virtuosi's request to assemble at all times, and professes to be one himself.[122] In October, 1661, we learn that Sir Robert Moray, who was president of the society until its incorporation in 1662, thanked the King for offering to become a member. It is Evelyn, however, who makes most clear the royal interest in the project. In *A Panegyric* addressed to Charles on April 23, 1661, he speaks of the honor shown the society by the King's

curious enquiries about the *Load-stone*, and other particulars which concern *Philosophy*; since it is not to be doubted but that so magnanimous a Prince, will still proceed to encourage that Illustrious Assemblie; and which will celebrate and eternize Your memory to future Ages . . . . when for you is reserv'd the being Founder of some thing that may improve practical and Experimental knowledg, beyond all that has been hitherto attempted, for the Augmentation of Science, and universal good of Man-kind, and which alone will consummate Your Fame and render it immortal.

Later in the year he is even more extravagant in his praise of the King and the society, which, he thinks, will continue to enjoy royal favor, and which, for the first time that I have discovered, is called the Royal Society.[123] Evelyn's dedication is an important document relative to the formation of the Society. While it is clearly a bid for royal patronage, it also acknowledges how much the King had already favored them. Furthermore, it breathes the highly optimistic and confident spirit which animated the group. It shows that to all intents and purposes the Royal Society was already founded, though it had to wait almost a year for its charter. It praises the members to the sky, expatiates upon the worthiness of the enterprise and the extent to which the King's fame will be increased by favor shown them, speaks gratefully of Clarendon's affection for the organization and his inclination to promote and protect it, praises the experimental in contrast to the scholastic philosophy, and, as Sprat later did in the *History of the Royal Society*, exalts the society as far above the French Academy as things are above words and the study of nature above the study of language. On August 13, 1662, the first charter was read to the organization, which thus became the Royal Society *de jure* as it had been for some time *de facto*.

THE RESTORATION

CHAPTER VIII

THE DEFENCE OF THE EXPERIMENTAL PHILOSOPHY

> This is the Age wherein (me-thinks) Philosophy comes in with a Spring-tide; and the Peripateticks may as well hope to stop the Current of the Tide, or (with *Xerxes*) to fetter the Ocean, as hinder the overflowing of free Philosophy: Me-thinks, I see how all the old Rubbish must be thrown away, and the rotten Buildings be overthrown, and carried away with so powerful an Inundation. These are the days that must lay a new Foundation of a more magnificent Philosophy, never to be overthrown: that will Empirically and Sensibly canvass the *Phaenomena* of Nature, deducing the Causes of things from such Originals in Nature, as we observe are producible by Art, and the infallible demonstration of Mechanicks: and certainly, this is the way, and no other, to build a true and permanent Philosophy.
>
> Henry Power, *Experimental Philosophy*, 1663.

With the establishment of the Royal Society, protected by a royal patent and comprising among its members some of the most important men in church and state, one might naturally surmise that the new science had been placed beyond the reach of detractors and the need of defence. But such was not the case, for the seventh decade of the seventeenth century witnessed the most vigorous defence ever carried on in behalf of experimental science. The conservative elements, undoubtedly assisted by the reaction against the Puritans, were too strong in the church, the universities, and the professions to permit the scientists to abandon propaganda and devote all their time to investigation. Furthermore, the comparative newness of the experimental idea and the enthusiastic faith with which it had been embraced by the Baconians inspired the latter to push its claims and publish its virtues beyond the bounds of necessity. Yet indifference and opposition to it undoubtedly existed, and would have been much more in evidence had the experimental philosophy not enjoyed royal favor. It is idle to speculate what might have happened had not Charles II been favorably disposed toward science, but one thing we may be sure of, the groups of experimenters which have been described in the preceding chapter would have found much rougher going. The innate conservatism of the English and the favor shown to Baconianism by the Puritans are factors that cannot be ignored in following the growth of science. Even the royal sanction, the quiet face-about of Puritan scientists, and the presence among the scientists of ardent royalists were not entirely sufficient to remove the Puritan stigma. But the King's example must have done much to stifle outspoken criticism, at least, on that score.

Another important consideration must be kept in mind. The new science did not embody only the announcement of important discoveries made through direct investigation of nature and exerting a profound influence on human thought; it was distinctly an intellectual movement, established upon definite philosophical principles and inspiring, on the one hand, ardent faith and allegiance in progressive minds, and on the other, opposition and resentful reaction in conservative spirits. So it had to make its way in the same manner as any great thought movement, such as protestantism, democracy, and the like. The controlling ideas had to be spread, defended, and enthusiastically recommended. It was necessary to conduct offenses against the old and maintain defences of the new. The idea of science thus assumes in importance a higher position than science itself. To this idea even great discoveries are placed in the subordinate position of props, and scientists bend their efforts in its support with even more enthusiasm than they conduct their experiments. Today science is so securely entrenched that the intellectual concept upon which it is based is almost ignored, and discoveries alone are valued, but in the seventeenth century the concept, new and in need of protection, was the greater concern. This chapter attempts to follow through five active years the numerous efforts to establish firmly the idea of the new science.

The Royal Society, then, having been founded, it became necessary to defend the faith upon which it was based. This faith had to do for the most part with the *method*, the method of experiment and observation. And it was constantly thrown into sharp relief against the traditional habit of reliance upon the ancients. In fact, again and again we are impressed with the central problem that confronted the beginnings of modern science, namely, the break with antiquity and the adoption of a method of direct investigation. This problem necessitated not only a vigorous espousal of the experimental method but also an equally vigorous

attack upon scientific ideas derived immediately or ultimately
from antiquity. So the great discoveries of Copernicus, Vesalius,
Gilbert, Kepler, Galileo, Harvey, and others, while frequently
cited to prove the fruitfulness of the new method, are more
usually employed to point out the falsity of traditional beliefs,
and thus to weaken the authority of the classics so as to gain
support for the new method. Furthermore, the hostility to the
ancients was strengthened a great deal by the mechanical philosophy, the first seeds of which are to be discovered in Galileo,
and which was much promoted by Gassendi's revival of the
ancient atomical philosophy and perfected by Descartes, with
whom, indeed, it came to be almost entirely associated, though
many of his ideas were rejected. The atomical, the Epicurean,
the corpuscularian, the mechanical, and the Cartesian philosophy
were terms which possessed in common the fundamental idea
that all physical phenomena are the result of matter and motion,
matter consisting of minute atoms or corpuscles. This philosophy, as developed by Descartes, was first embraced in England
by the Cambridge Platonists, who, however, were animated less
by scientific than religious motives, in that, eschewing Bacon's
separation of science and religion, they wished to reconcile the
two, or rather to bend the mechanical philosophy to support the
Christian religion as platonically interpreted by them. Thus
Descartes's own antagonism to ancient authority and the opposition of his philosophy to Peripateticism inspired in the English
Platonists and other Cartesians the same attitude toward Aristotle which characterized the experimentalists.

During the Restoration the mechanical philosophy, under the
influence of Descartes and especially Hobbes, was being more
and more widely accepted not only by the scientists but also by
the gentry and the court. But it is a mistake to think that Cartesianism inspired the scientific movement in England. That inspiration was the experimental method derived from Bacon.
The mechanical philosophy was considered by the scientists a
hypothesis, the truth of which was gradually being revealed to
their eyes through experimental verification, but experiment and
observation as the proper method for the discovery of natural
truths represented a faith, to doubt which was heresy, and which
was common ground for all members of the Royal Society. By
no means did all fellows of the Society embrace Cartesianism;

Glanvill says that only some did, and Henry More draws a definite distinction between the mechanical and the experimental
philosophy, upon which alone the Society was established.[1] So it
is not surprising to find that the official history of the Society,
written by Sprat, constantly harps on the experimental and all
but ignores the mechanical philosophy, deifies Bacon and almost
forgets Descartes. It is interesting to note that More's abandonment of Cartesianism in *Enchiridion Metaphysicum* coincides
with very high praise for the experimental philosophy,[2] and
that after his admission into the Royal Society, Glanvill's enthusiasm for Descartes wanes considerably and his admiration
for Baconianism reaches great heights. In general, the virtuosi
were free to accept or reject the mechanical philosophy, but not
the experimental method. Of course, there was nothing contradictory between Cartesianism and Baconianism. Both opposed the authority of the ancients, both insisted upon first-hand
examination of nature, and the mechanical philosophy offered
the clearest explanation of the data furnished by the Bacon-
inspired method of experimentation and observation. Thus it is
not strange to find the two discussed together in treatises of the
period.

In fact, the first work to be considered in this chapter, Simon
Patrick's *A Brief Account of the new Sect of Latitude-Men.
Together with some reflections upon the New Philosophy*, 1662,
is primarily devoted to a defence of the mechanical philosophy
and the Cartesian Platonists of Cambridge.[3] Patrick notes with
approval their substitution of a new philosophy for Peripateticism and their rejection of the principle of authority as a test
of truth, though he admits that he never had much "skill" in
the mechanical philosophy, and was too old to learn. This hesitation to pronounce upon the truth of the new hypothesis is
further revealed in a passage which defines the purpose of philosophy in mechanical terms.

Then certainly it must be the Office of Philosophy to find out the process of
this Divine Art in the great automaton of the world, by observing how one
part moves another, and how those motions are varied by the severall magnitudes, figures, positions of each part, from the first springs or plummets,
as I may say, to the hand that points out the visible and last effects; This
physicians have taken the boldness in part to do, in those little watches, if
I may so call them, the bodies of animals, and *Descartes* hath proceeded
farthest in the like attempt, in that vast machin, the Universe, with how good

successe, others are better able to Judge; But this I dare boldly say, it was
a noble effort, and if he had wholly failed in it, he would have been more
pitied and less envied.[4]

What Patrick is really arguing for is revealed in his expression
"new and free Philosophy," that is, the freedom of science to
throw off the shackles of ancient authority and to receive the
truth which observation and experiment reveal. Toward this
end he points out how all the new discoveries made by the aid
of the telescope, Gilbert's magnetical experiments, Columbus'
geographical discoveries,[5] and Harvey's discovery of the circulation of the blood have proved the erroneous nature of Aristotelian
science. He further comments on the ancients' ignorance of
"that most Luciferous though obvious experiment of the sling,"
the prism, and numerous truths revealed by the microscope. He
mentions "many Chymical experiments" and "the discoveries of
the Air-pump lately exhibited to the world by that Noble Gentleman [Boyle] . . . . [and] the ingenious experiments of *Galileo*,
Lord *Bacon*, and many others."[6] In short, Patrick's position is
that modern discoveries and experiments have overthrown the
old philosophy erected upon too narrow foundations, and have
revealed a world of new data, the explanation of which necessi-
tates the formation of new hypotheses, and that philosophy
should be free from the domination of antiquity to form such
hypotheses. While he is primarily concerned with the mechanical
philosophy because it offered the best explanation of phenomena,
he is indirectly praising the experimental philosophy and enthusiastically supporting the demand of the scientists for the
liberty of freely philosophizing. Patrick also notes the relationship between the revival of learning, the Reformation, and the
new scientific spirit in denying that the last is inimical to true
Christian doctrine, in stressing the service performed by Erasmus
and Melanchthon in overthrowing scholasticism, and in denying
that the Christian religion was bred in the Peripatetic school.
Finally, he reveals the ardent enthusiasm and unquestioning op-
timism of an age clearly conscious of its own character. It is as
impossible to stop the new philosophy as to hinder

the Sun from rising, or being up, from filling the whole *Horizon* with light;
Learning and Knowledge will break forth like fire, and pierce like Lightning
through all impediments; . . . . the inquisitive *Genius* of latter years, like a
mighty wind hath brushed down all the Schoolmens Cobwebs. There is an

infinite desire of knowledge broken forth in the world, and men may as well hope to stop the tide, or bind the Ocean with Chains, as hinder free *Philosophy* from overflowing: it will be as easie to satisfie mens corporal appetites with chaff and straw, as the desires of their minds with empty words and terms.[7]

Perceiving the deficiency of traditional learning as revealed by the early discoveries of modern science, and conscious of the vast amount of new data which demanded explanation, he attacks the ancients in order to overthrow their authority, urges freedom of thought, and insists upon experiment and observation.

Belief in the mechanical philosophy did not save the Cambridge Platonists from a vigorous attack by an ardent Baconian, Samuel Parker,[8] who, however, arraigns them only on the score of their Platonic ideas. In his dedication to Ralph Bathurst, president of Trinity College, Oxford, a fellow of the Royal Society, and formerly a member of the Oxford group of experimenters of the preceding decade, at whose request Parker says he is writing his treatise, the author is profuse in his gratitude for Bathurst's rescuing him from the "Chains and Fetters of an unhappy Education," apropos of which he eulogizes "Freedom and Ingenuity of Mind." In the same way in which others had treated Aristotelianism Parker brings the touchstone of material reality, sense-observation, and experiment to the Platonic philosophy of the seventeenth century. The result is that he condemns Platonism, with the exception of Plato's moral principles, which he praises highly, for being merely a notional philosophy, which deals with ideas unverifiable by the senses. The Platonists, he says,

by rejecting the Testimony and Judgment of sense in matters of Philosophie, ... do but involve and perplex the Principles thereof, under the pretext of a more abstracted and intellectual discovery of things: For hereby the minds of men are taken off from the *native Evidence* of plain and palpable Truths, and are fain to ground all their knowledge upon nice and subtle Speculations, whereby, at least, clear and unquestionable *Truths* are resolved into *Principles* infinitely more uncertain and disputable then themselves.[9]

He inveighs against archetypal ideas, innate ideas, and abstracted essences because they are not susceptible of scientific examination; it is impossible to apply to them the test of experiment and observation. Parker, who had to a certain degree come under the influence of Hobbes, may be described as one who believed only what he could see. He brings out in clear relief the materialistic nature of Bacon's insistence upon the "thing" and the similar nature of the mechanical philosophy. He seems to have embraced completely the narrow scientific view that would throw out of the count all that cannot be weighed, measured, and perceived. The only way, he says,

to be fully satisfied of their Truth and Sincerity, is to examine them [Platonic ideas] by a wary and discreet Experience, the Test whereof will remove all ground to doubt for the future of their Integrity. And if so, to what purpose do *Connate Principles* serve: for before I have made Tryal, I cannot use them, because I have no Reason to trust them, till I can be certain of their Veracity; which I cannot be, but by Experience, which yet makes them useless; because Experimental knowledge is of all others the safest and most unquestionable, and therefore must needs render all lesser evidence vain and unnecessary. At least when our knowledge proceeds in an Empirical way 'tis solid and palpable, and made so undoubtedly certain from the plain and most undoubted Testimony of Sense and Experience, as undeniably to convince *Scepticism* of a pitiful and ridiculous Obstinacy. But when we begin our knowledge from Notions within our selves, besides that 'tis a difficult and nice dispute to prove that the mind of man is furnished with any such innate *Prolepses*, and that we are destitute of any sure κριτήριον to discern Natural Anticipations from Preconceptions of Custome and Education (unless we bring them to the Touchstone of Experience) 'tis doubtless that Generalities are not capable of so palpable and convictive an Evidence, as singular and particular Observations.[10]

This last sentence brings out clearly Parker's firm belief in, and earnest advocacy of, the experimental method, which here, as we have noted in other instances, is opposed to theories and hypotheses. And the origin of this antithesis between authentic data and suspected theory is revealed in the next sentence, which says, "And therefore my *Lord Bacon* has well noted it as none of the least obstructions to the advancement of knowledge, that *Men have sought for Truth in their own little Worlds*," and not in nature.

Perhaps more clearly than anyone else, Parker expresses dissatisfaction with any theory that goes beyond the sensuously obvious. To him the experimental philosophy laid bare to the senses the true nature of things, beyond which it was both dangerous and unlawful to go. Curiosity which seeks only a knowledge of what the senses may reveal "is a gallant and heroical Quality . . . . but when it aspires after the knowledge of things placed above its Reach, it degenerates into a vain and fruitless Ambition, or

rather an unnatural lust of the mind after strange and extravagant Notions." And in the latter class he would place most
hypotheses. But it is the idea of abstracted essences which most
displeases him, and in the discussion of which he most clearly
reveals the influence of Hobbes and manifests his idea of what
science should concern itself with:

But yet further, we are so far from attaining any certain and real knowledg
of *Incorporeal Beings* (of an acquaintance with which, these *Visionists* so
much boast) that we are not able to know anything of *Corporeal Substances*
as abstract from their Accidents. Ther's nothing can more perplex my
Faculties, then the simple *Idea* of naked matter. And certainly, it was never
intended that meer Essences should be the Objects of our Faculties. And
therefore the truly wise and discerning Philosophers do not endeavour after
the dry and sapless knowledge of abstracted Natures, but only search after
the Properties, Qualities, Vertues and Operations of Natural Beings; the
knowledge whereof may be acquired by Observations and Experiments; but
there are no certain means or rational methods (that I could ever yet meet
with) to investigate the mysterious Ideas of bare and abstracted Essences.
Besides, all Beings are either Objects of sence, or not; now to go about to
discover the nature of the former by metaphysical definitions, would be
ridiculous, seeing they are far better understood by our senses .... And besides to abstract sensible things from materiality, is to abstract them from
themselves, because their very Essences are Material. And then of them
that take upon them to describe the Natures of Beings that are not obnoxious
to sence I demand by what ways and methods they came to that knowledge.
For tis not enough to prove that this or that is the Idea of any thing, because some fanciful men are able to make pretty *Hypotheses* concerning it,
but if any man have attain'd any certain knowledge thereof, he is able to
give a rational account of the way and method, by which he proceeded in
his Enquiry. But this these bold definers neither have, nor can do; but if
you will be so civil as to take their words, they will requite your Civility by
acquanting you with more strange and stupendious Mysteries.[11]

The essentially concrete and materialistic nature of the Baconian science of the seventeenth century is clearly revealed in this
treatise.[12] Experiments and observations, which reveal the undoubted truth of the physical world, are the true concern of
science and stand in sharp distinction to hypotheses which go
beyond the concrete data. The experimental philosophy represents the road to truth, and to it Parker is wedded. Toward
hypotheses and theories he is sternly critical, though he is somewhat inclined toward the mechanical philosophy.[13] Yet even as
regards the latter he is uncertain, believing that too much conjecture and too few observations and experiments had gone into

its making.[14] It was probably the fact that ancient theories were too far divorced from material reality which inspired Parker to depreciate ancient learning. He was imbued with the modern spirit which was warring upon antiquity, but though he was of the opinion that the ancients were pigmies in knowledge compared with the moderns, he maintained that the world was much beholden to the former for having been "the first founders and discoverers of that knowledge, which after Ages have but improved." As is true of most of the critics of antiquity, Parker is interested in destroying the hold which its authority had upon his age.[15]

The hypothetical nature of the mechanical philosophy as seen by seventeenth-century eyes is apparent in the title of the next work to be considered, Henry Power's *Experimental Philosophy, In Three Books: Containing New Experiments Microscopical, Mercurial, Magnetical. With some Deductions, and Probable Hypotheses, raised from them, in Avouchment and Illustration of the now famous Atomical Hypothesis,* 1664.[16] Though Power distinguishes in his title very clearly between the experimental philosophy, which was no hypothesis but an accepted fact, and the mechanical, he is a firm believer in the latter. Among the claims which he enthusiastically makes for the microscope is the greatly exaggerated one that it enables men to "see what the illustrious wits of the Atomical and Corpuscularian Philosophers durst but imagine, even the very Atoms and their reputed Indivisibles and least realities of Matter, nay the curious Mechanism and organical Contrivance of those Minute Animals, with their distinct parts, colour, figure and motion, whose whole bulk were to them almost invisible," so that if Aristotle were alive, he would write a new history of animals. Though he thinks the atomical hypothesis will receive illustration from the microscope, he does not lose sight of other philosophical possibilities:

if our *Microscopes* arise to any higher perfection: and if we can but, by any artificial helps, get but a glimpse of the smallest Truth, it is not to tell what a Fabrick of Philosophy may be raised from it; (for to conclude with that Patriark of Experimental Philosophy, the learned Lord *Bacon*) The Eye of the Understanding, saith he, is like the Eye of the Sense; for as you may see great Objects through small Cranies or Levels; so you may see great Axioms of Nature, through small and contemptible Instances and Experiments."[17]

But however much he may have subscribed to Cartesianism, it is the experimental philosophy and its "Patriark" that come nearest to his heart. He eulogizes Descartes, it is true, but the motto on the first title page is taken from Aphorism 39 of the second book of the *Novum Organum;* the section entitled "Mercuriall Experiments," begun, he says, in 1653, is headed by a passage from Aphorism 109 of the first book of the same work; and he quotes "the learned *Verulam*" regarding the necessity of studying the little and invisible things, which principally govern nature.[18] The treatise itself is largely occupied with a description of the new world revealed by the microscope, but he refers to the "noble Experiment of *Torrecellius,*" praises Galileo, "to whom all the Common-wealth of Learning are exceedingly oblieg'd, because thereby he has excited the greatest modern Wits to higher and nobler Experiments," and he speaks of a "noble Experiment . . . . the first hint [of which] was given us, by those acute and singularly accomplished Gentlemen of *Townley-Hall* in *Lancashire,* who were as Judicious as Honourable Spectators of these our *Hydrargyral* Experiments; and whose mechanical Prognosticks seldom failed, but were still made good by the future event of the Experiments." Elsewhere he speaks of experiments carried on by him at Townley-Hall during the years 1660 and 1661. He contrasts "solid and Experimental Philosophy" with litigious Peripateticism, and characterizes the virtuosi as "true Lovers of Free, and Experimental Philosophy."[19]

Power's unqualified espousal of the experimental philosophy is seen in his glowing appreciation of the society that soon became the Royal Society, and which was founded upon the experimental method alone. "But these, and a hundred more Experiments of this nature," he says with all hope, "are every day excogitated and tried by our Noble Society of *Gresham Colledge,* which in a little time will be improved into far nobler Consequences and Theories, than can possibly be done by the single Endeavours of any Person whatsoever," an echo of Bacon's faith in the efficacy of co-operation. After speaking of the great scientific possibilities of the future, he exclaims,

But these are the Reaches that are beyond all those of the *Stagyrite's* Retinue, the Solutions of all those former Difficulties are reserved for you (most Noble Souls, the true Lovers of Free and Experimental Philosophy) to gratifie Posterity withall.

impossibilities) though but in an Ingenious Hypothesis: And, certainly, there is no Truth so abstruse, nor so far elevated out of our reach, but man's wit may raise Engines to Scale and Conquer it.[23]

The conclusion of his treatise, from which this quotation has been taken, clearly reveals abundant hope of the future, unalloyed joy at having escaped the tyranny of the past and over the prospect of future progress, and consciousness of living in an intellectually stirring time, a consciousness which first dawned in the Puritan era. It is not strange that there was a let-down when the extravagant dreams of science were not realized. Part of the following passage has already been quoted, but it is worth repeating:

And this is the Age wherein all mens Souls are in a kind of fermentation, and the spirit of Wisdom and Learning begins to mount and free it self from those drossie and terrene Impediments wherewith it hath been so long clogg'd, and from the insipid phlegm and *Caput Mortuum* of useless Notions, in which it hath endured so violent and long a fixation.

This is the Age wherein (me-thinks) Philosophy comes in with a Spring-tide; and the Peripateticks may as well hope to stop the Current of the Tide, or (with *Xerxes*) to fetter the Ocean, as hinder the overflowing of free Philosophy: Me-thinks, I see how all the old Rubbish must be thrown away, and the rotten Buildings be overthrown, and carried away with so powerful an Inundation. These are the days that must lay a new Foundation of a more magnificent Philosophy, never to be overthrown: that will Empirically and Sensibly canvass the *Phaenomena* of Nature, deducing the Causes of things from such Originals in Nature, as we observe are producible by Art, and the infallible demonstration of Mechanicks: and certainly, this is the way, and no other, to build a true and permanent Philosophy: For Art, being the Imitation of Nature (or, Nature at Second-Hand) it is but a sensible expression of Effects, dependent on the same (though more remote Causes;) and therefore the works of the one, must prove the most reasonable discoveries of the other. And to speak yet more close to the point, I think it is no Rhetorication to say, That all things are Artificial; for Nature it self is nothing else but the Art of God. Then, certainly, to find the various turnings, and mysterious process of this divine Art, in the management of this great Machine of the World, must needs be the proper Office of onely the Experimental and Mechanical Philosopher.[24]

Here is a clear instance of the way in which the experimental and mechanical philosophies were combined. The former furnished the method ("Empirically and Sensibly canvass[ing] the *Phaenomena* of Nature") of securing data from which the latter could be deduced. Yet in the eyes of the experimentalists, especially Boyle, the Cartesian philosophy was not so certain that it did not require support and verification by experiments.

In a direct and straightforward manner the preface of the next work[25] to be considered announces as the book's purpose the defence of the experimental philosophy. Its author, Robert Hooke, declares that what is needed for a reformation in philosophy is not so much imagination or contemplation "as a *sincere Hand,* and a *faithful Eye,* to examine, and to record, the things themselves as they appear . . . . a man so qualified, as I have indeavoured to be, only with resolution, and integrity, and plain intentions of imploying his *Senses* aright, may venture to compare the reality and usefulness of his services, towards the true Philosophy, with those of other men, that are of much stronger and more acute *speculations,* that shall not make use of the same method by the Senses."[26] Here again we find experiment opposed to hypothesis or theory, for Hooke goes on to say that science has for too long a time been the work of "the *Brain* and the *Fancy,*" and needs above everything else to "return to the plainness and soundness of *Observations* on *material* and *obvious* things." Hooke, like Bacon, feeling that many data must be gathered before a foundation for theories can be laid, says he will be content if he can contribute to "the large stock of *natural Observations,* which so many hands are busie in providing," and which will show the uncertainty of generalizations.[27] Though distrustful of the theories which depend too much upon reason, he is by no means ignorant of the part the latter must play in experiment and observation, for, he says, it must examine, arrange, and dispose data, and must distinguish between "the *sober* and *well collected heap*" and "the *extravagant Idea's,* and *mistaken Images,* which . . . . it may sometimes light upon." The primary duty of science, as Hooke, Boyle, and the majority of the experimental philosophers see it, and as Bacon had pointed out, is through observations and experiments to collect a large enough body of authentic data to permit the safe establishment of theories, in short, a universal natural history. If this method were conscientiously followed, *"Talking* and *Contention of Arguments* would soon be turn'd into labours; all the fine *dreams* of Opinions, and universal metaphysical natures, which the luxury of subtil Brains has devis'd, would quickly vanish, and give place to *solid Histories, Experiments* and *Works."* And elsewhere he says that the remedy for the too free use of reason in the creation of theories lies in "the *real*, the *mechanical*, the *experimental* Philosophy." In his dedication to the Royal Society, he compliments the latter for having adopted the best rules for philosophical progress ever practised, "And particularly that of avoiding *Dogmatizing*, and the *espousal* of any *Hypothesis* not sufficiently grounded and confirm'd by *Experiments*." This, claims Hooke, is the best way to preserve philosophy and natural history from their former corruptions.[28]

Hooke expressly states that his book was undertaken with the purpose of promoting the design of the Royal Society, the members of which

having before their eys so many *fatal* Instances of the errors and falshoods, in which the greatest part of mankind has so long wandred, because they rely'd upon the strength of humane Reason alone, have begun anew to correct all *Hypotheses* by sense, as Seamen do their *dead Reckonings* by *Cœlestial Observations;* and to this purpose it has been their principal indeavour to *enlarge* and *strengthen* the *Senses* by *Medicine*, and by such *outward Instruments* as are proper for their particular works.

He maintains that as a result of observation and experiment old theories have been discarded, and men are inclining to the mechanical philosophy, although, as becomes one who speaks concerning a hypothesis, he expresses a very cautious approval of that philosophy.

By this means they find some reason to suspect, that those effects of Bodies, which have been commonly attributed to *Qualities*, and those confess'd to be *occult*, are performed by the small *Machines* of Nature, which are not to be discern'd without these helps, seeming the meer products of *Motion*, *Figure*, and *Magnitude;* and that the *Natural Textures*, which some call the *Plastick faculty*, may be made in *Looms*, which a greater perfection of Opticks may make discernable by these Glasses; so as now they are no more puzzled about them, then the vulgar are to conceive, how *Tapestry* or *flowred Stuffs* are woven.

Hooke, like Power, entertained the hope that microscopes could be so developed as to make visible the "corpuscles," to use Boyle's term, the motion of which, according to the mechanical philosophy, was responsible for natural phenomena. But after all, Cartesianism only satisfied the curiosity of the mind, and for this reason ranked lower than the method which could establish man's dominion over nature for his own utilitarian ends. "And the ends of all these Inquiries," says Hooke,

they [members of the Royal Society] intend to be the *Pleasure* of Contemplative minds, but above all, the *ease* and *dispatch* of the labours of
mens hands. They do indeed neglect no opportunity to bring all the *rare*
things of Remote Countries within the compass of their knowledge and
practice. But they still acknowledg their *most useful* Informations to arise
from *common* things, and from *diversifying* their most *ordinary* operations
upon them. They do not wholly reject Experiments of meer *light* and
theory; but they principally aim at such, whose Applications will *improve*
and *facilitate* the present way of *Manual Arts.*

The materialistic and utilitarian function of science finds even
more definite expression in what he says about the encouragers
of the Royal Society. After stating that such undertakings as
the Society's meet with small encouragement because men are
generally more interested in theories than in "the real and solid
part of philosophy" (i.e., experiments for utilitarian purposes),
he says,

yet by the good fortune of their institution, in an Age of all others the
most *inquisitive*, they have been assisted by the *contribution* and *presence*
of very many of the chief *Nobility* and *Gentry*, and others, who are some
of the *most considerable* in their several Professions. But that that yet
farther convinces me of the *Real esteem* that the more *serious* part of men
have of their *Society*, is, that several *Merchants*, men who act in earnest
(whose Object is *meum et tuum*, that great *Rudder* of humane affairs)
have adventur'd considerable sums of *Money*, to put in practice what some
of our Members have contrived, and have continued *stedfast* in their good
opinions of such Indeavours, when not one of a hundred of the vulgar have
believed their undertakings feasable. And it is also fit to be added, that
they have one advantage peculiar to themselves, that very many of their
number are *men of Converse and Traffick;* which is a good omen, that their
attempts will bring Philosophy from *words* to *action*, seeing the men of
Business have had so great a share in their first foundation.[29]

It is the observational and experimental method in which
Hooke is interested. In his dedication to the King, he states that
not the least of the many felicities of the Restoration is "that
Philosophy and *Experimental Learning* have prosper'd under
your *Royal Patronage*," and he assures the monarch that several
members of the Royal Society "are now busie about *Nobler*
matters: The *Improvement* of *Manufactures and Agriculture*,
the *Increase* of *Commerce*, the *Advantage* of *Navigation*." After
making an acute observation about Nature's not being concerned
to repair the decays of old bodies but being very solicitous to
perpetuate the species with new, he asserts that the only way to

discover the truth of nature is "to begin to build anew upon a sure Foundation of Experiments." And if this true method is followed, there is nothing that lies beyond the power of the human mind. Even the great discoveries of Copernicus, Galileo, Gilbert, Harvey, and others, whose "discoveries seem to have been the products of some such method, though but imperfect," may be surpassed. The optimistic view of the future, inspired by faith in a perfected experimental method, he shares with his age. The great discoveries in anatomy, astronomy, and physics, especially the air-pump of Boyle, "the *Patron* of *Philosophy*,"

and the now seemingly great *obviousness* of most of their and divers other Inventions, which from the beginning of the world have been, as 'twere, trod on, and yet not minded till these last *inquisitive* Ages (an Argument that there may be yet behind multitudes of the like) puts me in mind to recommend such Studies, and the prosecution of them by such methods, to the *Gentlemen* of our Nation, whose *leisure* makes them fit to *undertake*, and the *plenty* of their fortunes to *accomplish*, extraordinary things in this way. And I do not only propose this kind of *Experimental Philosophy* as a matter of high *rapture* and *delight* of the mind, but even as a *material* and *sensible Pleasure*. So vast is the *variety of Objects* which will come under their Inspections, so many *different wayes* there are of *handling* them, so great is the *satisfaction* of *finding* out *new things*, that I dare compare the *contentment* which they will injoy, not only to that of *contemplation*, but even to that which most men prefer of *the very Senses themselves*.[30]

The Royal Society and experimental philosophy possessed no more loyal and vigorous defender than Joseph Glanvill, whose first book, *The Vanity of Dogmatizing*, discussed in the preceding chapter, represented an attempt to combine an enthusiastic avowal of Cartesianism with a scepticism which would destroy all certainty. Two reasons for this scepticism are apparent. One is that Glanvill held to Bacon's idea that the belief of having attained certain knowledge was a great obstacle to the further development of science. The other is the fear that the mechanical philosophy, especially as developed by Hobbes, would lead to atheism, an undesirable end which might be defeated by distrust of the very reasoning that produced the philosophy. But in the years immediately following the appearance of his book, Glanvill must have developed a desire to affiliate himself with the Royal Society, for, when in 1665 he published a second edition under the title *Scepsis Scientifica*, he prefixed to it a long "Address to the Royal Society," which is highly eulogistic of that body.[31]

Thereafter Glanvill transferred much of his allegiance from
Descartes to Bacon, from the mechanical to the experimental
philosophy. The "Address," which continues his attack on Aristotelianism, is largely concerned with attacking both those who
employed the mechanical philosophy to promote atheism and
those who opposed philosophy in behalf of religion, and with
asserting the reconcilability of science and religion. He still
reveals a close affinity with the Cambridge Platonists and maintains his scepticism, which he considered the surest path to scientific knowledge.

Glanvill, however, both expresses and implies his advocacy of
the experimental method, and also furnishes further evidence of
the distinction between the two philosophies which we have been
discussing. In attacking the dogmatism of the upholders of
traditional learning, he says the "great and instructed Spirits"
of the Royal Society "think we have not as yet *Phænomena*
enough to make as much as Hypotheseis; much less, to fix *certain
Laws* and prescribe *Methods* to Nature in her Actings," and he
states that the Royal Society is engaged upon a history of nature,

without which our *Hypotheseis* are but *Dreams* and *Romances,* and our
Science meer *conjecture* and *opinion.* For while we frame *Scheames* of
things without consulting the *Phænomena,* we do but *build* in the *Air,* and
describe an *Imaginary World* of our *own making* . . . . And tis possible
that all the *Hypotheseis* that yet have been contrived, were built upon too
narrow an *inspection* of *things,* and the *phasies* of the *Universe.* For the
advancing day of *experimental knowledge* discloseth such *appearances,* as
will not lye *even,* in any *model* extant . . . . *So* that little can be looked
for towards the *advancement* of *natural Theory,* but from those, that are
likely to mend our *prospect* of *events* and *sensible appearances;* the defect of
which will suffer us to proceed no further towards *Science,* then to *imperfect guesses* and *timerous supposals.*

This, he asserts, may be expected from the Royal Society, which
can be counted upon to render truthful and impartial relations
of observations and experiments.[32] The purpose of the Society,
then, is the promotion of experimental philosophy, "real and
useful Knowledge," whereby correct data may be gathered, either
for furnishing the foundation of hypotheses, though that day
might be far off, or for the discovery of ways and means of
promoting man's material welfare. Regarding the latter, Glanvill
is emphatic, for he claims that whether the virtuosi succeed in
explaining phenomena or not, it is pleasant "to behold the *shifts,*

windings, and *unexpected Caprichios* of distressed *Nature*, when pursued by a *close* and *well managed Experiment*," and he emphasizes the useful service which they can render to the improvement of agriculture, mining, and the like. Already Glanvill is shifting the emphasis from the mechanical hypothesis to the useful experimental philosophy.

Perhaps the most significant and influential defence of the experimental learning besides Sprat's *History of the Royal Society* issued from the pen of that earnest propagandist of the new science who figured prominently in the preceding chapter, Robert Boyle.[33] As the title, *The Usefulnesse of Experimental Naturall Philosophy*, indicates, Boyle bases his plea for experimental science squarely upon its utility, though he does not ignore its intellectual value. He begins his treatise with a comparison of the new philosophy with that taught in the schools, which, he claims, is little other than a system of the opinions of Aristotle, a barren and litigious philosophy easily learned by a perusal of a few current authors. Experimental philosophy, on the other hand, is difficult and laborious, imposing, as it does, upon its followers continuous physical activity. We have already noticed how the Puritans emphasized the industry of the experimental scientists as contrasted with the sloth and laziness of the Aristotelians, and the charge continues to be made throughout the Restoration. He next stresses the pleasure of experiments,[34] even in matters possessing as much horror as the dissection of dead bodies, asserting that he had "often spent hours much less delightfully, not onely in Courts, but even in Libraries, then in tracing in those forsaken Mansions, the inimitable Workmanship of the Omniscient Architect." He continues to elaborate upon his willingness to deal manually with repulsive material, and not to scruple to handle with naked hands clay and charcoal, in defiant answer to the charge of dealing with low material things, which had been and was to be preferred against the new science. The second essay upholds experimental philosophy on the ground that it incites men to devotion, since experiments and observation, especially with the aid of such instruments as the telescope and microscope, reveal the power, wisdom, and goodness of God as seen in his marvellously contrived creations. The third essay continues the argument by showing that one ignorant of nature cannot perceive the divine attributes

mentioned above. The fourth essay, styled "a requisite Digression," touches upon a problem which was becoming more and
more acute for the scientists. Though Descartes had tried to
reconcile his mechanical philosophy with religion, his theory
hardly left any place for God in the universe, and the fact that
some of the ancient propounders of atomism all but denied the
existence of God accentuated the atheistic implications of the
philosophy. These implications were fully revealed in Hobbes,
who based his explanations of all phenomena, human and physical, upon matter and motion. His philosophy proving popular at
the court, which was rapidly revealing its atheistic and de-
bauched nature, there arose a natural inclination to attribute
that deplorable condition to the philosophy.[35] Since so many of
the scientists had subscribed to it, and were finding in it the best
explanation for the data furnished by their experiments, it became in the popular eye definitely associated with the experimental philosophy. Thus Boyle rightly considered his discussion
of the matter "requisite," but he also considered it a "digression,"
because his treatise had to do with the experimental philosophy,
which he held to be distinct from Descartes's. The logical infer-
ences arising from the latter, that matter was not created, that
all material things represented a fortuitous concourse of atoms,
even though obeying mathematical laws, and that motion was
absolute, depending on no mover, left no place for God. Boyle,
though embracing the theory as regards the physical universe,
insists on God's creation of the world, His original imparting of
motion to matter, and His concourse with the operation of the
machine of the world.[36] This leads him to attack Aristotle's
philosophy, because certain Aristotelians excluded God from the
world on the basis of Aristotle's belief in the eternity of the
universe, and he shows Aristotle's errors in magnetism, nature's
abhorrence of a vacuum, and the like. The fifth essay returns
to the theme discussed before the digression, and insists that a
knowledge of the world as revealed to a scientist must convince
one of the existence of the Deity. In general the first part of
the treatise is concerned with justifying the experimental philosophy in the eyes of religion.

The first section of "The Second Part. Of its Usefulnesse to
promote the *Empire of Man* over things *Corporeal*" attempts to
show the contribution which the experimental method may make

to medicine, and the first essay is devoted to the usefulness of the method to "the Physiological part of Physick." At the very outset Boyle makes plain, as he does in most of his writings, his pronounced utilitarian view of science, an attitude which is the outstanding characteristic of the scientific movement in seventeenth-century England. "I should not," he says,

have neer so high a value as I now cherish for Physiology, if I thought it could onely teach a Man to discourse of Nature, but not at all to master Her; and served onely, with pleasing Speculations, to entertain his Understanding without at all increasing his Power. And though I presume not to judge of other Mens knowledge, yet, for my own particular, I shall not dare to think my self a true Naturalist, till my skill can make my Garden yield better Herbs and Flowers, or my Orchard better Fruit, or my Fields better Corn, or my Dairy better Cheese than theirs that are strangers to Physiology. . . . . Me thinks, it should be a disparagement to a Philosopher, when he descends to consider Husbandry, not to be able, with all his Science, to improve the precepts of an Art, resulting from the lame and unlearned Observations and Practice of such illiterate Persons as Gardeners, Plow-men, and Milk-maids.

He proceeds to make the utility of the new method a strong talking point against the uselessness of the traditional philosophy, which indeed constituted the chief and most frequently mentioned charge brought against it, and in doing so he echoes loudly the words of the great Chancellor.

And indeed, *Pyrophilus*, though it be but too evident, that the barren Philosophy, wont to be taught in the Schools, have hitherto been found of very little use in humane Life: yet if the true Principles of that fertil Science were thorowly known, consider'd and apply'd, 'tis scarce imaginable, how universal and advantagious a change they would make in the World: For in Man's knowledge of the nature of the Creatures, does principally consist his Empire over them, (his Knowledge and his Power having generally the same limits) . . . . Theological inquiries excepted, there is no Imployment wherein Mankinde is so much and so generally concern'd, as 'tis in the study of Natural Philosophy.[37]

Many of the values and attitudes that became prominent during the Puritan supremancy continue in their full vigor throughout the Restoration: hostility to the ancients and traditional philosophy, worship of Bacon, regard for the experimental philosophy, and, above all, the strong spirit of utility. From the latter sprang the feeling that man's chief concern was his material welfare, and that to contribute to it was more important than all other activities. Thus science is raised to a place of

importance higher than all other human enterprises, for it
reaches to the interest of all mankind. Boyle says, "those great
Transactions which make such a noise in the World, and establish
Monarchies or ruine Empires, reach not to so many Persons with
their influence, as do the Theories of Physiology."[38] He asserts
that natural philosophy is no "barren speculative Knowledge,"
but that physic, agriculture, and very many trades, such as those
of tanners, dyers, brewers, founders, and the like, "are but
Corollaries or Applications of some few Theorems of it." It is
his purpose, he says to show how much may be contributed to
such arts and trades by "the Philosophical Experiments and Observations" of experimental science.[39]

So in his first essay Boyle, accepting the Cartesian idea that
the human body is but an engine in which all actions common to
man and animals are performed mechanically, proceeds to point
out how much light the experiments of vivisectionists may throw
on human anatomy, and he exposes various errors in current
anatomical beliefs, which he had discovered through observations and experiments. Naturally at this point Harvey's great
discovery comes in for notice. Boyle also points out that chemical
experiments may reveal the nature of fluids in the body, and
may discover better ways of preserving bodies for observation.
From anatomy he passes in his next essay to pathology, in discussing which he has occasion especially to mention the chemical
doctors. Here, as in other treatises, Boyle expresses distinct disgust with the boasting claims of the "vulgar chymists" and
strong suspicion of the truth of their theories, but he does think
that if "Chymicall tryals were skilfully varied and judiciously
applyed to the illustrating of Pathologicall *Phænomena,* the
former might be made conducing to the better explication of the
latter: especially if the business were mannag'd by a Naturalist
well vers'd both in Chymicall Experiments, and in Anatomy,
and the history of Diseases, without being too much addicted
either to the Chymist's notions, or the receiv'd opinions of
Physitians."[40] Boyle tries to steer a middle course between the
warring factions in the medical world, holding on the one hand
that the chemists scorn too completely traditional theories, and
on the other that Galenists cling too tenaciously to outworn
theories and ignore modern experiments too much. He especially
preferred the optimistic outlook of the chemists to the dishearten-

ing belief of the Galenists in the incurability of many diseases. Boyle could not help leaning somewhat to the side of the chemical doctors because of their addiction to, and faith in, the experimental philosophy and because of their forward-looking attitude. He also perceived the value of experiments to pathology in the discovery of the circulation of the blood, which had led to new explanations of diseases. The next two essays discuss with less animation the usefulness of experiments in improving prognostics, in the preparation and preservation of drinks, foods, and meats, and in determining the healthfulness of a climate. In the last essay, devoted to the contribution which the experimental philosophy may make to therapeutics, Boyle again emphasizes the importance of chemistry to medicine, and though he condemns the chemists for railing at the Galenists without recognizing the worth of some of their ideas, he blames the latter for blindly holding to unprofitable and outworn theories, learned from the schools and based on barren Peripateticism, to the great hindrance of the advancement of medicine. He even goes so far as to say that he has known not only empirics but even ladies and old wives who were more successful in treating diseases than learned physicians. Boyle's attitude toward the chemical doctors is at times somewhat equivocal, so that it is not strange to find, in the clash between chemists and Galenists, both citing him as an authority, but in general we find him distinctly more sympathetic with the former than with the latter, for as has been said, they subscribed to a method which looked forward to the advancement of medicine, and not backward to the stagnation of science.[41]

In two works[42] published soon after the *Usefulnesse of Experimental Naturall Philosophy*, Boyle continues, somewhat indirectly, to put forward the claims of the experimental philosophy. Here, as elsewhere, he is particularly interested in making clear the function of experiments and their relationship to theories. In short, he says his purpose in these two works is to furnish experiments toward a history of cold and colors. Here again he is one with Bacon in stressing the danger of drawing up hypotheses upon inadequate data, and he is careful to disclaim any intent to announce a theory of colors, but only "the beginning of a History of Colours, upon which, when you and your Ingenious friends shall have Enrich'd it, a Solid Theory

may be safely built . . . . an *Apparatus* to a sound and comprehensive Hypothesis."[43] To the seventeenth century the ultimate
function of experiments was to furnish a solid basis for true and
useful hypotheses, though inventions and knowledge that could
be turned to practical uses might much sooner result from them.

In no field of science during the seventh decade of the century
were the ancients and their followers more bitterly attacked or
the experimental and observational method more enthusiastically
espoused than in medicine. The controversy between ancients
and moderns finds full expression in the relentless war which
the chemical doctors were waging against the Galenists and their
"method." The chemists inveighed against them for their blind
following of ancient dictates, their espousal of a medical theory
based upon the erroneous idea of the four elements, and for the
inefficacy of their remedies. The chemical doctors insisted that
the authority of the ancients should give way to the experimental
method. For the most part, they confined the contribution chemistry could make to the advancement of medicine to the preparations of new and better medicines, and their faith was based on
the knowledge of the parts and principles of mixed bodies gained
through experiments with fire. They were more fortunate in
what they condemned than in what they supported, but they enjoyed two arguments, both in part derived from Bacon, which
gave them a distinct advantage over their opponents: their hostility to ancient authority and their faith in the experimental
philosophy. Furthermore, they were in full sympathy with the
greatest scientists of the age, and were imbued with the spirit of
the freedom of scientific investigation and thcught.

Though the Paracelsians had for a century or more striven
bitterly against the domination of the ancients, the struggle
entered upon an acute stage with the publication of Marchamont
Nedham's *Medela Medicinæ*, 1665,[44] and raged with great intensity. Nedham was the most earnest and intelligent advocate
of chemical medicine. He drew much of his inspiration and many
of his ideas from the chief scientific thinkers of the seventeenth
century, especially Bacon and Boyle. He is found in the van of
those who were ardently fighting for a new freedom in scientific
investigation. His chief purpose in writing the book was "to
plead for a freedom for such as labour in the secrets of Nature,
and that no discouragement be given to such laborious and in-

genious Inquirers."[45] It is this unfaltering loyalty to the ideal of intellectual freedom that chiefly animates his severe arraignment of the subserviency of traditional medicine to the authority of antiquity, though he is convinced of the erroneous nature of traditional physic and of the fallacious natural philosophy upon which it was based. After speaking of the way contagious diseases are spread through the dissemination of invisible atoms, he breaks out with "This fine way of Communication suits not with the Conceit of a Brain, that measures every thing by the gross Philosophy which *Aristotle* ties men to in the Schools." He makes much of the fact that new diseases have arisen, of which antiquity was ignorant, and that "many things lie hid in the nature of things which were unknown to the Antients, and to such also as have lived of late years, which the extraordinary Sagacity of the present times hath discovered by the benefit of the *Microscope,* and laid before our eyes." He was fully alive to the promise which the new philosophy held for the improvement of medicine. He contrasts with enthusiasm the sloth and indolence of the classical doctors with the industry of the experimentalists, and boldly declares, as Boyle had before him, that he was humble enough to learn from empirics and even old women, who frequently gave him good reasons for what they did, though not "cloathed with such delicate Terms of Art, as pass current among the Schools."[46] Again and again he belabors the utter inadequacy and innumerable errors of the philosophy and physic derived from the ancients and entrenched in the universities.[47] He is not without hope, however, for his inquisitive age, in which "the later and more excellent Brood of Physicians do every day run under the wings of *Chymistry,* and addict themselves only to Experimental and more rational Learning." It is upon the experimental method that Nedham bases the superiority of chemical to traditional medicine. In his eyes men must learn the nature of diseases and their remedies through direct observation and experiment. If, he says, the example of the chemists

were followed by some of the surly stately Sirs of the Faculty, sometimes to handle *Coales* and the *Bellows,* sometimes the *Pestle* and *Mortar,* and ever and anon to be sifting and observing the new natures of Diseases, and the nature of our own people, and not look upon them through the false Perspectives of old Authors, the Nation had not been now to seek remedies proper to its condition, nor would so many every day run away from their unprofitable Recipes, grounded upon old *Forein Gallimafries.*

He quotes with vigorous approval what Willis had said about how much medicine would be advanced if physicians betook themselves "to Observations only, and Experiments," and he insists that improvements in remedies for disease are to be expected only "from one that is qualified for the work by acquaintance with Mechanick and *Experimental Philosophy*."[48]

Nedham was in close touch with the great spirits that had battled and were battling against ancient domination and in behalf of modern science.[49] He was especially incensed with the treatment that new thinkers, departing from ancient paths, had received from the world: Paracelsus, who though guilty of some whimsies had opened up the most excellent part of physic; Tycho Brahe, who demolished the theory of solid spheres and was saved from punishment only by his noble descent; Galileo, who was sent to gaol for seeing more with his telescopes than others could see with their eyes; Descartes, against whom the Aristotelian professors at Utrecht stirred up the rabble; the profound Van Helmont; and Harvey, who complained of the ill usage he had received for departing from traditional anatomical ideas. He is fully cognizant of the importance of the scientific discoveries of the sixteenth and seventeenth centuries, and even more so of the courageous departure from the past which made these discoveries possible. He cites as opponents of antiquity Bacon, Helmont, Descartes, Boyle, More, Willis, and others. "Seeing men of greatest Learning and Abilities in the Profession, have judged the old Philosophy and Physick insufficient, and thereupon receded from it. And whereas in this prying, laborious Age, ... new discoveries of Medicine are every day wrought out of the Fire, and other ways, and more like to be discovered, for the benefit of man-kind . . . . it is the concernment of Nations to admit a greater latitude in the Profession and Practise of Philosophy and Physick, than the Interest (and many times the ignorance) of some men of the same Profession would be willing to allow." He exclaims that it would be well for the universities to follow the advice given by Willis and not "lie down . . . . under the shadow of a great name, and make an Idol of it; We should rather reject what runs to superstition, and not pin the Faith of mankind upon the sleeve of Hippocrates, because others have done so."[50]

Though, of course, many of the views which Nedham expresses derive ultimately from Paracelsus and the Paracelsians of the preceding and his own century, and though he, like other chemical doctors of his day, leans heavily upon Van Helmont, yet the most casual perusal of his treatise reveals that it was Bacon who chiefly inspired him with faith in the observational and experimental method, and whose words Nedham looked upon as carrying the greatest weight. His citations of the earlier philosopher are frequent and his quotations from his works generous.[51] He emphasizes such ideas of Bacon's as the obstacle placed in the way of the advancement of science by subserviency to ancient authority, the paralysis of men's activity through the opinion that nothing has been left for the moderns to discover, the erroneous belief that ancient wisdom represented the childhood of knowledge, and the necessity of the sciences' being reformed from the very foundation. Especially does he stress Bacon's dictum that scientific knowledge cannot be founded upon mental conceptions but only upon sensible objects and experiments, in which connection he cites Willis' analysis of matter as an excellent example of Bacon's idea of raising a philosophy upon sensible experiments. Again and again Nedham expresses that distrust of the reliability of the mind in itself and of intellectual operations too remote from material data which is so characteristic of St. Albans, to whom indeed Nedham appeals, and he also reveals the same craving for a material basis and standard for truth. He asks rhetorically if the Galenists, who, ignorant of chemical experiments, "rest in mere Traditional Doctrines and phantastick opinions, wrap't up in *general Conclusions* and *Definitions* of *pretended Principles,* and of Diseases, the mere Products of wit and opinion, than which, as my Lord *Bacon* saith, there is nothing more *Polydædalous,* various and uncertain (the Brain of man being a wondrous winding Labyrinth of Conceptions)," are to be preferred to the "Chymists, who build their Philosophy or knowledge of Nature in man, and in Medicaments, and in Diseases, upon real operations and productions of things natural, which they *see, feel, and handle?*"[52] One needs no clearer evidence of the way Bacon's influence was pervading all branches of science than is revealed in Nedham's volume, which indeed only gives Bacon's general rules specific application to medicine.[53] The only other scientist who can vie with Bacon in Nedham's pages is Boyle, who also, as has been seen, considered himself an humble disciple of the great teacher.

Most of the theories of chemical medicine have gone by the board, and perhaps in no branch of science were there so many blind empirics, brazen charlatans, and mercenary quacks, but the chemists' appreciation of the potential value of chemical analyses, their faith in, and insistence upon, observation and experiment, and their severe condemnation of the enslaved followers of antiquity give them an importance which their discoveries in general do not warrant. Boyle, who is an important figure in the history of chemistry, could hardly help being disgusted with much that was vain, vulgar, and idiotic in the pyrotechnists, but we must not forget that he sufficiently sympathized with their tireless experimenting to enable them to quote him in their favor. They were facing in the same direction as the other Baconians, and, their theories aside, the attitudes they maintained were worthy of the scientific spirit of the age. Freedom of scientific thought and investigation finds no more enthusiastic advocate than Nedham. In no one does the Baconian spirit burn more brightly. No one perceived more clearly the vital necessity of breaking with antiquity and of adopting the proper method of scientific research. Keenly aware of the importance of the great discoveries of the Renaissance and confident of the powers of the new science, he optimistically faced the future, a future which to him held unlimited possibilities.[54] He was imbued with the spirit of progress, a progress made possible by the new method.

Nedham's book was hardly off the press when it was answered by Robert Sprackling, a member of the Royal College of Physicians.[55] The chief fault that Sprackling finds with Nedham, and he finds it with a good deal of abuse, is his lack of reverence for the ancients, whom he himself praises highly. Yet Sprackling expresses just as severe condemnation of submission to authority as his opponent, insisting that if ancient learning is still accepted, it is because reason and the experience of many ages have approved it. He readily admits that eminent discoveries in his own age had corrected many errors of antiquity, some of which he points out, and he is loud in his praise of Harvey, Glisson (to whom the book is dedicated), and Boyle, the latter two of whom were conspicuous in the Baconian tradition. He tries to show that chemical principles are the same as the four elements of the ancients, and that the latter comprehended chemistry by "mental abstraction," whereas the moderns have only proved it by experiments.[56] Yet after all, he heartily approves modern discoveries, honors those who made them, and certainly implies sincere commendation of the experimental philosophy and need of further inquiry to advance knowledge, though he makes a covert reference to the association of the scientific movement with Puritanism in the years preceding the Restoration. But he cannot bear to hear anyone speak irreverently of the ancients. The sceptre of power was being taken from antiquity, but its crown and regal robes were still tokens of reverence.

In the same year Edward Bolnest continued the attack on traditional medicine in *Medicina Instaurata* (a pretentious title suggestive of Bacon), dedicated to the Duke of Buckingham for his contributions to "experimental philosophy," and containing an "Epistolary Discourse" written by Nedham in answer to Sprackling. Nedham repeats his criticism of the unprofitable physic of the schools, and expresses pleasure at Bolnest's dedication to Buckingham, who, he says, maintained two laboratories, one at home and one at the court. More than in his earlier treatise he is engaging in ardent propaganda to establish the "whole frame of Physick upon a new foot of Operative and Experimental Philosophy." Eager to secure converts to the new religion, he issues a call to all the nobility, in emulation of the King himself, to accept and practice the new method. How powerful and confident the chemists were becoming is revealed in Nedham's frequent references to "our Society" and in their efforts to establish a chemical college, undoubtedly with an idea of securing the King's patronage. He speaks of the chemists' being given an audience at "the King's Council-Table" and of the consideration shown every one of them. "When we have settled," he says, "our publick place or College, with a grand Laboratory suitable to so worthy an Undertaking, we purpose (God willing) to turn the stream of Practice out of the *Galenick Channel.*" And elsewhere he speaks as if the college were all but established, at the same time hinting at the opposition of the traditional doctors to it.[57] Bolnest adds little that is new to the contest. He is chiefly interested in advertising some newly invented pills, the marvelous virtues of which he is not backward in proclaiming. He holds forth the experimental method as the only test of truth and the sole avenue to discovery. He not only praises chemical experiments for their purifying of medicines, a standard argument, but he realizes the wider possibilities of chemical investigation and the possible contribution of the science to such homely commodities as bread, beer, wine, and the like. In the usual fashion he attacks the futility of academic learning with its fruitless disputations, and declares that while we should be ready to acknowledge the little received from the ancients, we should not "set up our Rest, and *Herculean* stops, or a *non ultra*, to all farther inquisition and search after knowledg."[58] No one proclaimed freedom more loudly than the chemists.

The conservative view of medicine during this period is best expressed by John Twysden, a physician and fellow of the Royal College of Physicians, in another answer to Nedham.[59] Twysden's fundamental conservatism is revealed in his lament over "this scribling age," in which men have taken the liberty to present unheard of things, and to assault not only traditional medicine but law and divinity as well. This association of the free spirit of science with Puritanism is further revealed in his pointing to the Puritan rebellion as an example of how dangerous it is to shake foundations without erecting firmer ones.[60] The argument upon which he most heavily leans, and which has been the cry of conservatism since the beginning of things, is the test of time and consent of ages. It is better, he urges, to stick to the old and tried than to follow the new, which is not backed by the authority of even one age. Those medical principles that have met with the approbation of the most learned in many centuries, the elements, humours, qualities, and the like, in which there may be more truth than in the mechanical philosophy, are to be preferred, for men must go very slowly in accepting new truths in medicine. He, too, is outraged by the lack of reverence for the ancients manifested by Nedham, whose "disgraceful speaches" against Galen and Hippocrates he severely condemns. He hopes to strengthen the cause of antiquity by insisting that the scientific spirits of the seventeenth century whom Nedham had used to good advantage, Bacon[61] and Harvey, were not disparaging in their attitude toward the ancients, that, in fact, Harvey entertained a high regard for Aristotle. He further claims that in attacking the ancients Nedham is really assaulting the universities and the Royal College of Physicians. He, like most of the conservatives, denies that authority has any weight with the Galenical doctors, and somewhat inconsistently he advocates full liberty in examining traditional theories and in subscribing to them only as far as reason permits. He likewise has high praise for the discoveries of Harvey, Glisson, and other exponents of the new philosophy, but he derogates from this praise in claiming that scientific investigations in biology and anatomy, while quite permissible in those that have time for them, are unnecessary for a physician, and that such discoveries as the circulation of the blood have not contributed to the curing of diseases. Better accept the theories of the ancients, he advises, which, though perhaps not true, as he evidently felt compelled to confess, are useful in that they have led to the discovery of remedies. Like most of the Galenists of the time, he found himself in the unhappy situation of being compelled to recognize the fruitfulness of the experimental method as manifested in the great discoveries of his age, and at the same time to feel reverence for the ancients and their principles, the inadequacy of which these same discoveries had revealed.

One factor which tended to intensify the conflict between the two medical fraternities was the Royal College of Physicians, founded in the preceding century firmly upon Galenical principles and remaining a stronghold of Galenism to the very last. Yet such were the important discoveries in medicine and such the impression produced by the more moderate chemists that at this time the College presented a wavering front. A number of its fellows were members of the Royal Society and earnest upholders of the experimental method, as well as being favorably disposed toward chemistry, so that the College itself felt compelled to acknowledge the value of chemistry. Yet the more violent attacks of the pyrotechnists upon the ancients, and the reverence with which the latter were still regarded by the rank and file of the body, resulted in an interpretation of the attack on antiquity as an attack on the College. This fact is quite apparent in the controversy between George Thomson, M.D. of Leyden, and William Johnson, who describes himself as chemist to the Royal College of Physicians. In *Galeno-pale*, 1665, Thomson severely attacks Galenical principles and methods, and accuses the Galenists of being lazy, mercenary, and enslaved to the authority of antiquity.[62] On the other hand, he praises the chemical physicians, though admitting the injury done them by the boasting claims of quacks and ignorant empirics, and he bases their excellence squarely upon the virtues of the experimental method, which, he claims, is the only true way of discovery, and which he contrasts with the disputatious scholarship of the Galenists. The ignorance of the latter, he says, springs directly from a want of true experimental knowledge. Thomson, evidently fearing the charge of hostility to the universities which had been brought against his kind, insists that he is not opposed to learning, and inveighs as loudly as any against ignorant empirics. The very same year Johnson answered him in *Aγνωστο-Μάστιξ. Or, some Brief Animadversions upon two late Treatises*, in which he accuses Thomson of attacking the Royal College of Physicians, declares that the members thereof are chemists, and maintains that the division between chemists and Galenists is unjustified since both Van Helmont and Galen are honored. Yet his bias is soon perceived when he states that Galen has by no means been buried by later discoveries and that discreet physicians of all ages have received Galen's theories as oracles. One matter of interest in the treatise is his extended comparison of a chemist or, as he states, a fanatic in chemistry with a fanatic in religion, and his clear recognition of the association of experimental philosophy with the Puritans. The chemical physicians, he maintains, "could never have gotten such a Repute in the World, but that they Politickly made an advantage of the Factious Principles then abounding in the Common People of our late Unruly times, when the Common Interest was to be carried on by crying down Humane Learning; then these Illiterate Fellows spit in the face of all the Liberal Arts and Sciences."[63] Throughout this whole period the former relationship between the new science and Puritanism, discussed in a previous chapter, receives more than passing notice from the upholders of the ancients.

Thomson immediately retaliated with an answer in which he accused Johnson of being only a tool of the College, and firmly denied that the Galenists were lovers of chemistry. But the most interesting feature of the volume is an "Epistolar Discourse" contributed by George Starkey, who inveighs sharply against the "rotten Foundations, ruinous Arches and Pillars, mouldering and tottering Walls, a leaky and almost falling Roof"

of the ancient edifice of traditional medicine, and especially
against the obstacles which the Galenists place in the way of
those who would advance medical knowledge. One of the most
commendable features of the spagyrists was the way in which
they faced the future rather than the past, and their ardent desire for progress in medical science. The reason why hope for
it does not lie in the Galenists is that "they think they have already attained it [truth], and therefore conceive there is no need
of further learning, and disdain to be taught, or reproved, but
will abuse him that shall attempt it, and load him with opprobrious contumelies."[64] As for the Galenists who make pretensions to chemistry, and who have taken to themselves the high
sounding name of "Rational Chemists" to distinguish themselves
from "Melancholy Operators,"[65] he speaks with utmost scorn of
their foolish belief that they "can by depth of reading, and profound Reason" discover medicines or their proper application. It
is upon experiment and direct observation of nature that Starkey
lays all emphasis. It is the method and the method alone that
will lead to truth, and that is the strongest argument for the
chemical doctors. " 'Tis as possible and probable," he says,

to be an expert Chemist, without seriously, sedulously, indefatigably, and
constantly attending the Mechanical part of *Pyrotechny*, making new
[experiments] daily, pondering also, and amending future Operations by
past experience, as to be an expert Anatomist without knowing the use of
the Knife, by bare Rational speculation. 'Tis the Mechanist informs, in-
structs, and confirms the Theorical speculator, and not on the contrary.[66]

The pyrotechnists were as ardent believers in the supreme importance of experimentation for the advancement of science as
can be discovered in the seventeenth century. They were just as
sincere followers of Bacon as more noted scientists.

The upholders of chemical medicine may in general be divided
into three groups. The first, and lowest group, was composed of
quacks, charlatans, and ignorant empirics, men of no education
or standing, who had no other resources than their hands and
eyes, who clung tenaciously and vociferously to the skirts of the
experimental method, and who were more concerned with making
money than advancing science. It was with this group that the
second was especially anxious not to be confused. The latter was
composed of educated men who had received the traditional
medical training, but who had turned from it to embrace the

chemical faith based upon experimentation and to subscribe to the chief chemical theories of the day, especially those of Van Helmont. Nedham and Thomson were of this class. And finally there was a group of men who believed firmly but by no means exclusively in the achievements and potentialities of chemistry, who did not confine to chemical remedies the benefits which the experimental and observational method would bring to medicine, who subscribed to mechanical rather than to Helmontian principles, and who, though committed whole-heartedly to the new science, were more respectful to antiquity and tradition as well as less severe upon the established physicians. To this class belong the next three men to be discussed, fellows both of the Royal Society and the Royal College of Physicians. Their view is essentially the view of Boyle, who applauded chemical experiments but condemned chemical theories and the extravagant claims of the Helmontians.

The first of these is Christopher Merrett, librarian of the Royal College of Physicians, who in *A Letter Concerning the present State of Physick, And the Regulation of the Practice of it in this Kingdom,* 1665, is seeking to persuade the members of the Royal College to adopt the experimental and observational method which he outlines. These men, he asserts in opposition to Nedham, are enslaved to no authority, neither that of Hippocrates and Galen nor of Paracelsus and Helmont, and although when chemistry first appeared, it was condemned, "in these more illuminated times," they now approve "The new and more exquisite way of *preparing Medicines,*" i.e., by chemical analysis. Though Merrett reveals at all times respect for the great medical names of antiquity, it is their methods and habits he recommends, not their theories. As regards the former, he insists that doctors return to the ancient way: exercises in anatomy, study of herbs, compounding of medicines, and visiting of the sick in company with an experienced physician, "nor," he says, "let any man think to disgrace this Method as Mechanical." The ancient principles, however, are barren and unfruitful, teaching nothing about diseases and resulting only in disputations and controversies of books, the exact view of the other chemists. The adherence of educated physicians to ancient theories, he says, has caused the method of observation and experiment to fall into the hands of those who are not capable of employing it properly.

Merrett is all for experiment, observation, and first-hand knowledge as opposed to books and traditional theories. In his eyes medicine as yet was not a science founded upon unquestionable truths, but "a mass of noble Experiments" to which must be added many more careful and laborious ones before physiology, at any rate, could be established. These experiments, he held, only the fellows of the Royal College were qualified to make for "the use of humane life and also to the advancement of Philosophy." If they would undertake "a *natural History both of Galenical and Chymical Medicines,*" it would prove

a more likely and hopeful way, then to put the business of *Chymistry* . . . . into the hands of a few men (for the most part) unlearned, and unprovided of such principles and helps as should make them able to reason and conclude intelligibly from that *brave stock* of *Experiments* which this Art affords: And for this Cause onely it has been (until of late) so unserviceable to Philosophy, and produc'd nothing but idle and extravagant Theories, such as those of *Paracelsus* and *Helmont;* who, if they had been as good *Philosophers* as they were *Operators* [experimenters], would have done a great deal more service to mankinde.[67]

Merrett runs true to the Baconian form in insisting upon the need of more experimental data before theories are devised, but he is equally insistent upon the necessity of putting the work of collecting them into the hands of capable men. He also sees a wider scientific field for chemistry than medicine, as well as the possibility of improving medicine in other ways than by chemical medicines. He insists upon doctors' preparing their own medicines (the strongest argument on the part of the chemical doctors), using the microscope, and carrying on their own chemical experiments. He has much in common with the men whom he rightly depreciates for their wild theories.[68]

The spirit of Bacon and the Royal Society animates Merrett's work. He refers to "that great Person" several times, but his indebtedness frequently appears without citation.[69] Yet, strange to say, he does not mention him in the thumb-nail sketch he gives of the development of modern science. After speaking of the peculiar genius of his age for scientific research, and of what medical investigations may contribute not only to the needs of life but also to science in the way of "so great a number of real and useful *Experiments,*" he claims that the true method of discovering natural truths is due to his age alone, for in the sixteenth century only Copernicus did anything of moment in science. Then Galileo was the first to introduce sound specula- tions by the application of geometrical theories to natural phe- nomena. This method was followed by Kepler and perfected by Descartes, whose works inflamed men's minds with the zeal of scientific investigation, even *"His Most Excellent Majesty, who has set on foot the greatest design for this end that ever any Nation saw, by Founding his Royal Society for the advancement of Natural Philosophy by Experiments,* which will certainly be as Immortal as his own Name and Fortunes."[70] This neglect of Bacon and others was not allowed to stand long, for Timothy Clerke, physician to Charles II, in his extended comment on Merrett's treatise says, "besides the famous men you mention for it [promoting useful knowledge drawn from experiments] *our own nation ought not to have her Verulam, and Gilbert For- gotten, nor our profession in its Harvey, Glisten* [Glisson?] *nor Ent;* to mention no others now."[71] Clerke possessed a firm faith in, and love for, the experimental philosophy, yet he was more dubious of its ultimate success than his *confrères,* because he found so many ignorant men, especially in chemistry, paying loud allegiance to a method that exacts unusual qualifications of mind and training, men who boasted much and promised more, and who deluded the world into expecting more from the Royal Society than could possibly be allowed. He is one of few to note the injury done the scientific movement by the ignorant pretend- ers to it. "I shall not now stand," he says,

to discourse how much more difficult a thing it is than usually considered to get experience from experiments: how few compare (or are indeed quali- fied to do it) all circumstances; and since a little one, may make great alterations in things, and the knowledge resulting from experience is, that actions, or causes so circumstantiated, always will produce such an effect; the circumstances not considered, or taken notice of, what can such an ex- perience amount to? How many impertinent Fooles do we meet with every day, that talk mighty big of experience, without knowing anything of it; if they did, these would neither pretend to miracles themselves, nor expect them from others, this I confess among other things, makes me despair of any advantage to be procured, for a real and true promoting the good our profession might bring to mankind, or of any great advantages from the Royal Society.[72]

Clerke declares himself a great lover of chemistry, and believes that it is indispensable to a physician, yet he thinks only educated physicians, such as the members of the Royal College of Physicians, are qualified to pursue it. He stoutly defends the fellows of the College against the charge of being enslaved to authority, and declares that since Harvey's great discovery no society of physicians can boast so many excellent anatomists. Furthermore,

they have long since embraced the more generous, and useful way of Philosophising freely; they cherish it, and encourage it in all their members, and I do assure you Sir, no man was ever discouraged by them, who gave a good account that he had judgment, and abilities for the practice of Physick, though he were deficient in the abstruser notions, and nice distinctions of the Schooles.[73]

Not by them but by "that lying and knavery which makes the very enquiring into it [chemistry] a reproach in the opinion of some sober men" is he induced to depart from the common opinion that his age was one for the advancement of science. The depreciation of reason in behalf of experimentation, he feared, would lead to the greatest barbarism and most profound ignorance. They who know only "enough to rally the ancient learning, and talk of experiments, will disturbe the gravest Societies with their impertinencies, silence the most learned, and think themselves the only dictators in universal learning, and if they are so unfortunate to be endowed with wit and good natural parts, *they turn everything that is sober into Ridiculous, and go out Doctors in Atheisme.*"[74]

One other tribute to the importance of chemistry to medicine is discovered in Jonathan Goddard's *A Discourse Setting forth the Unhappy Condition of the Practice of Physick in London,* 1670, written five years before publication and so falling within our period. The treatise, like Merrett's, was inspired by the apothecaries' practising medicine, and the remedy Goddard proposes is that physicians learn to make their own medicines by chemical experiments and thus render apothecaries unnecessary. The treatise is very reasonable and restrained. He shares the general antipathy to ignorant chemists, and advocates a doctor's receiving the traditional classical training, but the physician should then "apply his study to Natural Philosophy, such as is more real and solid in this Age, by many happy Experimental Discoveries in Nature: and lastly to the Art of Physick, and the knowledge of the Body of Man, with al the parts of it, by

Anatomical Administrations, Experiments and Observations."[75]
He admits that like all other human knowledge medicine is defective, since it has "too much abounded with Notions and Speculations wanting foundation in Nature and Experience," yet in view
of the great improvement arising from many happy discoveries
in his age, he thinks it can vie with any other science. Though
impatient with boasting quacks, he grants that because of their
experiments, "some Empiricks have stumbled upon very considerable and effectual Medicaments, wherewith in some particular
cases, they have outdone learned Physicians: and by the advantage of making their own Medicaments, they bear up, and will
do, till they be outdone in the same kind by such Physicians."[76]

Of the three doctors just discussed, Merrett and Goddard had
been members of the first London group of experimenters, and
Clerke was a charter member of the Royal Society and was appointed to its first council.[77] Thus having been in close touch
with Baconian science from the very first, they naturally took a
broader, saner, and more intelligent view of the relations of
chemistry and medicine. But their faith in the experimental
method and their allegiance to Bacon were no stronger than
those of men like Nedham and Thomson, who, though well educated, had fallen under the spell of Paracelsian and Helmontian
theories, and so were less restrained in their comments upon
traditional science. Because of exorbitant hopes for their medicines this latter group ran the risk of being confused with the
rabble of quacks, blind empirics, and charlatans who indiscriminately and with bad taste condemned everything savoring
of antiquity, at the same time loudly singing the praises of experiment and observation. Since the difficulties involved in the
true prosecution of the method, which Clerke pointed out, had
not become very apparent, and since even respected scientists
had at times favorably compared ignorant mechanics with men
of traditional learning, this last group seized with alacrity upon
a method which seemed to depend only upon the senses (they
could have adduced Robert Hooke as a witness), and to the
employment of which their ignorance appeared to offer no obstacle. They decried learning partly because they lacked it, and
shouted for experimentation partly because it offered opportunities to their limited capacities. Yet, though their boasting
claims for chemical medicines suggest a crass, mercenary motive

for their connection with science, undoubtedly true in most cases, one may at times detect a sincere and untainted, even though unintelligent, loyalty to the cause of experimental learning.[78] While to them, as well as to the respectable Helmontians, the experimental method as applied to medicine made only for improvements in medicines, men like Merrett and Clerke perceived the physiological and clinical importance of it. Both the doctors just mentioned insisted upon clinical training for medical students, who were urged to accompany experienced physicians in order to observe with their own eyes the nature of diseases.

It was in this last respect that a great contribution was made to medicine in the writings and example of Thomas Sydenham, a man of strong Puritan sympathies, who in the course of his education at Oxford must have established contact with the group of Baconians at Wadham College. Sydenham "laid for all time the foundation of the practice of clinical medicine."[79] His aim was to study diseases without any preconceived hypothesis, whether Galenical, chemical, or mechanical. In this attitude he was a true disciple of Verulam, in whose works, Dr. Payne says, the germ of his method is to be found.[80] To him Bacon was "that great genius of rational nature"; his motto was that aphorism of Bacon's, which Boyle employed several times, "Non fingendum, aut excogitandum, sed inveniendum, quid Natura faciat aut ferat."[81] Observation and induction were the keystone of his philosophy. As Sir George Newman says, "Without speculation he deals with what he sees and knows by observation. In method he is inductive; in philosophy he follows Locke and Bacon."[82] Maintaining that he would be most skilful in the cure of disease who most carefully and accurately observed its phenomena, he laid down the only two principles by which in his eyes medicine could be advanced: a faithful history of the disease, and a regular and exact method which must be "built upon a sufficient number of experiments, and has in that manner been proved competent to the cure of this or that disease." Elsewhere he argues the validity of his method because it has been tested by the infallible touchstone of "re-iterated experiments."[83] He was as ardent an advocate of inductive science as the chemists, but he applied it in a more fruitful way to the field of medicine.[84]

The climax of propaganda for the new science was reached in

Thomas Sprat's *History of the Royal Society*, 1667. Its importance lies not only in its being the most elaborate and comprehensive defence of the Society and experimental philosophy in this century, but more especially in its constituting an official statement on the matter. It was written at the instigation and under the auspices of the Society, was closely followed by the members during its composition, and when finished was heartily approved by them.[85] Furthermore, Oldenburg says that Brouncker, Moray, Wilkins, Evelyn, and others read the manuscript.[86] Sprat has given a rather definite hint that Wilkins was as much responsible for the work as he himself.[87] We are justified, then, in considering the views expressed by Sprat as representative of the whole body of Baconians. As such it becomes the most significant document in all the propagandist literature in behalf of the new science from Bacon to Wotton.

The *History* is divided into three parts. The first gives a brief survey of the rise of science in the East and its development through Greek and Roman civilizations down to Sprat's own day. His characterization of classical philosophy is pretty much the same as that which we have already discovered in earlier works by progressive thinkers. The Greeks exercised their wit and imagination upon nature instead of conducting sincere inquiries, since they disliked the drudgery and burden of observation, with the result that their conclusions were too sudden and supported by arguments rather than by proofs. From these talkative sects disputative philosophy was introduced into the early church. Next came the schoolmen, whom Sprat describes in a manner reminiscent of Bacon. They, too, possessed extraordinary strength of mind and quickness of imagination, but lacking matter to contrive, they relied on general terms which could lead but to disputing and "notional wars." Thus speculation became divorced more and more widely from material things, to which it should have clung. For this reason men in the past failed in their true duty of erecting a vast pile of experiments to be increased throughout all ages. As it was, their argumentative, contentious, and unnourishing philosophy, based upon those arts which Aristotle had drawn into method, proved useless to the good of mankind. (This utilitarian and materialistic standard is also apparent in Sprat's statement that true natural philosophy is intended primarily for mechanics and artificers.)

Fully convinced of the superiority of modernity over antiquity, he is willing to let the schoolmen, who, he regrets, are still esteemed by some as the only masters of reason, prevail in the schools and govern in disputations, but the Royal Society, he proudly asserts, will bow to no authority. Conscious, however, of the sensitive reverence entertained in many quarters for the great masters of the past, he politicly, though somewhat inconsistently, defends at length the virtuosi against the charge of irreverence, and declares that they should not try to destroy the ancients but should dwell on amicable terms with them.[88] Though he has as scant respect for antiquity as Webster, he is prudent enough to insist that it is subserviency to the ancients rather than the ancients themselves which the scientists abhor.

Modern philosophers he discusses under three general heads. First are those who, renouncing Aristotle, have revived the philosophy of some other ancient worthy such as Epicurus, Democritus, or Philolaus. This action is permissible only in so far as no tyranny is established over men's minds. It is hard not to believe that in this instance Sprat has in mind, besides the Copernicans, the enthusiastic followers of Gassendi, who revived ancient atomic philosophy and whose name was frequently on the tongues of English scientists. Another class are those who, though repudiating the ancients, seek to impose upon the world their own theories, defective through being based upon general conclusions arrived at before all the evidence was in. They only substitute one tyranny for another. Here he may have had Descartes in the back of his head, though he certainly shared the admiration of many of the scientists for the Frenchman. Descartes and the mechanical philosophy figure slightly in the *History*, though Sprat was sympathetic with the theory. He is careful to deny that soul and body constitute "one *natural Engine* (as some have thought)," yet he believes that through a study of the body "very neer ghesses [may be] made, even at the more *exalted*, and *immediate* Actions of the *Soul;* and that too, without destroying its *Spiritual* and *Immortal* Being."[89] In another passage he states that all alterations arise only from the meeting of little bodies of differing figures, magnitudes, and velocities.[90] His references to Descartes, however, are generally touched with criticism. He speaks of an experiment which proved the philosopher wrong in his theory regarding lines of force in a magnetic field; disagrees with the method described by the "excellent Descartes" of rejecting all previous impressions from experience and reading;[91] and ascribes the doctrine of motion "by geometrical demonstrations" to Christopher Wren with the explanation that though Descartes had begun it before, he based his ideas largely upon conjecture, so that some of his conclusions seem very questionable.[92] In view of the official nature of the *History*, these infrequent and somewhat indifferent references to Cartesianism, especially as contrasted with the praise heaped upon Bacon and the constant stress laid upon the experimental philosophy, are important in affording conclusive evidence of the fact that Baconianism was far more influential with the early English scientists than Cartesianism.

The third class of modern philosophers is the experimenters, who are considered under several heads and may be described as chemists, specialists, and co-operators. The first are subdivided into three groups: those who seek through chemistry a knowledge of nature in general; those who use the science for the purpose of discovering medicines; and those who are interested in the transmutation of metals and the grand elixir. From the first two Sprat expects science to receive the noblest improvements, and even from the last, whom Sprat characterizes as fanatics or enthusiasts, wonderful things might be looked for, provided they would be content with moderate expectations and ambitions. Experiments went a long way toward giving anyone a good reputation with the scientists. As we have seen, Boyle and others, though denying the Helmontian principles of chemical doctors and denouncing roundly the boasting claims of ignorant empirics, always express respect for their experiments. We may repeat again that the chief goal of science in the third quarter of the seventeenth century was the collection of authentic data, and experiments, however pursued, furnished such data.

The specialists, who in our eyes seem largely responsible for the history and advancement of science, receive high praise from Sprat. These men, conscious "of the vastness of the Design of an *universal Philosophy*," have made such discoveries in astronomy, geometry, anatomy, and the like that had all philosophical matters been handled so well, posterity "might have sate quietly down, and injoy'd the pleasure of the true *Speculative* Philosophy, and the profit of the *Practical*." Yet some, he says, have too hastily concluded upon axioms, and though others have done as much as any single wit could do, "they must pardon us, if we still prefer the joynt force of many men."[93] This, of course, is the spirit of Bacon, who, obsessed with the fear of generalizing upon insufficient evidence, opposed the Copernican theory and undervalued Gilbert's work. Let me repeat again that the chief task of science at this time was not the discovery of truth through specialized research, but the accumulation of a vast supply of observational and experimental phenomena which might safely furnish the basis for general laws.[94] For this reason the senses were emphasized and reason was distrusted. Naturally such an ambitious scheme made co-operation necessary, and it is on this score that Sprat espouses the third group, whom alone he calls experimenters.

The second part of the *History* is devoted to an account of the rise of the Royal Society, a description of its purpose and methods, illustrative experiments and observations, the charter, and extracts from the statutes. The Society proposed, as we might expect from what we have already seen, to gather genuine data rather than to discover laws.[95] These data were to be secured from the whole world, but wherever possible they were to be tested by experiments.[96] Experimentation furnished the supreme test; the method was to them the whole picture.[97] With remarkable modesty and self-denial they expressed willingness to leave to posterity the discovery of laws, being content only to gather in wholesale the data furnished and tested by experiments, with little or no attempt at correlation and deduction. Sprat explains that "as their purpose was, to heap up a mixt Mass of *Experiments*, without digesting them into any perfect model: so to this end, they confin'd themselves to no order of subjects; and whatever they have recorded, they have done it, not as compleat Schemes of opinions, but as bare unfinish'd Histories."[98] They experimented for data, not for theories, because hastiness in devising the latter seemed to them "the *Fatal point,* about which so many of the greatest *Wits* of all Ages have miscarried." This Baconian scepticism, however, he would distinguish from a "*speculative Scepticism,*"which would deny the right to venture a probable opinion or cause. While they permitted perfect freedom in the method of conducting experiments, prescribing no rules, they insisted on "the *critical* and *reiterated* scrutiny of

those things, which are the plain objects of their eyes," slowness
in assenting, and moderation in dissenting. This scrupulous, in-
credulous attitude Sprat considered the chief honor of the experimental philosophers.[99] In short, the members of the Royal
Society, true to the view which has been repeatedly emphasized,
hoped to advance science through the accumulation of material
secured by experiments carefully scrutinized and cautiously reasoned about (Bacon's natural history), so that an authentic basis
might be established for future generations to build reliable
theories upon.[100] Their faith was in the experimental method,
which alone could give them a sufficiently large body of authentic data.

Throughout the whole book there is constant stress upon material things, the evidence of the senses, the achieving of practical,
tangible, and material results as contrasted with the speculations of reasoning philosophers remote from a world of material
reality and producing mere notions instead of visible achievements. Hands not heads should be used, eyes not reason should
be trusted, works not thoughts should be sought. Intellectual
effort is discounted, and the world of human thought abandoned
for a world of sensuous impressions. The reason that in past
ages learning was for a time swept from the world, says Sprat,
was that it held too little converse with the senses, possessed no
utility for human life, and so was easily banished from mankind. The experimental philosophy, however, was no house
founded upon the sands, for

by bringing *Philosophy* down again to mens sight, and practice, from whence
it was flown away so high: the *Royal Society* has put it into a condition of
standing out, against the Invasions of *Time*, or even *Barbarisme* it self:
that by establishing it on a firmer foundation, than the *airy Notions* of men
alone, upon all the *works of Nature;* by turning it into one of the *Arts of
Life,* of which man may see there is daily need; they have provided, that
it cannot hereafter be extinguish'd, at the loss of a Library, at the overthrowing of a Language, or at the death of some few *Philosophers*: but
that men must lose their *eyes,* and *hands,* and must leave off desiring to
make their *Lives* convenient, or pleasant; before they can be willing to
destroy it.[101]

The last part of the *History* is expressly devoted to a defence
of the experimental philosophy, its purpose being to show that
increase of experiments is not injurious to traditional education,
the Christian religion, the Church of England, manual trades,

medicine, the nobility, the gentry, or the whole nation. In view of the energetic attacks made by the scientifically-minded Puritans upon the educational system of England, which have been discussed in a previous chapter, there was definite need to exonerate the Society in that respect. Sprat is compelled to admit that some "Asserters of new Philosophy" (whom he compares with the Puritans, and who in his eyes had prejudiced rather than advanced the cause of science) had advocated the destruction of the "ancient arts" and their "nurseries," but the only specific fault he finds with them is lack of prudence.[102] He says that though experimental studies were in the Restoration widely dispersed, they had their rise in the universities, not altogether a true statement, as we have seen. Furthermore, he maintains that the scientists at Oxford, whom he mentions by name,[103] saved that institution from Puritan frenzy, so that experimentalists evidently . could have no intention of sweeping away the honor due antiquity.[104] Acknowledging that there were some good men of severe and ancient manners who thought that a universal inquiry into things hitherto unquestioned would disturb the discipline and instruction of youth, he explains that men and not boys are engaged in carrying on experiments and that the new method of science, which consists in works, is proper for men, and the talkative arts for youth. So saying, he proceeds to show that the new science will not prove injurious to the "old talkative arts." Grammar and rhetoric will not be affected, since the same words and expressions will be used and new things will have to be named. Moral philosophy will have the same influence and authority as before. History cannot be impaired, since it covers an alien field. Mathematics will only be enlarged, for mathematical reasoning will be applied to other things. Metaphysics does not deserve to be considered. Logic will remain the same, though he cannot forbear remarking that men have been too busy with the productions of their minds instead of with the works of nature. It is true that natural philosophy will be changed, but all for the better, since men will lose only a few definitions, idle questions, empty disputations, and obscure Peripatetic terms, in place of which, "The Beautiful Bosom of *Nature* will be Expos'd to our view: we shall enter into its *Garden,* and tast of its *Fruits,* and satisfy our selves with its *plenty.*"[105]

Sprat's defence, except where he is admitting charges, is half-hearted for the simple reason that in his own mind he favored a radical change in the traditional methods of education. He differs from the Puritans in "prudence." It is true that in explaining why the Royal Society did not adopt the educational features of Cowley's *Proposition* he states that experimental science would consume too much of the student's time, that pedagogical dogmatism is antipathetic to scientific scepticism, and that old revenues could only with difficulty be diverted to new ends. Yet he lays the blame upon the age, which was not yet ready for the introduction of the experimental method into the schools. In one passage he so far forgets that prudence which he accused the Puritans of lacking as to advocate a scientific education:

To this purpose I will venture to propose to the *consideration* of wise men, whether this way of *Teaching* by *Practice* and *Experiments*, would not at least be as beneficial, as the other by *Universal Rules?* Whether it were not as profitable to apply the eyes, and the hands of Children, to see, and to touch all the several kinds of *sensible things*, as to oblige them to learn, and remember the difficult *Doctrines* of general *Arts?* In a word, Whether a *Mechanical Education* would not excel the *Methodical?*[106]

Visible things should precede doctrines and precepts, a more practical way to prepare the minds of youth for the world and the business of life. Here we have both the spirit and the letter of the Puritan treatises on education. Experimental science had not discarded the ideas which accompanied its infancy.

The next and perhaps most dangerous charge against which Sprat defended the Royal Society was the danger to religion, a problem coeval with the rise of science and probably destined to last till the end of time. A suggestion of the liberalizing influence exerted by science upon religion is seen in his approval of the Society's admitting men of all creeds and in the implication that there may be many ways of salvation. It was because of this avowed latitude that Henry Stubbe, as we shall see later, fell foul of the *History*. One compelling reason for such liberalism inheres in the earlier association of science with Puritanism, and though Sprat always expresses submission to, and belief in, the Church of England, he cannot refrain from recognizing the manifestation of the scientific spirit during the Puritan era. "The late times," he says, of *Civil War*, and *confusion*, to make recompense for their infinite calamities, brought this advantage with them, that they stirr'd up mens minds from *long ease*, and a *lazy rest*, and made them *active, industrious* and *inquisitive:* it being the usual benefit that follows upon *Tempests*, and *Thunders* in the *State*, as well as in the *Skie*, that they purifie, and cleer the *Air*, which they disturb.[107]

These words are a clear and perhaps necessary recognition of the part which Puritanism played in the rise of modern science, and represent the length to which Sprat dared go in paying tribute to the enemies of the King.

But it was not the Puritan association with early science which constituted the chief element in the controversy between science and religion. It was the crass materialism of the new science and the atheistic implications in the mechanical philosophy which seemed to damn the Society more than anything else. Hobbes had carried the theory of matter and motion to a degree which eliminated all spirit, and other Cartesians, ignoring the pious claims of Descartes's dualism, were adopting a purely mechanical explanation for all things. Soon Descartes himself became suspect. Henry More finds it necessary to turn his back on him; Baxter engages in a direct onslaught; Glanvill moderates to a great extent his former enthusiastic praise; and Boyle engages in much mental twisting and squirming to defend him. Hobbes's and Descartes's philosophy was the greatest obstacle the scientists had to overcome in seeking popular favor. Sprat recognized "what a tender matter" he was touching, and "in what a slippery place" he stood.[108] He boldly declares that experiments do not darken eyes, deceive minds, or deprave hearts (the debauchery of the court notwithstanding). " 'Tis true his [the experimenter's] employment is about material" objects, but this does not lead him to deny invisible things, for he knows that there is a subtle as well as gross matter which is not apparent to the sharpest sense. Spiritual things do not seem so impossible to one who by means of the microscope sees the numberless particles in the blood. (Sprat is consciously or unconsciously maintaining that spirit is matter not apparent to the senses, a defence as grossly materialistic as the idea he is denying.) He next makes use of what had been and what continued to be the stock argument in behalf of science's assistance to religion; namely, that from the former will come full knowledge of the beauty, contrivance, and order of God's wisdom. Nature is the instrument of God; the study of it is divine; and the praise of the knowing man is worth much more than that of the ignorant. The experimental philosophy, he asserts, is not hostile to the doctrine of the Gospels because the latter has been attested by miracles which can be appreciated only by those who know how far nature goes; nor to the doctrine of the primitive church, though it may be suggested that a knowledge of things will abolish this doctrine by insinuating into men's minds the thought that it cannot stand scientific scrutiny. His reply to this suggestion is that the scientist can as readily as anyone else submit his judgment to what lies beyond his reason; he can have faith in the supernatural elements in religion which no philosophy, Peripatetic or otherwise, can reach. An echo of Puritanism is heard in his statement that the plain things of religion need no philosophy; religion does not stand in need of reason, and should not be made the subject of disputations. If the experimental philosophy does not undermine doctrine, neither does it hinder the practice of religion. He admits that though the scientist explains authentic miracles and prophecies on the ground of God's infinite wisdom and power, he does apply critical scrutiny to pretended supernatural phenomena and often destroys modern pretences to them by explaining many apparent miracles by natural law and the power of man's fancy. If the experimenter removes the enthusiastic abuses of religion, he reveals honor enough for God in His guidance of creation along its wonted course of cause and effect; what he takes from prodigies he gives to the works of God. Furthermore, the scientific attitude has been a great boon in dispelling the melancholy fear produced by superstitious omens and prognostications during the recent plague and fire. To the charge that men cannot despise the world when they study it so much, that holiness suffers when so much time is spent on the labor of the senses, he replies that experiments reveal God's goodness, if not His justice, and if they do not contribute to godly sorrow or contrition, they give men more opportunities for charity, friendship, and generosity, "which are all of them *divine Graces,* as well as *Faith* and *Repentence.*"[109]

In his effort to show that experiments are not dangerous to the Church of England, he draws a parallel between it and the Royal Society by pointing out that the inquiring spirit of modern science was first produced by that liberty of judging, searching, and reasoning which effected the Reformation. Though seeking reformation in different fields, both institutions sprang from the same spirit of intellectual independence and both have been accused of having forsaken ancient traditions and of having ventured on novelties. This relationship Sprat, as we have seen, was by no means the first to perceive, and we may accept it as real. To show that experimental science is peculiarly the task of the Anglican Church he points out the difficulties with which it is beset in the Catholic and Reformed Churches. But conscious of the contributions the Puritans had made to the promotion of science, which he himself had noted on a previous page, he must have been in somewhat of a quandary as to how to show that experimental science "is an honor which seems reserved for it [the Church of England] alone." What he does is to hurry over the difficulty with a single misleading and guarded sentence: "From all the several sorts of *Enthusiasts*, I fear, there cannot much help be expected towards such *Works*, till they shall have left off to abhor them under the Title of *vain Philosophy*."[110] While undoubtedly there were those among the Puritans who would have characterized all philosophy in such a manner, Sprat certainly knew, what we have already shown, that the term "vain" was applied by the science-touched Puritans to the Aristotelian philosophy which had become involved with divinity.

The extremes to which enthusiasm for the new science went are quite apparent in the analogy which Sprat draws between science and the Christian religion. Miracles he calls God's *"Divine Experiments,"* and he identifies their purpose with that of the new philosophy, namely, to convince men through their senses. The Apostles would have made good scientists for they were men of honesty, trades, and business. Furthermore, Christ, like the experimenter, commanded his disciples to believe his works rather than his words. The qualities of an humble Christian are equated with those of a true experimentalist: he must judge himself aright, doubt his own thoughts, and be conscious of his own ignorance. Certainly the sceptical, scrupulous, diligent observer of nature is nearer to the modest, meek, severe Christian than the proud, speculative man. Moreover, science and Christianity join in a common humanitarian purpose, for Christ by feeding the hungry, healing the lame, and curing the blind showed that "it is the most honorable Labor to study the benefit of Mankind; to help their infirmities; to supply their wants; to ease their burdens. . . . All which may be calld *Philosophical Works,* perform'd by an *Almighty hand."* Science, religion, and humanitarianism stand in an interesting interrelationship here.[111]

What is, perhaps, even of more interest, in view of modern developments in naturalism, is the connection which Sprat tries to establish between science and morality. After declaring that the virtuosi have no intention of encroaching on the domain of morality, as well as that of politics and oratory, because since it is difficult to reduce it to any certain observation of the senses, they would run the risk of talking instead of working, "yet, when they shall have made more progress, in *material* things, they will be in a condition, of pronouncing more boldly" on that, too, for though soul and body are not one machine, "yet by long studying of the *Spirits,* of the *Bloud,* of the *Nourishment,* of the parts, of the *Diseases,* of the *Advantages,* of the accidents which belong to *humane bodies* (all which come within their Province) there [may], without question, be very neer ghesses made, even at the more *exalted,* and *immediate* Actions of the *Soul;* and that too, without destroying its *Spiritual* and *Immortal* Being." This evident proposal to place morality upon a sensuous and materialistic basis, and to make it a matter of natural rather than spiritual law, in the manner of Hobbes's philosophy, clearly reveals one influence of science, which in our own day has become conspicuous. In another passage, he insists, in a manner prophetic of Wordsworth, that the pleasant images of nature will purify the thought more than all the rigid precepts of the Stoics or empty distinctions of Peripatetic moralists.[112] Seldom do we find in this period such an unmistakable example of the antithesis between naturalism and humanism, coincident with the rise of science and important in the later development of the neo-classical period, as is afforded by this passage, which had received the sanction of more minds than Sprat's. Finally, the innocent, industrious, sensible delights of science will free the mind from vanities, and leave no room for dark and melancholy passions.[113]

The other points in Sprat's defence of the experimental philosophy need be only slightly touched upon. He has little difficulty in freeing the new science from the faults attributed to other sorts of learning, since, indeed, the latter were largely traditional and had been frequently and unfavorably contrasted with experimental knowledge. These charges are that learning makes men contentious, obstinate, impractical, idealistic, contemplative rather than active, unfit for the business of life, and prone to dote on the past and despise the present. He dilates upon the great benefit which experimental philosophy will confer upon the manual arts, in the course of which discussion he paints a glowing picture of the improvement of trades and mechanical arts, and what they will do for the world when method rather than tradition and chance determines discoveries and inventions. In an *ex parte* manner he shows why London is peculiarly suited to be a world-center of scientific activity, and why experiments are the proper study of the gentry and nobility. He speaks of the contribution science may make to literature in furnishing to wits and writers an inexhaustible supply of images from nature and works of art.[114] Again and again he stresses the utilitarian value of science, contrasting it with the inutility of classical philosophy.[115] In fact, "pure" science receives but the scantiest recognition from him, and so, of course, from the whole Society. Merchants, artisans, and mechanics come in for the highest praise.[116] This democratic spirit is further revealed in his elaboration of Bacon's idea that the method requires of its devotees neither learning nor genius. To allay the fears of artisans and tradesmen that new inventions and discoveries will injure their business, he enlarges upon the economic value of science in creating trades and multiplying demands. The part that science may possibly have played in the development of a mercantile nobility is suggested by what he says about the gentry's engaging in commerce and by his assurance that they do not debase themselves in so doing.[117] The new values of science were helping to make possible the social rise of the trading and industrial classes.

Almost every page of the *History* bears testimony to the great importance of Bacon in the development of science in seventeenth-century England. The frontispiece shows Fame placing a wreath on a bust of Charles II, on one side of which Lord Brouncker, president of the Society, is sitting, and on the other the Lord Chancellor, with the legend "Artium Instaurator." In the ode to the Royal Society, prefixed to the history, Cowley attributes the reformation in science to Bacon, comparing him to Moses, who pointed out the promised land, though he could not enter it himself. Early in the volume Sprat praises Verulam in a most fervent manner because he

had the true Imagination of the whole extent of this Enterprize, as it is now set on foot. . . . In whose Books there are every where scattered the best arguments, that can be produc'd for the defence of Experimental Philosophy; and the best directions, that are needful to promote it. All which he has already adorn'd with so much Art; that if my desires could have prevail'd with some excellent Friends of mine, who engag'd me to this Work; there should have been no other Preface to the *History* of the *Royal Society,* but some of his Writings.[118]

In another passage he gives Bacon credit for devising the Society. In view of such statements as these, the Royal Society may be considered the legitimate offspring of the great Chancellor.[119] Equally important, however, in showing the indebtedness of the virtuosi to him are the ideas of his which have been appropriated and elaborated upon. The pronounced emphasis upon experiment and sense-observation, the turning from books to nature, from theories to things, which represent the heart of the work, are directly traceable to Verulam. But just as apparent is Sprat's debt to more restricted ideas of his predecessor: the hindrance offered scientific discovery by the desire of mechanics to seize profit too soon; the idea of building experiments upon experiments; and the injury done the advancement of science by the despair of achieving anything.[120] Bacon's caution against devising theories upon insufficient evidence and reducing science to systems before sufficient experimental data have been collected receives even greater emphasis from Sprat than from anyone else. Bacon's hardly complimentary conception of the intellectual qualifications requisite for a scientist figures more prominently than we should suspect. Learning and genius are not necessary for the pursuit of scientific investigations; as soon as men "have the use of their *hands,* and *eies,* and common *sense,* they are sufficiently furnish'd to undertake them [experiments]."[121] But excepting only the experimental method itself, no idea of Verulam's receives so much stress or elaboration as the view that science must be a work of co-operation and of many ages. This was one of the most fertile of all Bacon's ideas and was largely responsible for the Society. It was Sprat's realization that the co-operative idea was the most distinctive feature of English science which prompted him to argue at great length the need and advantage of joining forces, and to prefer such a method to the specialized endeavors of individual men, no matter how important.[122] The virtuosi were so ambitious in the goal they set for science in the way of accumulating huge stores of data that co-operation, and that through many years, seemed essential for success. Bacon's conception of a universal natural history was their unchanging objective.

Sprat was fully conscious of the character of his times. In "this Learned and Inquisitive Age," he says, *publick Faith* of *Experimental Philosophy*" is manifest, and "the Genius of Experimenting . . . . much dispersed."[123] Yet he makes it perfectly plain that there was no inconsiderable opposition to the new science. While one purpose of his *History* was to recommend and promote the experimental method, an equally important one was, as we have seen, to defend it against educational, religious, and other interests and prejudices. He speaks of the "Enemies of Real Knowledge" and "the present contemners of *Mechanic Knowledge*," who reveal much prejudice toward discoveries in nature. Some possessed of an excessive censuring spirit, he admits, prophesy that the Society will be diverted from its stated purpose, that private men will not impart their secrets to it, while others ask, "What have they done?"[124] But worse than the cavils of the idle and malicious and the jealousies of private interest were the "wits and railleurs," whose ridicule, owing to the temper of the age, was more dangerous than the opposition of serious adversaries.[125]

The period covered in this chapter represents only five years, but how many ardent words were spoken in behalf of the new faith! In the Royal Society the virtuosi possessed a tangible expression of the new science, which raised them to the highest pitch of enthusiasm and filled them with confident visions of the future. Reassured by the favor of the King and by the association of many of the most important men of the realm, they carried on a vigorous campaign to establish their new philosophy on a firm basis. Certainly some defence against the upholders of antiquity and other enemies was necessary, but the scientists are on the offensive more frequently than on the defensive. Riding buoyantly on the swelling tide of a new age, they are filled with the spirit of battle, and are confident of success. Yet there was one more battle to fight, the most intense of all, before victory was assured.

CHAPTER IX

THE "BACON-FACED GENERATION"

No *Law* ever made him [Bacon] *our Dictator,* nor is there any *Reason* that concludes him *Infallible:* Nay, it is manifest that he was frequently *deceived.* And, since the *Gardiners* have *protested against him,* and that *justly:* Since the *Chymists,* and the *Mathematicians* disclaim him: Why may not a *Physician* refuse to be tried *by him?* . . . . 'Tis by *his great Example* that the *Baconical Philosophers* are such *Plagiaries, and Relators of false and defective Experiments: Contemners of the Ancients,* and *opinionated concerning themselves* . . . . And the *Root* of all our *present Distractions* was planted by *His Hand.*

      Henry Stubbe, *The Lord Bacons Relation of the*
              Sweating-sickness Examined, 1671.

The appearance of Sprat's *History* was an important event in the scientific world. Besides being an official pronouncement of the Royal Society, it presented the strongest possible case for the experimental scientists, expressing in a comprehensive, assured, and unequivocal manner the principles upon which they took their stand. In a way it represents a turning point in the scientific movement, for it revealed to the world the secure position in which the Baconians were entrenched and the strength of their cohorts. Furthermore, being written in an attractive, clear style, it must have reached far into the social world, furnished the topic of many conversations, and been the subject of numerous arguments. Certainly the most violent episode in the controversy between the supporters of antiquity and those of modernity had its rise in such an occasion. This last significant onset of the forces of intellectual conservatism furnishes the material for the final chapter of the present study.

In 1666 Joseph Glanvill, who has appeared in previous chapters, became rector of the Abbey Church at Bath. His hostility to the scholastic philosophy was well known from his earlier writings, and in his new domicile he maintained a vigorous interest in the Royal Society and the new philosophy.[1] As it happened, not far away lived Robert Crosse, vicar of Great Chew in Somersetshire and a confirmed Aristotelian with a wide reputation for learning. According to Anthony à Wood, some neighboring scholars of a mischievous turn of mind wished to see the two men meet, and sometime in 1667 conducted Glanvill to Crosse's house. The latter, knowing that Glanvill was a member of the Royal Society, spoke violently against it. Taken by surprise, the virtuoso said very little at the time, but later declared his opinions in so emphatic a manner that a considerable quarrel arose. Henry Stubbe, or Stubbes, a Warwick physician with a summer practice at Bath, took Crosse's part and encouraged him to write a book against his opponent, which, however, was so full of vituperation that it was refused a license both at Oxford and at London.[2] Thereafter Crosse contented himself with writing ballads against Glanvill, in which he was followed by the wags at Oxford, who were amused by the quarrel.[3]

Glanvill's experience with Crosse inspired him to publish a book, the title of which, *Plus Ultra,* employs an expression which had proved popular with the new thinkers since the time of Bacon.[4] The author proposes as the main purpose of his treatise to free the capable and the ingenious from a dull acquiescence in the discoveries of former times by comparing modern with ancient achievements in science and by showing the advantages which his age possessed over former times for the promotion and increase of knowledge. He adds, however, "I confess, I had a principal *eye* upon the *Royal Society.*"[5] This consideration probably moved him more than the first, for he elaborates upon it with the statement that in order to remove the fear of the Royal Society entertained by some pious men, "I have . . . . given a *succinct* Account of the *Reason, Nature,* and *Designs* of *that Establishment,* for the information of such as have not met with their *Excellent History.*"[6] In fact, the Royal Society had become a sign and symbol of modernity, and the cause of one was the cause of the other, though, as we shall see, Stubbe made an attempt to distinguish between the two.

After an introductory chapter, in which the author gives the circumstances of his quarrel with Crosse, and maintains that the virtuosi acknowledge the excellence of the wit of the ancients but reject their authority inasmuch as their method was barren compared with the fruitfulness of the modern method, he launches on his thesis. He divides the field in somewhat the same manner as Hakewill, to whom he refers, and in each division shows the superiority of modern discoveries and means of

discovery to those of antiquity.⁷ His account presents a fair
catalogue of modern achievements in chemistry, anatomy, mathematics, astronomy, optics, and geography. He emphasizes the
importance of the microscope, telescope, thermometer, barometer,
and Boyle's air-pump, mentioning particularly Power's and
Hooke's discussions of the first, and he cites the oft-used inventions of printing, the compass, and gunpowder as proof of the
superiority of his own times. In answer to the accusing query
which was being raised regarding the achievements of the Royal
Society, he stresses the fact that the organization had set as its
goal the compilation of a natural history of such magnitude as
to require ages to achieve. He describes, however, Boyle's
published and unpublished works as examples of what the members of the Society had already accomplished and as earnest of
what they hoped to do. He is careful to note discoveries or
inventions made by members of the Royal Society, for he was
more interested in it than in any other product of modernity.
His treatise is much more a history of science than had appeared: Columbus, Magellan, Copernicus, Brahe, Galileo, Kepler,
Viete, Napier, Harvey, Torricelli, Gassendi, Mersenne, Descartes,
Boyle, Ward, Merrett, Evelyn, and others appear in it.

It is Bacon, however, to whom the highest praise is given. He
is "illustrious," "immortal," "famous," "deep and judicious,"
and "glorious." In speaking of the advantages for knowledge
derived from modern improvements in natural history, which
he considered *"fundamentally necessary* to all the Designs of
Science," he gives due credit to his predecessor, declaring that
"all the main Heads of *Natural History* have receiv'd *aids* and
increase from the famous *Verulam,* who led the way to *substantial Wisdom,* and hath given most excellent *Directions* for
the Method of such an *History* of *Nature."*⁸ Like Sprat he
ascribed the founding and purpose of the Royal Society directly
to the conception of a natural history originated and stressed
by Bacon, who, he says, noticing that philosophy was only a
combination of general theories and notions, rashly drawn from
too few particulars and productive of nothing but disputes,

made the *complaint,* represented the *defects* and unprofitableness of the
Notional way, proposed *another* to *reform* and inlarge *Knowledge* by *Observations* and *Experiment,* to *examine* and *record Particulars,* and so to rise
by degrees of *Induction* to *general Propositions,* and from *them* to take

directions for *new Inquiries,* and *more Discoveries,* and other *Axioms;* that *our Notions* may have a *Foundation* upon which a *solid Philosophy* may be built, that may be *firm, tite,* and close *knit,* and suted to the *Phænomena* of things: So that *Nature* being *known,* it may be *master'd, managed,* and *used* in the Services of humane Life.

This was a *mighty Design,* groundedly *laid, wisely exprest,* and *happily recommended* by the *Glorious Author,* who *began nobly* and *directed* with an *incomparable conduct* of *Wit* and *Judgment:* But to the carrying it on, It was necessary there should be many *Heads* and many *Hands,* and *Those* formed into an *Assembly,* that might *intercommunicate* their *Tryals* and *Observations,* that might joyntly *work* and joyntly *consider;* that so the *improvable* and *luciferous Phænomena,* that lie scatter'd up and down in the *vast Champaign of Nature,* might be *aggregated* and brought into a *common* store. This the *Great* Man *desired,* and formed a *Society* of *Experimenters* in a *Romantick Model;* but could do no more: His time was not ripe for such Performances.

These things therefore were consider'd also by the later *Virtuosi,* who several of them *combined* together, and set themselves on work upon this *grand* Design; in which they have been so happy, as to obtain the *Royal Countenance* and *Establishment,* to gather a great Body of *generous* Persons of all *Qualities* and sorts of *Learning,* to overcome the *difficulties* of the *Institution,* and to make a very encouraging and hopeful *progress* in their pursuits.[9]

Glanvill is just as insistent as Sprat that the true function of the Royal Society was not to set up new scientific theories, but to discover and report accurately "how things are *de facto* . . . . to free *Philosophy* from the vain *Images* and *Compositions* of *Phansie,* by making it *palpable,* and bringing it down to the *plain objects* of the *Senses* . . . . to erect a well-grounded *Natural History,* which . . . . ties it [imagination] down to *sober Realities.*" Toward this end, he says, "We must *seek* and *gather, observe* and *examine,* and *lay up* in *Bank* for the Ages that come after."[10] There could be no clearer evidence than these passages afford of the stimulating influence which Bacon's idea of a universal natural history exerted upon the seventeenth century. In Glanvill's eyes the greatest advantages which the moderns possessed over the ancients lay in the method of compiling such a history, which had been outlined by Bacon, and in a numerous company of wary, diligent, and eagle-eyed philosophers for carrying on the work.[11] Even with this advantage, it is foolish, he maintains, to expect too much from the Society since one age can do little more than remove rubbish, lay in materials, and put things in order for the building.[12]

The last chapters of the treatise engage in an attack on Aristotle, in the course of which the author denies Crosse's claim that Aristotle possessed greater opportunity for gathering data by virtue of his travels in Asia, on the ground that he received his information from unreliable people and chose his observations to suit his theories. On the other hand, the moderns have much the advantage over the Stagirite because of wider geographical knowledge gained by the compass, more astronomical facts revealed by the telescope, and the large amount of data secured by the Royal Society. He explains the addiction to Aristotle on the ground of the force of education, and severely censures those who would set up his philosophy as the pillars of science, beyond which no one could go. The last chapter answers the charge of atheism which Crosse in a rage had hurled at him on the close of their argument, and the conclusion seeks to shield the virtuosi in general from that accusation, at the same time praising their tolerance, modesty, and critical temper.

In the *History of the Royal Society* Sprat had contented himself with describing the method of the virtuosi in contrast to that of the Aristotelians, citing modern achievements more for illustrative purposes than as arguments for modern superiority. But Glanvill, though emphasizing the method and painting a glowing picture of what it would accomplish in the future, takes a lesson from Hakewill and proudly holds up to view most of the important discoveries of his and the preceding century, thus giving concrete evidence of the greater knowledge and advantages making for knowledge enjoyed by the moderns. The latter had, in his own opinion, produced more "works" in the preceding half-dozen years than the Aristotelians in eighteen hundred, for the disputatious philosophy of the past had not yielded as much beneficial, practical knowledge as would help toward the cure of a cut finger, a statement which enraged his adversaries.

Glanvill's book re-enforced in no small degree the impression created by Sprat, and the two became a common target of those who resented the attack upon antiquity. Among the latter was Meric Casaubon, son of the great classical scholar, and prebendary of Canterbury, who in a letter addressed to Peter du Moulin[13] offered the most intelligent criticism of the new philosophy advanced in this century. His treatise sprang out of arguments with du Moulin over the experimental philosophy and the

possibility of its undermining traditional learning, in one of
which arguments the latter had given him Glanvill's book to read.
What Casaubon especially criticizes in *Plus Ultra,* and in Sprat's
History, too, for he discusses them together, is the characterization of the old "humane" learning as useless and the new as useful and beneficial. He is severely critical of the materialistic
standard by which Glanvill and others would measure utility. If,
he says, usefulness were found only in what affords the necessities and conveniences of life, brewers and bakers, smiths and
veterinarians would have to be considered equal or superior to
those who have been regarded as the great lights of learning.
But man's soul is more important than his body, for all true
happiness depends upon it, and those men have proved the
most useful who have paid attention to it, and have rescued
men from the domination of the body and the senses. For this
reason he gives highest praise to the *Ethics* of Aristotle, and
condemns Glanvill for ignoring it. Casaubon is arguing a higher
utility for those things which foster man's spirit, and rightly
senses the materialistic and physical basis upon which the new
science would place life. He points out the danger of the new
science's leading to atheism, in that it fixes men's minds too
much on secondary causes, making them forget the spiritual
world and discredit supernatural operations. This danger had
been conspicuously revealed in Hobbe's philosophy, and it is undoubtedly with the latter in mind that Casaubon accuses the
experimental philosophy of leading directly to a denial of God
and to disbelief in the immortality of the soul. He, like many
others, recognized this as the most dangerous tendency of the
new philosophy, and he considered the scientists' contempt for
all non-scientific kinds of learning, especially metaphysics, which
was all the more divine for being abstracted from the senses, an
omen of future disaster.

Casaubon touches upon the difference between the moral and
scientific worlds, ridiculing the belief that experimental science
could "moralize" men, and laughing at Gassendi's story about
Peiresc's claim that he learned to control his passions by observing how all the blood of a louse when angered ran into its tail.
Can a louse, he asks, do what philosophers and the Bible cannot? He clearly perceived the danger of placing morality upon
a naturalistic basis, by which, he says, reason is prostituted to

nature instead of ruling nature, and as an example he cites the
justification of sexual freedom on biological grounds. He is
particularly concerned about the effect of such an attitude upon
boys, whom some would have trained in scientific study. In
modern phraseology Casaubon is insisting upon the law of man
as opposed to the law of the thing, and is manifesting his fear
of what will happen when the two are confused. In him humanism squarely fronts naturalism.

His essentially humanistic view is revealed in his comment on
Durie, whom he couples with Glanvill and Sprat, thus showing
the relationship between Restoration science and the Puritans'
espousal of science. After quoting Durie's attack on literary
studies for contributing nothing to the discovery of real truths
in science, he exclaims that when the reading of classical literature ceases, barbarism and the grossest ignorance will follow.
He expresses wonder at the new kind of poetry, founded on a
knowledge of nature and experiments and excluding ancient
mythology, which Sprat advocated, and which in Casaubon's
eyes meant the abolition of ancient poetry, an incredible conception. His humanistic spirit is also revealed in the exception
he takes to Sprat's invitation to the nobility to engage in experiments, on the ground that noblemen can be of more service to
their country by reading history and by military exercises than
by tending furnaces or raking in the entrails of man or beast.
To him the physical world could furnish few truths to the world
of man.

Casaubon is a true conservative in his reverence for antiquity.
In his eyes religion, morality, education, and art were so closely
associated with the past that a contempt for antiquity, generated
by an overweening confidence in, and by an exaggerated emphasis upon, the superiority of modern science over all other
learning, would tend toward the destruction of the church, the
corruption of education, and the brutalization of man. In his
view the lamp of civilization had been entrusted to antiquity.
Aristotle was the wonder of the ages; his philosophy was based
upon experimental knowledge and reduced to general rules,
the only useful form of knowledge. There was no art or faculty
in which the moderns were not inferior to the ancients, partly
because the industry of the latter had been stimulated by greater
rewards than any known to the moderns. Painting, sculpture,

and music reached their greatest height in antiquity. He is not opposed to experimental science, when properly controlled; he manifests pleasure in Glanvill's account of modern discoveries, though somewhat sceptical of their utility; and he expresses sincere regard for many members of the Royal Society as well as high praise of the King for being its patron. But he resents the attack on traditional learning, with which he thought man's highest interests were indissolubly bound, and the overvaluation of modern science, which insists on its monopoly of usefulness, and which would interpret morality and religion itself by scientific principles.[14]

Casaubon's treatise is the most important of all the criticisms of the new science. It contains the germ of much that has been said by later generations. Though he was almost as much under the domination of ancient authority as the blindest upholders of antiquity, he realized more intelligently than others the significance which the classics possess for man. He perceived clearly the dangers inherent in a materialistic philosophy, especially as revealed in Hobbes and also, perhaps, in the court of Charles II. He noted the perils, as regards man's moral and spiritual interests, of a naturalistic philosophy based upon the new science. Though encumbered with many traditional prejudices, he supported the cause of a humanistic culture against the aggressive demands of a utilitarian and mechanistic science, and though he is found on the losing side, one that should have lost, he has some things to say that have awakened echoes from his age to ours. His brief pamphlet is one of the significant utterances of the day.[15]

A much more severe critic of the Royal Society, who, indeed, came to entertain an almost insane hatred of the institution, was Dr. Henry Stubbe, previously mentioned as an instigator of Crosse. He was drawn into the controversy by hearing in conversation someone denounce Galenical medicine for possessing no usefulness whatsoever, and cite Glanvill as authority for the opinion. Taking this as a reflection upon his profession, which he considered based upon antiquity, he lost no time in replying.[16] According to his own statement, Stubbe had been passionately devoted to the experimental learning earlier in his career, and had tried to introduce it among the gentry and youth, but the enthusiasm with which the Baconians had exalted it above all other knowledge, and especially the contemptuous attitude they maintained toward the ancients drove him into active opposition. He says, however, that "were the thing *well modelled,* managed by *discreet persons,* and the *Question* rightly stated, there would be *no controversie* betwixt us."[17] He raises no objection to the experimental method; he gives ample evidence of having practiced it himself.[18] But he was incensed by the attitudes and spirit of the Royal Society, as represented by Sprat and Glanvill, and so he makes that body the object of his attack. Yet in upholding the ancients and in denouncing the Society, the embodiment of the modern spirit, he could not but find himself at times a supporter of antiquity against modernity, and thus was forced to oppose Glanvill in a controversy over the relative merits of the two ages. Stubbe is really confused about the issue; at one time he upholds the superiority of an ancient to a modern, and at another, lauds a modern over some member of the Society.

Since, however, he answered Glanvill in order to vindicate a medical profession based upon Galenical principles, he was compelled to take up cudgels in behalf of ancient medicine, and thus came to fight valiantly for the honor of the past in opposition to his own age. He defends Aristotle against the charge of teaching only notions and the art of disputation by pointing out that the criticism applies only to his dialectic and metaphysics.[19] On the other hand, "it is manifest that the *Ancients* (especially the *Aristotelians;* . . .) did with some curiosity examine the *Phænomena* of *nature,* and *regulated* their *opinions* by *sensible experiments;* and this was the *practice* of most of the eminent *Physicians* and *Naturalists* of old."[20] He supports Crosse in the claim that Aristotle had unusual advantages for observing nature because of his travels in Asia, and he discounts a similar claim for the moderns, based on the recent discoveries of new lands, for the reason that the narratives brought back to the virtuosi were inaccurate and unreliable.[21] He maintains that Aristotle possessed knowledge claimed by the virtuosi as modern, and he advances antiquity directly against modernity in asking if the atoms of Gassendi and the principles of Descartes are not as precarious as the ideas of Aristotle and less subject to the evidence of the senses.[22] He denies Glanvill's statement that the ancients were unacquainted with chemistry, and asserts, what Boyle had denied, that modern chemical principles are as uncertain as Peripatetic.[23] He even dares risk the generalization that "The ancient *Philosophy* better agrees with the *Phænomena* of *Nature,* and carries us on with more assurance to work (as they[the virtuosi] phrase it) then any other."[24]

Toward the principles of Galenical medicine Stubbe maintains a purely pragmatic attitude. These, he insists, whether true or not, have resulted in successful cures, and "whosoever hath acted, or shall proceed according to *those notions* [elements and qualities] in compliance with the *Ancients,* shall not stand in need of any *novel Method* from the *Virtuosi* to salve a cut finger."[25] Physicians in general, his argument runs, have looked with indifference upon the principles of natural philosophy, being concerned not with their truth or falsity but only with the cures to which they conduce.[26] Nothing is to be relied upon in medicine, he says somewhat inconsistently, but an exact knowledge of medicinal physiology, founded upon observations instead of principles and upon known methods and tried medicines.[27] The application of this test, which contains a Baconian element, to his profession led him to uphold Galenical medicine against the attacks of experimental scientists. For the same reason he depreciated the practical importance of anatomical discoveries, which Glanvill had made much of as evidence of modern superiority. In short, traditional medicine is justified by the success of its cures, and the discovery of new scientific truths or the detection of old errors, both of which he seems tacitly to admit, can do nothing to advance medicine.

Stubbe was highly incensed by the mechanical education, which in his opinion Sprat and Glanvill would introduce into the schools in place of the traditional curriculum, though Sprat had purposely put the soft pedal on the matter, and Glanvill likewise had for the most part avoided the issue. The educational treatises of the Puritans were still making their influence felt, and were rightly associated with the new science. Stubbe scornfully contrasts such vocational and utilitarian subjects as the making of cider, the planting of orchards, the grinding of optic glasses, and magnetic and agricultural "curiosities" with the logic and moral philosophy taught in the universities.[28] Scientific training and knowledge, he holds, contribute nothing to the wisdom by which human affairs are regulated. The Aristotelian philosophy, on the other hand, and the humanistic studies of moral philosophy, law, politics, and religion fit men far better for life than the corpuscularian philosophy, which generally leads to atheism or religious indifference.[29] In him, as in others, the inevitable clash between a humanistic and a naturalistic philosophy appears. In his eyes metaphysics and school divinity were essential supports of the Christian religion for the quieting of conscience and the convincing of adversaries; so the substitution of a scientific for a classical education would lead to the destruction of the universities, religion, and the state itself.[30]

Stubbe conducts his offense on two fronts, one against the moderns in general and one against the Royal Society. In his attack on the latter he tries to rob the members of all credit for the invention or even improvement of the telescope, microscope, and the like, upon which Glanvill had laid much emphasis.[31] He depreciates Boyle's discoveries, states that Willis wrote an important treatise of his before he became a member of the Society, and asserts that those who had most improved chemistry were not members.[32] On the other hand, he fully participates in the controversy between moderns and ancients by emphasizing the unreliability of optic glasses, the unimportance of modern astronomical theories, and the knowledge of the ancients in chemistry and meteorology.[33] He also stresses the inferiority of chemical medicines, which the moderns had advanced with great ardor against the ancients, denies all utility to the discovery of the circulation of the blood, a strong talking point of the moderns which had no bearing on the Royal Society, and condemns the transfusion of blood and the injection of medicines directly into the blood, operations which the virtuosi practiced, but which were, as Stubbe points out, by no means confined to them.[34] Stubbe is so eager to discover mistakes in Glanvill and Sprat that he confuses the issue of ancients versus moderns, and frequently speaks beside the point. Thus he argues at great length in behalf of Caesalpinus' claim to the discovery of the circulation of the blood as opposed to Harvey's, an opinion which, if substantiated, would have been neither in favor of the ancients nor against the Royal Society.[35] He does not evince hostility to the experimental method, which he himself employed; he is provoked by certain characteristics of the experimental scientists.

In the preface to his treatise Stubbe explains that upon read-

ing the *Plus Ultra* he found Glanvill so involved with Sprat's
History of the Royal Society that he was compelled to read that
volume also. As a result he developed an even greater animosity,
if possible, against the historian than against Glanvill, which received expression in three treatises written at the same time as
the attack on his initial opponent though not published until
later in 1670.[36] One, *Campanella Revived,* pretended to discover
the origin of the Royal Society in Campanella's scheme to introduce popery into England and Holland, listed some experiments
which supposedly had been falsely claimed by the virtuosi, and
pointed out the danger which the Royal Society held for tradesmen. Another, *Legends no Histories,* is largely devoted to cavilling at mistakes in Sprat's book, which, he asserts, makes it
plain "that the *Antient Learning* (and not only the *Natural Philosophy*) is the *Rubbish they would remove*: This work they
have so diligently pursued, as if they had forgot *"Their first and
chief Employment, carefully to seek, and faithfully to report
how things are de facto."*[37] He also upholds against Sprat, just
as he had done against Glanvill, a humanistic philosophy based
upon the classics in opposition to the materialistic and mechanical principles founded upon the experimental philosophy. "What
contempt," he says,"is there raised upon the *disputative Ethicks
of Aristotle,* and the *Stoiques?* And those *Moral* instructions
which have produced the *Alexanders* and the *Ptolomeys,* the
Pompeys and the *Ciceroes,* are now slighted in comparison of
day-labouring."[38]

The third treatise[39] is devoted to an analysis of five or six
passages in the *History* for the purpose of showing errors or
dangerous implications, in the discussion of which there are
many learned quotations, much sophistry, and some good re-
joinders to Sprat. In answer to the assertion that one versed
in nature will render God more intelligent and acceptable praise,
he maintains that the Bible rather than nature is the source
of religious truth, and that "a *Psalm* of *David* . . . . in a *blind*
and *ignorant,* but *devout Christian,* will be better accepted than
a *Cartesian Anthyme."*[40] He brands as irreverent the recommendation that figures drawn from the Bible instead of from
classical mythology be employed in secular poetry, and offers as
horrible examples various quotations from Cowley's *Mistress.*[41]
In defending the old learning against the usual accusation, the

irate doctor insists that religion stands in need of syllogistic logic and philosophical distinctions in order to refute false religions and to substantiate the true one—a stock argument.[42] But the passage in which he finds the greatest danger is that which asserts that Rome's assumption of infallibility and sovereignty over faith was responsible for the Church of England's foregoing communication with Catholics and not showing them that respect which so ancient and famous a church might deserve. He reads in this broad-minded view a pernicious assault upon Protestantism and a covert desire to introduce Catholicism into England.[43] The difference in attitude revealed in this matter is typical of the difference between the two men. Sprat displays tolerance and freedom from prejudice, while Stubbe is narrow-minded, dogmatic, vehement, profuse with quotations, and much given to quibbling.

In his reply[44] to Stubbe, published in the following year, Glanvill persistently denies that the Royal Society cherished any designs whatsoever on the universities. So far were its members from wishing to meddle with the traditional learning that they had even made gifts to the universities. For further evidence he points to the attitude revealed in Sprat's *History*, which disclaims any desire to change the curriculum of the schools, though in almost the next breath the author recommends a scientific education.[45] The Puritan attacks on the established education had so dangerously associated the new science with hostility to traditional learning that it became necessary for the virtuosi, though at heart they must have been sympathetic with the educational ideas of the Puritans, to adopt a hands-off policy. Glanvill also denies that the experimental philosophy is inimical to philological studies, though he would have been hard put to it to prove his point.[46] The virtuosi, he explains, respect Aristotle but do not regard him as a dictator nor servilely bow to his authority.[47] He notes the confusion in Stubbe's argument due to his thinking at one time in terms of the moderns and at another in terms of the virtuosi only. His opponent, he says, "takes the Credit of the *Inventions* I mention from the *Authors* to whom I ascribe them, and gives it to others of the *same* Age, or not far from it; and so the Ancients are *readvanced*, and the *virtuosi* undone." He insists that the doctor's elaborate attempt to prove that Caesalpinus was the true discoverer of the circulation of the blood was beside the point, since the earlier scientist was as much a modern as Harvey.[48] While Glanvill believed that the Royal Society was the greatest advantage which his age possessed over the ancients, he consistently holds the controversy to the larger issue.

Glanvill was not the only one to rise to a defence of the Society. Stubbe had asserted that Henry More of Cambridge had left the organization and entertained a poor opinion of it.[49] In a letter to Glanvill, More indignantly denied that he had given up his membership, praised the experimental philosophy, referring for confirmation to the preface of his *Enchiridion Metaphysicum,* and declared that natural experiments were very serviceable in proving the existence of immaterial beings.[50] More was particularly interested in reconciling his abandonment of Cartesianism with his approval of experimental philosophy, and in freeing the Royal Society from the suspicion of atheism which the mechanical philosophy had aroused in various quarters, and which he declared was not the philosophy of the Society. The clash between Descartes's theory and the demands of Bacon's natural history is clearly expressed in his words:

But the Philosophy which they aim at, is a more *perfect Philosophy,* as yet to be raised out of faithful and skilful *Experiments* in Nature, which is so far from tending to *Atheism,* that I am confident, it will utterly rout it and the *Mechanical Philosophy* at once, in that sense which I oppose, namely, as it signifies a Philosophy that professeth, *That Matter having such a Quantity of Motion as it has, would contrive it self into all those Phænomena we see in Nature.* But this Profession cannot rightly be called the *Mechanical Philosophy,* but the *Mechanical Belief of Credulity.*[51]

Although he expresses the belief "that *Generation, Corruption, Alteration,* and all the *Vicissitudes* of *corporeal* Nature are nothing else but *Unions* and *Dissolutions* . . . . of little *Bodies* or *Particles* of *differing Figures, Magnitudes,* and *Velocities,*" he earnestly rejects what might be called the atheistic implications in Cartesianism. Furthermore, he does not think that the latter explains all phenomena, such, for instance, as those of light and color. Though the mechanical philosophy may lead to atheism, the experimental philosophy conduces to the contrary effect.

In the same year appeared another short and anonymous defence of the Royal Society,[52] which attempts in brief fashion to dispose of Stubbe's charge that the Society was undermining religion, the church, the monarchy, education, the universities, and the practice of medicine. Since, the author's argument states, the Society's purpose is only to increase knowledge and light, it can hardly be antagonistic to the Christian religion, nor can it be accused of seeking to introduce Catholicism, whose principle of authority is a direct contradiction to the spirit of the motto *nullius in verba* adopted by the royal body. To accuse it of endangering the monarchy is to reflect on its patron, Charles II, and the illustrious prelates and nobles who are members of it. He follows Sprat, Glanvill, and others in denying that the virtuosi advocated any such educational program as had been advanced by the Puritans, but he firmly asserts that if they should influence education, they would make it more manly and polite, since the Royal Society contains in its membership many men of the best families and breeding, who are devoted to the cause of useful science. Admitting that the experimental philosophy is not the chief business of the universities, he points out that, not being opposed to the traditional subjects, it is countenanced there and pursued by some of their most worthy members. Finally, the Society can hardly be prejudiced against medicine, since the latter is founded upon observations and experiments, such as are carried on by the organization, and since the Baconians are not concerned with medical practice but with medical knowledge.

Stubbe had so beclouded the atmosphere with his intemperate charges that it behooved the author to state in clear and unequivocal language the purpose of the organization; namely, "to discover truth wherever it be in Natural, Mechanical, and Mathematical things, to help Old Inventions and encourage New ones, and all for the raising and enobling the Dignity of Humane Nature, and the serving of Mankind." He, too, bears testimony to the remarkable influence which Bacon's idea of a universal natural history was bringing to bear on scientific progress, an influence which is one of the significant facts in the history of science.

And since they are perswaded with the excellent Lord *Verulam*, and many other judicious Moderns, that hitherto all sorts of Naturalists have been too hasty in establishing Systems of Natural Philosophy, before they had laid a solid and comprehensive Foundation to build upon; they have esteemed it a laudable and useful Undertaking to endeavour, that all industrious and sagacious Inquirers of Nature every where may conjoyn their

Researches, studies, and labours, to examine what hath been performed
hitherto, to retain what will endure the Test, and to add thereunto what
they can, in order to the composing a faithful History of Nature and Art,
that may contain a competent stock of Observations and Experiments,
frequently and carefully made by Intelligent and Cautious men, which may
serve for a Magazeen of Materials, of which hereafter, by duly considering
the whole, and comparing all parts together, may be raised (if possible) such
a Systeme of natural Philosophy, as may give a rational Accompt of the
Appearances and Effects of Nature, and enable men to infer from confronted Causes and Effects such consequences, as may conduce to the greater
benefit and ampler accommodations of Humane life.[53]

Still another anonymous writer in *A Letter to Mr. Henry
Stubbe,* 1671,[54] espoused the cause of the Society, but abandoned
the forthright, direct method of defence in favor of irony and
sarcasm. The Aristotelians, he says, do not need to engage in
arduous experiments, look through microscopes and telescopes,
or tediously tend furnaces. All they require is unintelligible
terms which sound brave to the ear, glib explanations based on
forms, qualities, elements, and the like,. and traditional theories
of astronomy and physics. In the usual fashion he also seeks
to discredit Aristotle by ascribing the corruption of "practical
Holiness" to the incorporation of Peripatetic philosophy in Christian doctrine.

Quickly replying to the various protests against his invectives,
Stubbe collected his answers and appended them to a second
edition of the *Censure.*[55] In these he further reveals his fanatical hatred of the Society and his obsession with the supposedly
wicked designs it harbored against church, schools, and monarchy. He also manifests a ludicrous suspicion of evil machina-
tions toward himself, expressly stating that he had been threatened with law suits, the bastinado, and assassination, and that
several virtuosi tried to have him whipped at the cart's tail,
charges that hardly need to be taken seriously. He viewed every
attack on his attitude as not the opinion of one individual but the
concerted efforts of many. A whole cabal of the Royal Society, he
says, consulted upon the *Letter,* and he never considered Glanvill's works the sole product of the author's labors. He looked
upon himself as a doughty warrior at death-grips with those who
were intent upon destroying religion and learning. Employing
every means to enlist on his side the services of those whose influence might tell more heavily against the Society, he even wrote
to Sir Robert Moray, of all people, about the vicious tendencies in

Sprat's and Glanvill's treatises.[56] He likewise wrote Boyle, expressing sorrow at his remaining a member of the organization, and warning him, as he valued his reputation, to have nothing to do with such an iniquitous body.[57] Casaubon's plea for the ancients brought joy to his heart and regret that years alone prevented the aged cleric's pursuing his task to the overthrow of the virtuosi.[58]

As we have already seen, Stubbe's attitude toward the ancients was somewhat inconsistent. Though he insists that the disrespect shown them was responsible for the controversy, he refuses to uphold the truth of ancient over modern science. In fact, he seems anxious to dodge the issue of superiority as Glanvill presented it. He again maintains that it is not the truth but the usefulness of Galenical and Aristotelian science which he is upholding. Galenical principles may be erroneous, but they have led to many useful medicines. He indignantly rejects the sarcastic attribution to him of belief in the Ptolemaic system and other ancient theories found in the *Letter*. Yet he depreciates the importance of the telescope and recent discoveries in magnetism, and tries to belittle the moderns in every way possible. As was true of other men whom we have discussed, modern discoveries and propaganda for the new method had made such an impression upon the world that the upholders of the ancients were paying respect to hallowed names rather than to a credited philosophy.

Acutely aware of the earlier association of science with Puritanism, Stubbe does not hesitate to employ it as a weapon. He styles the new scientists fanatics in religion and science, and maliciously calls attention to the erstwhile Puritan affiliations of various members of the Society.[59] Undoubtedly it was the lingering reverberations of the vociferous attacks on the universities by the educational reformers which intensified his concern over the baleful influence which the Baconians might exert upon education. Though Sprat, Glanvill, and others had definitely denied any intent on the part of the Royal Society to dabble in educational matters, the attitude of Baconianism toward the traditional learning and syllogistic logic, which had undergone no change during the Restoration, rightly made him apprehensive for the future of the schools. He perceived, though not so clearly as Casaubon, how possible it was for the utilitarian and materialistic values of the new science to dehumanize education. "This is the usual deportment," he says,

of several *Virtuosi*, they declaime in *general* against the *Peri[pa]teticks*, and *Notions:* & being pressed to *instances* of *their deficiency*, they fly to the old *Scholastick Theology*, or the *Aristotelian Physicks*, as if our *Universities* were so employed as in the days of *Sarisburiensis*. They crye that *Phylosophy* doth not fit them for *Action*: and if you demand what *Action*, what *world* they are not thereby prepared for: their reply only is, you cannot learne hence the *sophistication* of *Wines*, the *Art of Dying*, and such like *Mechanick* trades. Is not this *rationally* objected?[60]

The danger of science's professionalizing liberal education and shifting the emphasis from the world of man to the world of nature was as apparent then as now.

Stubbe speaks frequently of the odium under which the Royal Society was resting, and in a letter to Boyle he elaborates at great length upon it. Lamenting the fact that Boyle's association with the institution had done much to advance its interests, he states that the virtuoso's reputation has suffered much with sober men, and

What I believe you could never have apprehended, is come to pass; the body of the nation, and all learned men abet me: the two universities (especially that of *Cambridge*, which was most inclined to novelty) avow my quarrel; and they, that favoured experimental philosophy, judge those virtuosi unfit to prosecute it. The bishop of *Chichester* [Peter Gunning] reads against them, and intends to hold his lecture a year or two, and that for this reason, to overthrow the esteem of them. The concurence to hear him is such, as the university never saw before. The elect of *Bath* is as much as any for me, and against *Glanvill*. In sum, all men apprehend it now their common interest to oppose the R. S. I know not what any physicians may, as the mode is, tell you to your face; but except it be such as Dr. *Sydenham*, and young *Coxe*, I believe not one lives, that doth not condemn your experimental philosophy; and either the college, or Sir *Alexander Fraser*, and the principal of them intend me a letter of thanks for what I have done, with the liberty to print it; and perhaps others will follow the precedent; and how consistent these things are with the repute of the Society you adhere unto, judge you.[61]

Stubbe's words must be heavily discounted, but when all due allowance is made for his fanatical spirit, there still remains enough evidence to indicate a strong prejudice against the virtuosi. The glory with which ancient names were invested and the ingrown conservatism of the universities must have aroused genuine resentment against the critical attitude of the new

science. Even after all possibility of a convincing defence of
ancient authority had vanished, the feeling of reverence and
dislike of change paid homage to erstwhile powerful figures. In
the Restoration this resentment was keener than is apparent in
published works. Dr. Peter Gunning, to whom Stubbe refers in
the last quotation, and who had been famous as a royalist
preacher, would consent to the publication of a volume of poems
by Peter du Moulin only on condition that a poem in praise of
the Royal Society be omitted.[62]

It is not hard to understand the alarm at such a wide-sweep-
ing innovation as the new science. The fear that if one plank
were loosened in the intellectual edifice, the whole would crash
to the ground represents the most constant element in conservatism. Sprat's liberal attitude toward Catholicism created concern
for the Anglican Church, and the atheistic tendencies of the
mechanical philosophy bred anxiety for religion itself. But above
everything else the radical changes which the educational reformers of the pre-Restoration era would have introduced into
the schools, and which would have substituted a mechanical and
vocational training for a humanistic one, and the fact that the
scientific values of the Puritans in general appeared unchanged
in the virtuosi must have pressed heavily on the conservative
heart. Wary scientists like Sprat, though sympathizing with
most of what their predecessors had said, tried to free science
from the odium of its earlier manifestation, but with indifferent
success. John Wallis gives an account of a long oration delivered
by the popular preacher Robert South at the dedication of the
New Theatre at Oxford,

The first part of which consisted of satyrical invectives against *Cromwell*,
fanaticks, the Royal Society, and new philosophy; the next, of encomiasticks,
in praise of the archbishop, the theatre, the vice-chancellor, the architect,
and the painter; the last, of execrations against fanaticks, conventicles,
comprehension, and new philosophy; damning them *ad inferos, ad gehen-
nam.*[63]

Though South's oration has not survived, the coupling of the
new science with Puritanism in Wallis' report suggests strongly
that the speaker definitely associated the two.

Behind every defence of the Royal Society and the experimental philosophy stands the great figure of Bacon. Even when
he is not mentioned—a rare occurrence—his spirit is perceived.

Men had come to see centered in him all for which the new
science stood, and paid him that passionate allegiance which
did much to sustain the scientific movement. Clear evidence of
his pre-eminent position is revealed in Glanvill's next reply[64] to
Stubbe, in which the author defends himself almost entirely by
Bacon's authority, and seems to dare Stubbe to attack so great
an authority. Glanvill strenuously denies that he himself is an
enemy to the universities, giving high praise to the moral philosophy, anatomy, mathematics, languages, history, and divinity
taught there. He objects only to metaphysics and the old natural
philosophy, which he considers of no value, but which he thinks
should be studied because required by the statutes. Trium-
phantly he calls on Bacon as a witness to the sterility of Peripateticism. "What think you of my *Lord Bacon?* Either acquit
the *Virtuosi* with *him,* or condemn him with the *Virtuosi.*"[65]
He points out that in ascribing knowledge of anatomy and chemistry to the ancients and in denying the virtuosi credit for
various inventions Stubbe missed the chief contention, namely,
the improvements the moderns had introduced into the various
sciences by virtue of the experimental method. He further defines the issue as the usefulness of the experimental method in
contrast to the sterility of Peripateticism, and in maintaining
his position he quotes passage after passage from the *Novum
Organum,* at the same time apologizing for having spoken after
so great, learned, wise, profound, and celebrated a philosopher,
who was no enemy to physicians or learned men.[66] "And now
M. *Caviller,*" he cries,

did you ever *read* my *Lord Bacon,* or did you not? If *not,* for all your
boasts of great *reading,* You are not acquainted with some of the Authors
of greatest note; And 'tis a *shame* for you to write *against* the *Experimental
Philosophers,* and not to have *read* the *Founder* and one of the Chiefest Men
of *that Way.* If you *have read* him, you know he said all *this* that I write
the *Experimenters thought;* and why did you pass it by in *Him* who was
the *Author;* and insisted so largely upon it, and *so maliciously* censure it in
me, that spoke after him in two, or three transient Passages? If these
Periods in my Book were the *occasion* of the *Quarrel,* as you *say,* why was
not the Quarrel begun *before?* Are the *Physicians* more injur'd by my
writing those things, than by my *Lord Bacon's?* Is *my* saying what he
thought, a greater Affront to the *Ancients,* than his *declaring* to the World
the *same himself?* . . . . You knew that your *cavilling* could not injure my
Lord Bacon, but *thought* it might *expose me,* and other Friends of the
Experimental Way, to the displeasure of the *ignorant, envious,* and *misin-*

form'd; 'Twas not any *concern* for the Honour of the *Ancients,* and your *Faculty,* that ingaged *You;* For then you would have endeavoured their *Vindication* against my *Lord Bacon,* (if you had thought them *so* injured by those Sayings) But 'twas a *malevolent, envious* humour against the *Royal Society,* and its Friends, was the cause of your *impudent* Assaults.[67]

Glanvill repudiated with emphasis the charge of reflecting upon established physicians, acknowledging that ancient doctors used the method recommended by the Baconians, but he insisted that the notional and disputatious philosophy of substantial forms, real qualities, and *materia prima* was utterly unprofitable and led to no cures or other useful discoveries. The experimental philosophy, on the other hand, he represents as holding out promise of rich rewards, and to support his opinion he continually calls on his great predecessor.

Stubbe was not slow in perceiving that in order to undermine the experimental learning it was necessary to attack the knowledge and the authority of the man upon whom all the new scientists united to base it. To do this he selected none of Bacon's scientific or philosophical works, which at this time could hardly have been overthrown, but he seized upon his account of the sweating-sickness given in the *History of Henry VII,* and proceeded to show with much gusto how erroneous it is.[68] More than anything else, the realization that the chemical doctors had arrayed themselves under the banner of Bacon inspired Stubbe's attack, for in his dedication to the president and fellows of the Royal College of Physicians he expresses the opinion that it might be to the benefit of the institution, "if I descanted upon the miscarriages of the Lord Chancellor *Bacon,* and in *him* shewed what *incompetent Judges* they [chemical doctors] are in *Physick,* . . . . In representing of his [Bacon's] *faileurs,* I thought I might deterr others from the like attempts," for "the repute of my *Lord Bacon* is great in *this Age,* and made use of to the prejudice of *Our Faculty,* more wayes than *one.*"[69] Though he never excluded from his mind the virtuosi in general, whom he compared to the Goths and Vandals in their destruction of old structures and books, a comparison which had earlier been applied to the Puritans, his chief concern was with the chemists, particularly George Thomson, who had made loud use of Bacon's name in onslaughts on Galenical medicine. He characterized chemical doctors as liars, their medicines and laboratories as "delusory braggs"; he would gladly see burned "all those [books] of the Lord *Bacon*, the *Baconical* Philosophers, and *Van Hel-mont*."⁷⁰ He was not far wrong in pointing out that the Baconian philosophers had given distinct encouragement to empirics and quacks, much to the injury of established medicine, for as has already been stated, it was a polyglot mob who availed them-selves of Bacon's prestige and ranted about observation and experimentation.

Contained in the *Relation* is a definite attack on George Thom-son,⁷¹ which carries to greater lengths the depreciation of Bacon, and which argues without reservation for Galenical as opposed to chemical medicine. " 'Tis *malapertness* in this *Bacon-faced* generation," he says, to oppose Galenical physic. Old books and not the *Novum Organum* nor the writings of modern experi-menters will qualify a man for medicine. He clings to the typi-cally conservative view that "those *foundations* are most *sure*, which are laid by the *Most Men*, if they be *judicious* and *observ-ing*, and have endured the test of *more ages* and *tryals*." He is consistently contemptuous of all followers of Bacon, who, he regrets, have infatuated the nation, and who are more like "Cobblers, or *day-labourers*, than *Practitioners* in *Physick*. . . ; and as little doth the title of *Experimental Philosophers*, and *Verulamians* avail *them*." He bewails the sorry plight of medi-cine when "the reading of two or three *Books*, a *Comical Wit*, a *Bacon-face*, a *contempt of Antiquity*, and a *pretence to novel Experiments* (which are *meer* excuses for *Ignorance*, and *Indis-cretion*) are sufficient Qualifications" for a doctor.⁷² He widens his censure to include Sydenham, who had applied Bacon's phi-losophy to medicine in a clinical rather than chemical way, be-cause the famous physician derogated from the authorized practice of physicians by erecting practice on his own observa-tions. "Oh! what a misery," he sarcastically exclaims, "is *much study!* and how many *scruples* is this Doctor *Sydenham* freed from, by suffering our best *Writers* to remain untouched, un-consulted."⁷³ Running through all the pages is the injunction to hold to the old, shun the new, for discoveries in science can contribute nothing to medical practice.

Appended to the *Relation* is "A Preface to the Reader," which characterizes Bacon and puts forward a weird explanation of his interest in science. Granting that the Lord Chancellor was a man of good literature and great eloquence, Stubbe declares that being a lawyer by profession he was unacquainted with, or had but superficially looked into, the scientific matters of which he treated, and so should not be considered an infallible authority. He accuses Lord Verulam of stealing the *Novum Organum* from Aristotle, the *Advancement of Learning* from Vives' *De Causis Corruptarum Artium*, and Solomon's House from "the *Peripateticks* establishment at *Alexandria* in the Musæum of Ptolmæus *Layides* and Demetrius Phalereus." He also attacks Bacon's morality, citing his trial, and with sublime indifference to dates ascribes his scientific activities to a desire to wipe out the stigma of his disgrace:

The only judgment I can make of my Lord *Bacon's* Actings, is, that being so *Flagitious*, and so *Ignominiously degraded:* He determined to redeem the *Infamy* of his *past Life* by amusing the World with *New Projects;* and to gain a *Chancellourship* in *Literature*, when he was excluded from that on the *Bench:* And to revenge himself of the *Nation* whom *He* had exasperated, by diffusing *Heresies in Philosophy*, and creating in the Breasts of the *English* such a desire of *Novelty*, as rose up to a Contempt of the Ancient Ecclesiastical and *Civil Jurisdiction*, and the *Old Government*, as well as the *Governours* of the Realm: and the *Root* of all our *present Distractions* was planted by *His Hand.*

It is too much to saddle the whole rebellion upon Bacon, but certainly Stubbe had good reason to note the relationship between his works and the Puritans. To the same purpose he emphasizes, as he had done in former works, the earlier political sympathies of some of the virtuosi, whom he calls his "Olivarian adversaries," and again in the preface to *A Relation of the Strange Symptomes Happening by the Bite of an Adder*, 1671, he throws up to the Baconians their relations with the Commonwealth. The part which the Puritans played in furthering the values of science has so faded from view that it is hard to realize how conscious of the fact the Restoration was.

Stubbe continues in the "Preface" to uphold the ancients and university learning against the mechanical education associated with the virtuosi. Highly praising Aristotle, Theophrastus, and other classical writers, he proudly boasts that his physic is based on Aristotelian physiology and principles. At the same time that he denies that the universities made Aristotle a dictator and deprived themselves of liberty of thought, he advances the moral philosophy of Aristotle and Thomas Aquinas against the scientific education which he so heartily despises. He looked upon the old learning as indispensable for the preservation of religion and government. "I believe," he says,

that the *World* of *business* and *Action* is not such as *Eccebolius* [Glanvill] talketh of; that the *Mechanical Education* is the most *Extravagant Folly* imagineable: That, to *Preserve our old Religion*, 'tis absolutely necessary that we retain our *Old Learning*: that there is a greater value to be placed on *Controversial Divines*, then our *Virtuoso* do allow of in *their Letter* against *me*; and perhaps one *Jewel, Lawde, Andrews, Davenant, Whitaker, Chillingworth, Scaliger, Grotius, Selden, Casaubon*, or *Salmasius* may ere long be more *serviceable* unto [the] Monarchy than a *Fleet of Ships, Thirty thousand horse and foot*, or *Three hundred Thousand Virtuosi* . . . . and when the people have a *little more* observed them, the *Works* which they have atchieved in Six Years above all that ever the *Aristotelians* ever effected, will only conduce to the *Improvements of Raillery*; We shall perceive that we have been out of the *Way* all this while; that the interests of a *Lineal* and *Hereditary Monarchy* are different from those of any *Usurper*; and if my *Olivarian Adversaries* may understand what *Arts* are subservient to the *advancing*, or continuing of a *Cromwell*, a *Regicide*: I am content to dissent *once more* from them in behalf of the *present Monarchy, the Church* of *England*, the *Universities*, and my own *Faculty*.

In this way Stubbe turns Glanvill's attack on Aristotelianism into an attack on religion and monarchy. He proceeds to show that the old natural history and metaphysics, which Glanvill had singled out for special reprehension, are essential to the continuance of established institutions. In the last analysis his argument is based upon the utility of error, for he asserts that no new discovery made by the experimental method can counterbalance the evil which arises from depreciating the learning and rendering contemptible the notions of the fathers and writers of the church. He maintains that the monarchy rests only on opposition to Catholicism, and that Protestants to hold their own with Catholics must understand and use the terms known to their adversaries. In short, both sides must employ common principles which are based on the old natural philosophy.[74] Finally, he expresses the fear of change characteristic of conservatism, holding that it is better to bear the ills we suffer than fly to those we know not of.[75]

Glanvill did not reply to Stubbe's attempt to overthrow Bacon, but Thomson did and in no uncertain manner.[76] Exalting the philosopher as "the Honor of our Nation" and "a profound *Searcher* into *Nature*," he defends with enthusiasm his account of the sweating-sickness and rallies to his defence "all true-

hearted, virtuous, intelligent Disciples of our Lord *Bacon*."
"There hath not lived in *England*," he declares, "many Centuries
a greater *Natural Philosopher* than our Lord *Bacon*," who in
spite of his many other duties has

> given us such testimonies of his endowments, and *his indefatigable search*
> into the true *Fundamental Cause* of these *sublunary Essences*, withal set-
> ting us in the right way, even indigitating the means how we may acquire
> knowledg, so that we are obliged never to suffer his Name to be ill spoken of,
> or his works to be calumniated, unless we incur deservedly the sin of horrid
> ingratitude.[77]

He continually talks of experiments, calling all the Galenists in
the world to a contest of cures, and heaping upon them all
possible opprobrium. He speaks eloquently in defence of the
"intelligent virtuosi of the Royal Society," contrasting their experimental method with the lazy perusing of books. An unlearned laborious empiric in his eyes produces more real useful
knowledge than "the most Learned *Academick* in *Europe*, who
studies *Words* more than *Works*, a *Library* more than a *Laboratory*." And again he exclaims,

> 'Tis *Works*, not *Words*; *Things*, not *Thinking*; *Pyrotechnie*, not *Philologie*;
> *Operation*, not meerly *Speculation*, must justifie us Physicians. Forbear
> then hereafter to be so wrongfully Satyrical against our Noble Experi-
> mentators, who questionless are entred into the right way of detecting the
> Truth of things.[78]

So intense is his hostility to Galenical practices that he expresses
the wish that experimentalists like Merrett and Goddard, who,
however, approved of bleeding, were expelled from the Royal
Society, and he asks the former how he can declare himself "an
Experimentator, one of the Lord *Bacons Disciples*." Though
opposed to the Royal College of Physicians because of its Galenical sympathies, he was an ardent admirer of the Royal Society,[79]
and stoutly combated the charge that it was inimical to religion
or the universities.

A survey of Stubbe's writings seems at first to reveal little
else than contradictions. He confessed no opposition to the experimental method, frequently speaking of his own experiments
and observations, yet he waged relentless war against the Society which was the chief exponent of that method. His unqualified praise of the ancients was worthy of the blindest admirer of antiquity; his admiration of Aristotle was unbounded.

He supported Galenical medicine by the argument most frequently on the lips of the advocates of antiquity; namely, the
centuries through which it had been employed and approved.
Furthermore, he was incited to the controversy by the irreverence which Glanvill showed the ancients. Yet he indignantly
repudiated some of the most representative theories of the past,
such as nature's abhorrence of a vacuum and the Ptolemaic
system, and refused to defend the truth of Galenical and Aristotelian principles. Apparently the controversy between the ancients and moderns was entering its last stage, in which little remained for the supporters of the former to do but save the face
of antiquity by demanding reverence for names which had lost
their authority. But for the most part Stubbe's insane hatred
and frenzied fear of the Royal Society sprang from a deeper
source. State, Church, and School were so involved with ancient
learning that an undisguised attack upon it, without even the
semblance of perfunctory respect, seemed to his conservative
spirit, with its genuine fear of change, about to undermine all
established things. He also sensed, though not so clearly as
Casaubon, the beginning of a conflict between a scientific and
a humanistic education, between naturalism and humanism. We
may be sure this view was not peculiar to him. Men like Casaubon, South, and Gunning only expressed what many must have
thought, and had not so many powerful men, from the King
down, lent their support to the Royal Society, the Warwick
doctor might well have received in public the aid which he says
was privately granted him.

Stubbe and his enemies by no means monopolized the battle-
field of the ancients and moderns during this period. One very
lively skirmish occurred between a chemist and a Galenist, which
is worthy of record. In 1669 a William Simpson, who was described by his opponent as a proud, impudent young man without
university learning, published an enthusiastic plea for the experimental philosophy,[80] in which he took occasion to animadvert
upon a work by Dr. Robert Wittie. Simpson's volume is primarily devoted to critizing Galenical medicine in the usual
fashion and to eulogizing chemical remedies, but at times he
extends his view to include the new science in general. The importance ascribed to experimentation by Sprat and others is, if
possible, carried even further by Simpson. To him experiments

alone are the path to knowledge, the only satisfactory source of solid learning, and the test to which all opinions should be put. The speculative, argumentative, "putationary" medicine of the classical doctors, derived from the great volumes of Galen, Hippocrates, Dioscorides, and others, is empty and fruitless because its principles are not founded upon experiments. To the charge of empiricism, the standard Galenical argument, he replies that if physicians were more empirical, they would cure more diseases, because nature is not tied to any method, however spacious, which does not conform to her laws as discovered by observation and successful experimentation. For this reason, the bare experiments, if credible, of quacks, old wives, and other lowly persons, who come in direct contact with nature, frequently lead to more efficacious results than the splendid receipts of famed physicians.

Simpson's view of science, however, is not confined to the all-importance of the experimental method to medicine. He had come under the Baconian spell both as regards the application of the method to nature in general and as regards the absolute necessity of a universal natural history. If the Galenists, he says,

would more listen to a grand collection of experiments of all sorts, according to the Lord *Verulam's* advice, and totally cease from all manner of disputations; improving all other parts of Learning to the utmost, It would be the only way in tract of time, to establish a true solid Science of Philosophy; till then, we may expect no considerable improvement, in those places, of real Science.

Elsewhere he calls this plan "a direct road toward a real *Science of Experimental Physiology*," to which, he says, chemistry is of no small use. In the first step of "promoting an *Hypothesis* of experimental Philosophy (one large branch whereof would be this of Physick,) I think it would not be impertinent, yea, perhaps necessary, to lay aside all or most of our Books, excepting such, as by some Judicious persons, might be reputed faithful, in their communication of experiments." To the experiments gathered from this source by intelligent men, he would add experiments of all sorts, for which large laboratories, gardens of trees and plants, cattle for experimental purposes, and a free clinic for experimenting on the sick should be available. All experiments, then, according to his plan, would be digested, condensed, and entered in a volume entitled *Clavis Scientifica*, or *Philosophica*, from which could be derived a true natural philosophy. Seldom do we find Bacon's conception more clearly stated.[81]

Simpson squarely opposes the method to antiquity, direct investigation of nature to reliance on books. He shows scant respect for the ancients and even less for the philosophy derived from them, which students were compelled to waste their time in studying. He proposes that they give up this pedantic disputatious learning and take to experiments. It was especially this proposal that provoked Wittie to publish a defence of Galenism.[82] Wittie's attitude is hardly a fair example of what the conservative view was becoming. He disclaims any opposition to chemistry, which he considered "an excellent Appendix to the Noble Art of Physick," and in which he himself had engaged. But he looked upon chemical medicines as being an addition to, and not a substitute for, Galenical practices. Though he expresses approval of the confirming of notions by experiments where possible, he evidently considered experimentation limited in its proper operations. To Simpson's recommendation that all books not containing experiments be abandoned, he replies that if this were followed, the world would soon decline into a deplorable ignorance. The testimony of the ancients is still valid for him, and the learning which was derived from Galen and other "Princes in Physick" and which had proved successful for mankind's good, claims first place in his affections. His reverence for the past made him severely censure the impudent disrespect which Simpson had manifested for it as well as for learned men and the universities.

During this period there were those who, though not engaging in definite controversies, continued to pay public and sincere tribute to the new science. John Webster, author of *Academiarum Examen*, discussed in a previous chapter as one of the most important educational treatises inspired by Bacon in the Commonwealth, published in 1671 a volume entitled *Metallographia*, which chiefly consists of treatises by other men on traditional chemical problems such as the philosopher's stone and potable gold, most of them now on their last legs. Though Webster views with favor the greater part of these hallucinations, he is eager for "the advancement of Experimental Philosophy," and quotes with praise Boyle, Hooke, Harvey, and the *Philosophical Transactions*. In the dedication to his volume he eulogizes the Royal Society as "one of the happy fruits of His Majesties blessed and miraculous Restauration, ... which will speak him glorious to all succeeding Generations, beyond all his Royal Progenitors." Long after the necessity for them had been removed, except perhaps in academic circles, he embarks upon an attack on dogmatism and a plea for freedom of thought, "that Mens judgment may no longer be fettered in Scholastick Chains, nor kept always in the Prisons of Academick Opinions." The extent to which the Restoration had restained the attacks of science on the universities is shown by the difference between the educational views revealed in this book and those in the *Academiarum Examen,* a change characterized by a spirit of prudent conciliation. Now he expresses approval of the established education, with the one proviso that it does not hinder the increase of natural knowledge by the experimental method.[83] He repeats, however, Bacon's idea that a great hindrance to knowledge lies in men's trying to force nature to fit theories rather than fitting theories to experimental observations, deduction rather than induction.

In the same year Boyle brought forth the belated second volume of his *Usefulnesse of Experimental Naturall Philosophy,*[84] which had for the most part been composed as early as 1658. He probably published it at this time to counteract Stubbe's influence, for in the "Preamble" he says that he was told that "it could scarce come forth more seasonably to recommend the whole Designe of the Royal Society." The work lays claim to a threefold purpose: to furnish materials for a history of nature; to promote the practical or operative part of natural philosophy; and to enable gentlemen to hold commerce with tradesmen and artisans to their mutual benefit. The influence of Bacon's conception of the importance of a natural history remains the most powerful incentive of the new science. In this work Boyle appears again as a popularizer and propagandist of the experimental philosophy, for he hopes to incite men to a study of it by describing experiments which may be easily tried and are of immediate use. He is chiefly concerned with pointing out the various ways in which science can be of use to tradesmen and artisans through the discovery of principles which may be directly applied to the needs of life. Impressed with the importance of invention, he foresaw more clearly than any of his contemporaries the development of machinery, and he based his claim

for science not so much on the stage it had reached in his own
day as upon his vision of a mechanical future, which has been
fully realized in our times. Throughout the whole treatise the
democratic note is strong. He decries the superciliousness which
would prevent scientists' holding commerce with tradesmen and
mechanics, who though possessing no fine language or clothes,
are more diligent, industrious, inventive, and familiar with
material things than scholars, and from whom naturalists may
gain much knowledge. "I learn'd more of . . . . the Nature of
Stones," he says, "by conversing with two or three Masons, and
Stone-cutters, than ever I did from *Pliny* or *Aristotle*, and his
Commentators."[85] Boyle's worship of Bacon continues undimmed
in this volume. To the latter he ascribes the rise of the scientific
movement, for his wisdom and intellectual independence have
enabled men to make discoveries and remove hindrances to
discovery. Among the last he includes despondency, scholastic
philosophy, want of a natural history, want of the experimental
method, and lack of associated endeavors. Here in a nutshell
is the heart of Bacon's message to the seventeenth century and
evidence of the fertility of his conception of natural history and
the need of co-operation to achieve it.

My study of the idea of experimental science draws to a close
with the end of the quarrel between Stubbe and the virtuosi.
Boyle continued throughout the century to publish works extolling the Baconian philosophy and the Royal Society, one of his
main themes being the reconcilability of science and religion.
Open attacks on the new philosophy begin to disappear, though
there is evidence that opposition still existed in a more or less
latent state.[86] Wit and satire seize upon the virtuosi as their
proper victims, and indignation is turned into ridicule.[87] But
propaganda for the experimental method and attacks on the old
learning continue to appear, though not by way of open controversy, until the last decade of the century.[88] One such production is of special interest, for it caused the final conflict between
modernity and antiquity. In 1681 there appeared in Latin two
volumes by Thomas Burnet entitled *Telluris Theoria Sacra*,
which flauntingly wave the banner of the new philosophy.[89] An
English translation appeared in 1684, and five years later two
more books were added. It was this work that in 1690 inspired
the publication of Sir William Temple's *An Essay upon the An-*

cient and Modern Learning, which maintains the superiority of ancient philosophy and science, and which is only a revival of those attacks on the new science and the Royal Society that characterized the Restoration.[90] The title and content of the essay confirm this view. It is further confirmed by the fact that the Royal Society, in whose memory still lingered the assaults of its enemies a score of years before, engaged William Wotton, a young man of extraordinary learning, to answer Temple.[91] Wotton himself says, "The Society which I principally had in my Eye to defend, whom he more than once contemptuously calls, *the Men of Gresham*, were many of them every way as considerable for Birth and Quality, for Parts and Wit as Sir *W. Temple* himself."[92] Wotton's book, *Reflections upon Ancient and Modern Learning*, 1694, though covering more than science, is full of praise for the virtuosi, draws heavily upon Glanvill's writings, and inveighs sharply against the old philosophy. Temple's reply, *Some Thoughts upon Reviewing the Essay of Ancient and Modern Learning*, published posthumously, proves conclusively the scientific nature of the controversy. This controversy, re-enforced on one side by Bentley and on the other by Charles Boyle and the Christ Church crowd, led directly to Swift's *Battle of the Books*.[93]

CONCLUSION

The unceasing war which the Baconians waged against the ancients in the seventeenth century, as well as the obsequious submission to the authority of antiquity in the sixteenth, make it clear that the classical obstruction had to be removed before science could find a place in the sun. In his savage attacks on the ancients, Gilbert plainly showed his realization that as long as freedom of thought was restrained by internal or external coercion, truth warred in vain against error. It is not strange that during the first forty years of the seventeenth century, when his influence was at its height, the ideal of liberty of thought and discussion made marked progress, and prepared the way for the fierce onslaught upon antiquity in the time of the Commonwealth. Bacon's influence, of course, re-enforced and later overshadowed his, but the part Gilbert played in vigorously asserting his right to disagree with the ancients is an important fact in the history of the liberation of the mind. In general, this ideal, which has been pursued so ardently by subsequent generations, received more powerful and consistent support from the scientific spirit than from any other source.

The chief foundation of the blind reliance upon antiquity was the debilitating theory of nature's decay. The antithesis of the idea of progress, which colored and brightened the thought of the last century, it laid its enervating hand upon human powers, and bred a despairing resignation to an apparently inexorable decree of fate. In the light of this theory, one can easily understand why Bacon considered it so important to instill courage and confidence in the hearts of men. But unlike Hakewill he did not squarely meet the issue. He may have subscribed to the theory himself; thinking the obsession too deep to be eradicated, he may have considered a concession to his age expedient; or he may have made use of the idea to enhance the importance of his method. But the fact remains that he nowhere clearly states that the moderns equal the ancients in genius. On the other hand, he seems tacitly to admit the inferiority of the former. He justifies the confidence with which he wished to inspire his readers, in part, by the incremental nature of the growth of knowledge, which, together with his paradox that the moderns are the ancients, supported their claim to superiority in learning. But above everything else, it was his method which he impressed upon men as sinking the balance in the favor of modernity. He confidently opposed his method to the genius of the ancients.

The Commonwealth was a very significant period in the history of science. It firmly installed Bacon as the dominant influence of the day. The materialism, utilitarianism, democracy, social-mindedness, humanitarianism, and anti-authoritarianism, explicit or implicit in his writings, developed rapidly in the congenial atmosphere of Puritanism. Perceiving the practical worth of the experimental philosophy, the reforming Puritans seized upon it with eagerness, and made it the companion of their fanatical religion. At first sight this appears an odd couple, indeed, but upon closer scrutiny the strangeness disappears. Certainly no part of the population was more interested in the increase of wealth and the improvement of their material welfare than the "godly men." Bacon had pointed out only too clearly the relationship between his philosophy and such desirable benefits. Furthermore, there was nothing in his philosophical conceptions to antagonize the Puritans, for the potentialities which his views possessed for discrediting religion had not yet appeared, and he himself had clearly separated theology and science.

The Puritans chose what would seem to have been the most direct and fundamentally efficacious method of establishing the new science, namely, education. It is true that the first group of experimenters met in London, but before long its more imporant members were transplanted in Oxford, evidently for the purpose of introducing the experimental philosophy into the university, which thus remained for a decade the center of scientific activities in England. Most of the Puritan propaganda for Baconianism is found in educational treatises, which attack the established curriculum with virulence and support the new method with enthusiasm, and which gave rise to those values that have plagued our own educational world. In them professional, vocational, and scientific education is directly opposed to the study of subjects in part humanistic and cultural. Science and the classics fought their battle then as they did in the nineteenth century. Philistinism made its first appearance.

In the Restoration this attempt at educational reform retained its earlier associations more clearly, aroused more fear and animosity in conservative hearts, and suffered more from the general reaction against the Puritans than any other element in the scientific movement. The Puritan scientists as a whole made their way safely into the royalist camp, and science fortunately rested under the protection of a lenient King. Though the Baconians continued to manifest and recommend most of the values and attitudes of Commonwealth science, they carefully avoided advocating any disturbance in established academic subjects. Sprat, though personally approving a scientific education, disclaimed for the Royal Society any educational schemes. Glanvill followed him in this respect, and when Casaubon, Stubbe, and others, with an eye on the earlier treatises, accused the Society of designs on the universities, he strenuously denied the charge. Many of the writers of these treatises were not received into the scientific fold. Hartlib drops suddenly and completely out of sight, and John Webster tones down his former scathing criticism of the schools to an innocuous proviso. Though Boyle remained at Oxford until shortly before the Restoration, and a definite attempt was made to secure support and sympathy for the experimental philosophy there, the reaction against the Puritan ideas was too strong, and the university, which had been staunchly royalist while Charles I was alive, ceased to play any significant part in the scientific movement. The characteristic feature of Baconianism in the Commonwealth was educational; in the Restoration this feature all but disappeared.

Another reason for the abandonment of the educational aspect of the new science is found in the demands of a universal natural history. In the intense propaganda for, and defence of, the experimental method which followed the establishment of the Royal Society, this conception of Bacon's rose to a pre-eminent position, and became almost the sole motive and goal of scientific activities. The task which confronted the virtuosi was the accumulation of as much data as possible through observation and experiments, and all their energies were concentrated on the work. This pressing demand caused them to be less interested in the teaching of the method than in its application to natural phenomena. Their chief concern was to defend their new philosophy against the upholders of antiquity, who still sought their knowledge in the lore of the past, and to incite as many men as possible to observe and carry on experiments. They were more

interested in recruiting mature men to assist in carrying on the
work than in teaching the new science to youths. Thus they were
willing to relinquish that element in the scientific movement
which had most outraged conservative thought.

But the naturalistic tendencies of the new science, which had
lifted an ugly head in Sprat's *History,* together with the Philistinism of the Puritan treatises, intensified the opposition to ex-
perimentalism. The Stubbian period marks the first clear realization of the inevitable clash between naturalism and humanism
and of Bacon's relation to it. Aside from the fact that conservative thought naturally viewed with concern the weakening of one
element in the established social order, so that religion, education,
government, and the professions seemed threatened, men began
to sense clearly just what implications and potentialities lay in
the experimental philosophy and its extravagant demands. The
humanistic spirit was aroused, and engaged in an energetic defence of its creed, by which the world of man is sharply distinguished from, and opposed to, the natural world. And humanism planted itself squarely upon the classics. This fact, together
with the association of the new science with the mechanical
philosophy, which was falling more and more under the suspicion
of atheism, presented the most formidable obstacle to science.

The reaction against the Puritans helped in keeping the universities humanistic, though the new science succeeded in getting
a foothold in them, especially in Cambridge. Likewise, literature
remained humanistic in theory if not always in practice. The
influence of Ben Jonson and the French critics succeeded in establishing the ancients as literary arbiters, so that Aristotle, who
had suffered badly at the hands of the scientists, maintained his
prestige in criticism. Not that science did not exert influence on
critical ideas. It was one of the chief liberalizing forces playing
upon neo-classical dogma. Dryden's vacillation may in part be
explained by the influence, on the one hand, of classical and neo-
classical critics, and, on the other, of the new science, in which
he was interested. It is true that other poets of the time were
members of the Royal Society, but for the most part they were
little influenced by it. Restoration comedy, containing a definite
romantic element and reflecting a philosophy of life which had
been determined to some extent by the materialism of the scientific spirit, clashed with neo-classical criticism and was defeated.

Though Sprat had definitely tried to bring science and literature together, humanism won the day in the literary world, and later directed much of its satire against the "men of Gresham." So we have a neo-classical rather than romantic period in literature.

Thus science, having to thread its way between a humanistic literature and humanistic universities, found its growth retarded in spite of Newton's great discoveries, or perhaps partly because of them, but its values were by no means extinguished. They persisted especially in that element of the population which derived from the Puritans. When this class rose to greater power in the eighteenth century, and the dogma of neo-classical criticism began to disintegrate, partly owing to science, romanticism began to take form, and those attitudes which flourished in the third quarter of the seventeenth century—utilitarianism, humanitarianism, democracy, and the like—resumed their onward march.

NOTES

NOTES

CHAPTER I

[1] W. C. D. Dampier-Whetham, *A History of Science and its Relations with Philosophy and Religion*, 1930, p. 65. Where no place of publication is given, it is understood to be London.

[2] Quoted in J. L. McIntyre, *Giordano Bruno*, 1903, p. 21.

[3] Thus it became easy for anyone versed in the classical languages, no matter what his vocation in life, to become an authority on medicine. For example, we find one worthy clergyman attacking ignorant quacks and at the same time upholding the propriety of a preacher's writing on medicine on the ground that his book "contayneth the right Methode and way of healing, practised by the auncient fathers in Phisick, *Hippocrates, Galen* and *Auicen*." Epistle Dedicatorie to *The English Phlebotomy .... Collected out of good & approued authors.* By N.[ichols] G.[yer] 1592.

[4] Second dedication prefixed to *A profitable Treatise of the Anatomie of mans body: Compyled by that excellent Chirurgion, M. Thomas Vicary*, 1577. The first edition, 1548, has been reprinted by the E. E. T. S.

[5] See Philip Barrough's *Methode of Phisicke*, 1583, p. 1.

[6] Edward Jorden's *Briefe Discovrse*, 1603, fol. 3ᵛ, and "Briefe Declaration" (Nov. 7, 1566) prefixed to Thomas Gale's *Certaine Workes of Galens*, 1586.

[7] *Practicæ Medicinæ liber*, 1596.

[8] See *The second parte of William Turners herball*, 1552, Dedication.

[9] For example, Anthony Askham, *A litle Herball*, 1550.

[10] *The difference betwene the auncient Phisicke, first taught by the godly forefathers, consisting in vnitie, peace and concord: and the latter Phisicke proceeding from Idolaters, Ethnicks, and Heathen: as Gallen, and such other consisting in dualitie, discorde, and contrarietie. And wherein the naturall Philosophie of Aristotle doth differ from the trueth of Gods worde, and is iniurious to Christianitie and sounde doctrine*, 1585. R. B. says (p. 64) that the work was inspired by his hearing at a bishop's house an attack on "Paracelsicall Phisicke," as the cleric called it, but wrongly in the eyes of R. B., who holds that it is much older than Paracelsus. Evidently there was considerable discussion over the two kinds of medicine. The book is unpaginated, but I have used the pagination inserted in the British Museum copy.

[11] Cf. pp. 70-71.

[12] Pp. 99-100. " 'Sulphur' did not mean the particular substance of definite atomic weight and chemical properties which we call by that name, but that part of any body which caused it to be combustible and disappeared on burning. 'Mercury' was any part which distilled over as liquid, and 'Salt' the solid residue." W. C. D. Dampier-Whetham, *op. cit.*, p. 125.

[13] See Thomas Hill, *The profittable Arte of Gardening*, 1568, with its list of authorities headed "The Authours, out of which this woorke of Gardening is gathered"; E. Fenton, *Certaine Secrete wonders of Nature. . . . Gathered out of divers learned authors as well Greek as Latine*, 1569; and the "authorities" given at the beginning of John Maplet's *A greene Forest, or a naturall Historie*, 1567, interesting in the light it throws on Lyly's unnatural natural history. The expression "gathered," used constantly to express the Elizabethan's manner and method of scientific investigation, may be taken as the embodiment of the spirit of the age. For another example see the

letter from a St. Bredwell physician, prefixed to Gerard's *Herball*, 1597, which speaks of "The first gatherers out of the Ancients."

14 See *The profittable Arte of Gardening*, fol. 40.

15 See the preface to *The pathway to Knowledg*, 1551, and to *The Castle of Knowledge*, 1556.

16 William Cuningham in *The Cosmographical Glasse*, 1559, fols. 3-4, mentions among the astronomical controversies of his day the theory that the earth moved and was not the center of the universe. He did not, however, look upon the theory with favor, for he is typically Elizabethan in his reverence for authority. In discussing Galen's lunar theory of tides, the scholar says there seems to be a discrepancy "betwixt his authoritie, & dayly experience," but the teacher is careful to state that such a trivial matter "nothing infringeth Galen his authority." He believed "that men in our age, are not comparable in any thyng to those of times past. For how many sondry Artes, secrete Sciences, and wonderfull Ingens, through well spending of tyme, did the auncient Philosophers in their dayes invente?" See fols. 2, 145-6. Cf. fols. 66, 169.

17 *The Castle of Knowledge*, pp. 11-12, 104-5, 165. It is rather difficult to understand Recorde's inconsistent position, unless we attribute it to his fear of publicly espousing such a novel theory. At first he attributes the hypothesis to "grosse ignorance," and styles its supporters "contemners of philosophy," but later on in the book he speaks of the strong arguments, great learning, wide experience, and wonderful diligence revealed by the advocates of the idea, and when his scholar·speaks slightingly of it, he reprimands him sharply, and declares that when the young man learns enough to understand Copernicus' proofs, he may be ready to accept the theory.

18 See Dorothy Stimson, *The Gradual Acceptance of the Copernican Theory*, 1917, pp. 42-3.

19 Another reference to the theory, like Recorde's somewhat enigmatic, is discovered in Thomas Hill's *Schoole of Skil*, 1599. Though in one passage (pp. 49-50) Hill produces arguments in support of the Ptolemaic system, in another (p. 42) he seems to slight the Copernican only because its newness might be troublesome to young students: "*Aristarchus Samius*, which was 261 yeares, before the byrth of Christ, tooke the earth from the middle of the world, and placed it in a peculiar Orbe, included within *Marses* and *Venus* Sphere, and to bee drawne aboute by peculiar motions, about the Sunne, which hee fayned to stande in the myddle of the worlde as vnmoueable, after the manner of the fixed stars. The like argument doth that learned *Copernicus*, apply vnto his demonstrations. But ouerpassing such reasons, least by the newnesse of the arguments they may offend or trouble young students in the Art: wee therefore (by true knowledge of the wise) doe attribute the middle seate of the world to the earth, and appoynte it the Center of the whole." Both Recorde and Hill claim that the theory is not suited to young minds, but how far they would recommend it to adult intelligences presents a perplexing question. They seem, however, to consider it a live enough hypothesis to speculate upon.

20 See F. R. Johnson and S. V. Larkey, "Thomas Digges, the Copernican System, and the Idea of the Infinity of the Universe in 1576," *The Huntington Library Bulletin*, No. 5 (1934), pp. 69-117. The authors, to whom we owe a distinct debt of gratitude, have not only given a very intelligent discussion of Digges's importance, they have also reprinted entire his short treatise, *A Perfit Description of the Cælestiall Orbes according to the most aunciente doctrine of the Pythagoreans, latelye reuiued by Copernicus and by Geometricall Demonstrations approued*, appended to his edition of Leonard Digges's *A Prognostication euerlastinge*, 1576. It is almost incredible that this significant treatise should have been so consistently slighted by historians of science.

21 F. R. Johnson and S. V. Larkey, *op. cit.*, pp. 71-2.

22 *Ibid.*, pp. 79, 80, 101.

23 *Ibid.*, pp. 99, 111.

24 *The Cosmographical Glasse*, fols. 66, 169.

25 Richard Hakluyt, *Voyages*, 1589, pp. 611 ff.

26 "Epistle to the Reader" in *A treatyse of the newe India*, 1553.

27 See dedication, dated 1574, to a translation of Joannes Taisnerius' *Of continuall motions* in E. Arber's *The First Three English Books on America*, 1885, p. xlvi.

28 *Pilgrimage*, 1613, pp. 26, 42, 603.

29 William Barlow in his *Nauigators Supply*, 1597, sig. A, A4, calls the compass a "merueilous and diuine Instrument . . . . (being one of the greatest wonders that this World hath)," and he says, "Experience testifieth, that this [compass] began to be in common vse about the time that Printing was inuented, and the making of Gunnes. Both which, although they are of very excellent vse and great wonderments to the world, yet doeth this farre excell and exceede. For all things performed by them, are marshalled within the limittes and bounds of humane reason; and therefore their causes being knowen, their wonder ceaseth. But this being incomprehensible vnto humane reason, carieth it away captiue vnto the astonishment thereof, and leadeth it to the admiration of him, whose wisedome comprehendeth all things. . . ." In 1600 Gilbert claimed that no invention had done so much good to mankind as the compass. *De Magnete*, the Gilbert Club translation, 1900, p. 4.

30 In speaking of the many erroneous ideas about magnetism William Barlow says, "But many, being now at length weary to take such counterfeit shewes for good payment, doe seriously seeke to haue some sound knowledge of the things themselues, and not to depend vpon other mens extravagant disputes, but to be their owne iudges in matters that concerne themselues, and are subject to their owne senses and reasons, and especially because some of them must aduenture their whole estates, yea and also their liues vpon the right vnderstanding of some *Magneticall* conclusions." (*Magneticall Aduertisements*, 1616, preface.) For an enthusiastic eulogy of the compass see p. 86 of the same treatise.

31 Gilbert says that certain facts regarding the action of the compass "have been pointed out to me and confirmed by our most illustrious Sea-god, Francis Drake, and by another circumnavigator of the globe, Thomas Candish." (*De Magnete*, Gilbert Club translation, 1900, p. 117.) Some years later in arguing with William Barlow over some doubtful point in the movement of the needle, Mark Ridley hopes to settle the matter by entreating "those Masters and Marriners that trauel to the East-Indies, that they would make triall thereof by hauing two Inclinatory needles fitted for their instrument . . . . and thus we doe commit this point to be determined by these mens reports at their returne." *Magneticall Animadversions*, 1617, p. 30.

32 *The newe Attractiue, Containyng a short discourse of the Magnes or Lodestone, and amongest other his vertues, of a newe discouered secret and subtill propertie, concernyng the Declinyng of the Needle, touched therewith vnder the plaine of the Horizon. Now first founde out by Robert Norman Hydrographer. Herevnto are annexed certaine necessarie rules for the art of Nauigation, by the same R. N.*, 1581.

33 *The newe Attractiue*, "To the Reader."

34 Pp. 4, 6, 15, 18, 21. I am loath to finish this discussion of Norman without calling attention to the manly humility, sincerity, directness, and sweetness of spirit which characterize this small volume.

[35] In the Epistle Dedicatorie of his *Newe Attractiue* Norman says he found his discoveries "byee making sundrie experiments of the Magnes stone. . . . . Wherefore to further the noble studie of Nauigation and Hydrographie, and to giue occasion to industrious and skilfull trauailers by sea and by lande, to make diligent observation of these effects in sundrie places, whereby some generall conclusion mai bee inferred, I have heere set downe whatsoeuer I could finde by exacte triall, and perfect experimentes."

[36] A different attitude is discovered in William Barlow's *Nauigators Supply*, 1597, in the dedication to which the author tries to minimize the seriousness of his lack of experience as a sailor by saying that "in the minde onely, pure and true Arte, refined from the drosse of sensible or experimentall knowledge, is to be found," yet in one of the prefatory poems experience is accorded due importance:

> Triall may bring as much authoritie,
> As newnesse hinder it of due respect.

[37] See the Biographical Memoir prefixed to Mottelay's translation of the *De Magnete*, 1893, which notes that J. F. W. Herschel, Henry Hallam, Dr. Whewell, Thomas Thomason, and Dr. C. W. Cooke considered Gilbert the first to proclaim induction in England. Sir William Hale-White, in *Bacon, Gilbert and Harvey*, 1927, p. 45, thinks that since there was no likelihood that Bacon influenced Gilbert, the latter may justly claim "the glory of being the first in this country, at any rate, to break through the bonds of authorities and to properly use experiments and induction."

[38] See the preface to the *De Magnete*. All references to the latter are to the translation issued by the Gilbert Club of London: *William Gilbert of Colchester, Physician of London. On the Magnet, Magnetick bodies also, And on the great magnet the earth; a new Physiology, demonstrated by many arguments & experiments.* 1900.

[39] Pp. 35, 216, 228, and preface.

[40] Pp. 3, 6, 109-11.

[41] Preface and pp. 50, 169.

[42] See an article by the present writer entitled "Science and Language in England of the Mid-Seventeenth Century," *Jour. of Eng. and Ger. Philol.*, XXXI (1932), 315-31.

[43] Pp. 47-8.

[44] P. 28.

[45] Edward Wright, who sincerely admired Gilbert, and who contributed an "Encomiastic Preface" to the *De Magnete*, claimed to have written the chapter in the book containing the table of fixed stars, and also remarked that Gilbert was not well versed in the Copernican theory. Mark Ridley informs us that Wright was "most perfect in *Copernicus* from his youth," and that a Doctor Gissope, "a great scholler in the Mathematick," who was a close friend of Gilbert's, even lodging in the same house with him, wrote whole sheets of demonstrations out of Copernicus (*Magneticall Animadversions*, 1617, pp. 9-10). Surely the Copernican theory was more widely accepted in private than is apparent from published opinions. Even in cases where it was rejected, it made a deep impression. Thomas Blundeville, though holding to the Ptolemaic system, after stating the new theory says, "by helpe of which false supposition he [Copernicus] hath made truer demonstrations of the motions & reuolutions of the celestiall Spheares, then euer were made before, as plainely appeareth by his booke. . . ." (*M. Blundeville His Exercises*, 1594, p. 181.) See also his *Theoriques of the seuen Planets*, 1602, in which he puts forward as conscientiously as possible the arguments for both theories.

[46] See Dorothy Stimson, *The Gradual Acceptance of the Copernican*

Theory, 1917, p. 87, and the *De Magnete*, p. 224: "The Earth, then, which by some great necessity, even by a virtue innate, evident, and conspicuous, is turned circularly about the Sun, revolves."

47 See S. P. Thompson, *Notes on the De Magnete*, 1901, p. 47. In a letter to William Barlow, Gilbert says that he has received a letter from a Venetian mathematician, Johannes Franciscus Sagredus, who said that among the learned men of Venice and Padua there was a "wonderful liking" of the *De Magnete*. The letter is published at the end of Barlow's *Magneticall Aduertisements*, 1616.

CHAPTER II

1 J. B. Bury also finds such a theory suggested in Augustine. (See *The Idea of Progress*, 1920, p. 21.) Cyprian and 2 *Esdras*, frequently cited in support of the theory, furnished some of the most popular arguments. In fact, they played such a part in the matter that one opponent of the idea found it necessary to explain their position. "But it is objected, that *Esdras*, the Apostle, and *Cyprian* a Martyr, did intimate the decaying of things, and confirme that Principle, *Every thing the more it is removed from its beginning, the more it faints and fails.* It is true, and in expresse words in *Esdras. Consider also that you are of lesse stature than those that were before you, and those that shall come after you, will be lesse then you are, for that the Creatures now grow old, and are past the strength of their youth.* [In the margin reference is made to l. 4, c. 5, v. 54.] .... We cannot deny but these are *Cyprians* words, *You must know in the first place, that the world is now grown old, that it stands not so strongly as it did stand, nor is it so vigorous as formerly it was, etc. In Winter, there is no such plenty of showers to nourish the seed, nor in Summer is the Sun so beneficiall to ripen the corn; Nor are plants in the Spring so prosperous by reason of temperate weather; nor are Trees so fruitful with fruit in Autumn. There are fewer Marble stones dug forth of the Mountains that are worn out; they afford lesse quantity of Silver and Gold; Metals are exhausted, and the slender Veines daily grow lesse and decrease. The Husbandman failes in the Fields, Concord in friendship, skilfulnesse in Arts, Discipline in Manners, etc."* John Jonston, *Constancy of Nature*, 1657, p. 8. Jonston's treatise was first published in Latin in 1632.

2 Such for instance as Psalm 102, which declares that the heavens and earth shall wax old like a garment.

3 *A blazyng Starre or burnyng Beacon, seene the 10 of October last,* London, 1580, sig. Aiiii-Av. The Epistle Dedicatorie states the author's purpose: "to disswade the worlde, from freezing in the dregges of their synnes, by admonisheyng them of the finall dissolution of the Engine of this worlde, and seconde commyng of Christ in the cloudes, whiche by many manifest and ineuitable reasons I gather, can not bee farre of."

4 "You may say if you will that the worlds change and decay in old age, or that the earth seems to grow grey with years, and that all the great animals of the universe perish like the small, for they change, decay, and dissolve." See J. Lewis McIntyre, *Giordano Bruno*, 1903, p. 221. Obviously animistic philosophy lent itself to the idea of decay. Bruno believed that form alone changed; the atoms remained indestructible. For another continental reference to the theory, see the preface to Otho Casmann's *Marinarum Questionum Tractatio Philosophica Bipartita*, Frankfurt, 1596.

5 John Wood, *Practicæ Medicinæ liber*, London, 1596, Ad Lectorem. "What then shall wee now expect," cries another pessimist, "in these *doggedayes* of the world's declining age, wherein malitious detraction is esteemed the quintessence of wit, and an ingenuous acknowledgement of mens good deseruing, accounted too great a courtesie?" (Nathaniel Car-

penter, *Achitophel, or the Picture of a Wicked Politician,* 1629, p. 10.)
Spenser attributed modern degeneration to the fact that "when the world
woxe old, it woxe warre old." Cf. *Faerie Queene,* Book IV, canto viii,
stanzas 29-33, and the opening stanzas of Book V.

⁶ *Purchas his Pilgrimage,* 1613, p. 42.

⁷ Sermon XXXVI of the edition of 1640. See also the first paradox of
Donne's *Paradoxes and Problems.* He dares state, however, that "in the
Worlds early *Infancy,* there was a time when nothing was *Euill,* but if this
World shall suffer *dotage* in the extreamest *Crookednesse* thereof, there
shalbe no time when nothing shall bee *good.*" See the fourth paradox.

⁸ *The Fall of Man, Or the Corruption of Nature, Proved by the light of
our naturall Reason. Which Being the First Ground and Occasion of our
Christian Faith and Religion, may likewise serue for the first step and
degree of the naturall mans conuersion. First Preached in a Sermon, since
enlarged, reduced to the forme of a treatise, and dedicated to the Queenes
most excellent Maiestie.* By Godfrey Goodman, her Maiesties Chaplaine,
Bachelor in Diuinitie, sometimes a member both of Trinitie Colledge in
Cambridge, and of Saint Peters Colledge in Westminster. 1616. Two years
later another edition, differing in no respect from the first, appeared under
the title *The Fall of Man, Proued by Reason;* and thirteen years later a
third, with the title *The Fall of Adam from Paradice. Proued by Natural
Reason, and the Grounds of Phylosophy. Published some twenty years since
and now set fourth by the same coppy.* From this edition the dedication to
the queen and the "To the Reader" were omitted, and in their place were
substituted "My Lords Letter to R. P." and "R. P. his letter to the Stationer,"
dated, respectively, Feb. 28 and Mar. 6, 1629. For this last edition, R. P.,
whoever he was, assumed responsibility, for his letter indicates the difficulty
he experienced in securing Goodman's permission to reprint the book, and
then only with the understanding that Goodman's name be withheld from
the title page. In his letter the Bishop of Gloucester, for Goodman had
gained the office since the second edition, says that he would suppress the
book if he could, since it was written when he was but thirty-three years of
age and only a country parson in order not to forget the philosophy he had
learned at the university, and since it contained many things that he as
bishop condemned. He was careful to add, however, that he could still
"make good the scope and intent of the Booke." R. P.'s desire to reissue the
book, as well as Goodman's apologetic tone, was undoubtedly inspired by
Hakewill's *Apologie,* which had appeared two years before.

⁹ P. 214.

¹⁰ P. 357.

¹¹ Pp. 359-62.

¹² Pp. 361-62. This figure served the cause of both the ancients and the
moderns. Whereas the conception of the dwarf upheld the theory of decay
and thus argued the superiority of antiquity in genius, his position on the
shoulders of the giant pointed to modern superiority in knowledge. Glanvill,
a defender of the new science, seems both to attack and defend the idea of
decay. (See *Scepsis Scientifica,* 1665, p. 176, and *Lux Orientalis,* 1662, p.
188.) Sprat, though he makes use of the figure to emphasize the superiority
of the moderns in knowledge, seems to oppose the idea of corruption.
Newton, the greatest exponent of the new science, and Richard Bentley, who
sympathized with it, are said to have employed the comparison. Sir William
Temple, an admirer of antiquity, subscribed to the figure, but tried to deprive the moderns of the small consolation which it provided. How old the
figure is, I cannot say. Joseph Du Chesne in *The Practise of Chymicall, and
Hermeticall Physicke,* tr. T. Timme, 1605, sig. B1ᵛ, ascribes it to "learned
Guido," perhaps Guido de Cauliaco, who flourished late in the fifteenth and
early in the sixteenth century. But the conception certainly goes farther

back, for John of Salisbury, in the *Metalogicus*, c. 1159, says, "Bernard of Chartres used to say that we, like dwarfs on the shoulders of giants, can see more and farther not because we are keener and taller, but because of the greatness by which we are carried and exalted." (Quoted in C. S. Baldwin's *Medieval Rhetoric and Poetic*, 1928, pp. 167-68.) The conception of the superiority of the ancients goes far back before the Renaissance, and, like the theory of decay, was not the product of the revival of the classics.
    13 Pp. 365, 378-79.

    14 P. 268. Subserviency to the authority of the ancients goes hand in hand with belief in progressive decay. Goodman thinks it a shame "that Colledges and ancient foundations, should giue harbour to such a wretched brood; who as they are first nourished vp in obstinancy and wilfulness, so they will proue to bee firebrands in the Church; who neglecting the fathers, and all ancient orders and decency, shall with their owne furious and witles conceits, set the whole world in combustion." (*Loc. cit.*) He seems to be unaware that his own belief in the destructibility of the heavens is diametrically opposed to Aristotle, as he might have learned from Shackelton, for he tells us, "I doe ingeniously, and truly confesse, that whatsoeuer I haue spoken, for the proofe of mans fall, and of natures corruption, I haue only borrowed it from the grounds and foundation of his [Aristotle's] learning; so that the whole treatise, may not improperly bee ascribed to him." P. 389.

    15 *An Apologie of the Power and Providence of God in the Government of the World. Or an Examination and Censure of the Common Errour Touching Natures Perpetuall and Universall Decay, Divided into Foure Bookes: Whereof the first treates of this pretended decay in generall, together with some preparatiues thereunto. The second of the pretended decay of the Heauens and Elements, together with that of the elementary bodies, man only excepted. The third of the pretended decay of mankinde in regard of age and duration, of strength and stature, of arts and wits. The fourth of this pretended decay in matters of manners, together with a large proofe of the future consummation of the World from the testimony of the Gentiles, and the vses which we are to draw from the consideration thereof.* By G. H., D. D. Oxford, 1627.

    Three years later a second edition was published, in which the body of the text was increased by fifty pages, and sixty-four pages were added at the end in the way of "Advertisements," comprising articles and letters bearing on questions raised in the book, and indices of subjects, authors cited, and quotations from the Bible. In 1635 appeared the last edition "revised, and in sundry passages and whole Sections augmented by the Authour; besides the addition of two entire bookes not formerly published." The new books, consisting of 378 pages and devoted to answering objections raised since the second edition, swell the number of pages in the folio to almost one thousand. The progress of the work through these editions reveals how keenly Hakewill was interested in the problem. In the third edition the table of authors cited lists some eight hundred names ranging from earliest antiquity to his own time. It could almost serve as an index to Renaissance learning. Furthermore, in the "Advertisements" added to the second edition and in the additions to the third, he reveals remarkable assiduity in gathering new material from his contemporaries in England and from continental writers with whom he was not familiar when he first produced his treatise. Among the latter were Johannes Gerhardus, a professor at the University of Jena, Canonherius, an Italian, Carolus Gallus, professor of divinity at Leyden, and Besoldus, a German who, Hakewill says, proposed the question, "Commode hic quaeritur an mundus decrescat ac ratione naturalium et moralium indies deterior evadat," and reached the conclusion, "Mundus non solvetur fatigione, non enim instrumenti consumptione, vel diminutione, vel depravatione, multo minus finem habebit in justa proportione quæ intercedere potest inter motum et motorem." But the

man in whom he was especially interested was Secondo Lancellotti, a
Perugian, who in the same year in which the *Apologie* appeared, published
at Venice a work entitled *L'Hoggidi, Overo Il Mondo Non.peggiore ne piu
Calamitoso del passato*. In his first "Advertisement" Hakewill lists translations of the headings of all Lancellotti's chapters, "that it may appeare
in how many points hee accords with mee," and at the conclusion he remarks, "These are his heads: by which it appeares that though hee handle
not the question touching the state of the world in general, nor yet touching
the heavens, the elements, the plants, or other insensible & vnreasonable
creatures in speciall; yet by so much as he hath expressed, it cannot well
be but hee should accord with mee in the whole." Hakewill fails to notice,
however, that the Italian says practically nothing regarding the mental
abilities and achievements of man.

[16] *An Apologie*, third ed., 1635, p. 1, and preface. All references are to
this edition.

[17] *Ibid.*, p. 20.

[18] *Ibid.*, p. 259. In this connection it is interesting to note that Bacon
considered one of the obstacles to learning this very idea of the "floods and
ebbs of the sciences, that they grow and flourish at one time and wither and
fall off at another." (*Novum Organum*, Book I, Aphorism 92.) Without
holding to the view of the decrepitude of nature, Bacon believed that science
advanced by increments, and looked upon the idea of circular progress as
likely to persuade men that progress could not be made. Hakewill, steering
as widely as possible from the theory of decay, fell upon this other idea,
which he supports with some success by pointing to the eras of the Chal-
deans and Egyptians, the Greeks, the Romans, the Dark Ages, and the
Renaissance. It must be noted, however, that the Epistle Dedicatory of the
Apologie makes use of Bacon's paradox *saeculi antiquitas, mundi juventus*,
and also expresses Bacon's idea that the moderns cannot equal or excel the
ancients, if the latter are considered perfect. (See note F on Part I of
Dugald Stewart's *Progress of Metaphysical, Ethical, and Political Philosophy* in *A Supplement to the fourth, fifth, and sixth editions of the En-
cyclopædia Britannica*, 1824, pp. 157-8.) By his theory of compensation,
Hakewill shows that the universe does not partake of the nature of its parts,
for what is lost by one part is added to another.

[19] Epistle Dedicatory.

[20] See pp. 312, 316. These inventions, which loomed large in the Renaissance mind, are cited again and again throughout the seventeenth century.

[21] *J. Jonstoni . . . . Naturæ constantia: seu diatribe, in qua . . . . mundum,
nec ratione sui totius, nec ratione partium, . . . perpetuo in pejus ruere,
ostenditur*. Amsterdam, 1632. The title of John Rouland's translation, to
which all citations are made, is as follows: *An History of the Constancy of
Nature. Wherein, By comparing the latter Age with the former, it is maintained that the World doth not decay universally, in respect of it Self, or
the Heavens, Elements, Mixt Bodies, Meteors, Minerals, Plants, Animals,
nor Man in his Age, Stature, Strength, or Faculties of his Minde, as relating
to all Arts and Science*. 1657.

[22] See pp. 73-81. Jonston is quoting Du Chesne.

[23] See p. 84.

[24] Jonston bears testimony to the deep-rooted hold which the gloomy
doctrine had secured upon his age, and also reveals its direct bearing on the
controversy over the relative merits of the ancients and moderns: "Reader,
it is the common opinion, and is in every mans mouth, that this world, and
all things therein contained, do run towards their end, by a universall and
perpetuall declining to worse and worse; and that there is nothing now in
being, or can fall out, that may, I say not be prefer'd, but be compared to

the dayes of our Forefathers. That neither the uniformity of the Heavens, is the same as it was of elder times, nor the fruitfulnesse of the Earth, nor the same vigour is in Plants and Animals, nor is the stature or age of man so strong. *The excellencies of Arts, are, by reason of the worlds growing old, exhausted, and can do but little that is wonderfull, and but in few things.* Astraea *is gone to the gods, & there is no man, in common society, that may be compared with* Aristides *or* Socrates; *in* Policie *with the* Fabii, Scipio's, *or* Marcelli; *in* Houshold Government, *with* Æneas, *the wives of the* Myniae, Manlius, *or* Junius; *in the Church, with* Ambrose, Hierome, Augustine: *In the Schools, with* Aristotle, Demosthenes, Cicero, Galen. In brief, All things run downward, and they come not back again." (See pp. 1-6.) Such a view, he declares, dishonors God and hinders the endeavors of men.

25 See *Modern Philology*, XXXI (1934), 283.

26 *An Humble Motion*, 1649, p. 6. Cf. the anonymous poem *An Apology for Bishops or, A Plea For Learning, Against those lying, rayling, and scandalous libels so frequent, in despight of Learning and Learned Men. Dedicated to all the Lover[s] of Learning and Vertue for the Reformation of the Abuses and Corruptions of these Times.* 1641.

27 Cf. "To the Reader," prefixed to *Reflections upon Monsieur Des Cartes's Discourse of a Method*, 1655.

28 *Vanity of Dogmatizing*, 1661, p. 240, and *Scepsis Scientifica*, ed. John Owen, 1885, p. 209. See, however, his *Lux Orientalis*, 1662, p. 188.

29 Henry Power, *Experimental Philosophy*, 1664, pp. 183 ff., and preface.

30 *A Miscellany*, 1659, p. 102.

31 *Medela Medicinæ*, 1665, p. 10.

32 *Plus Ultra*, p. 13.

33 In 1677 John Webster says that Hakewill's book, a "profound Piece," met with considerable opposition in its own day, but that nature does not decay is "a truth now sufficiently known, and assented to." *The Displaying of Supposed Witchcraft*, 1677, p. 3.

34 For a more detailed account of this relationship see the present writer's "Background of the Battle of the Books" in *Washington University Studies*, Humanistic Series, vol. VII, no. 2, 146-48.

35 As early as 1633 (?) Henry Reynolds, a friend of Drayton and translator of Tasso's *Aminta*, applied the theory of nature's decay to poetry in his *Mythomystes, wherein a short survey is taken of the nature and value of true Poesie, and depth of the Ancients aboue our Moderne Poets*. He laments the senility of the world in the following fashion: "I haue thought vpon the times wee live in, and am forced to affirme the world is decrepit, and, out of its age & doating estate, subiect to all the imperfections that are inseparable from the wracke and maime of Nature, that the young behold with horror, and the sufferers thereof lye vnder with murmur and languishment. Euen the generall Soule of this great Creature, whereof euery one of ours is a seuerall peece, seemes berid, as vpon her deathbed and neere the time of her dissolution to a second better estate and being." (*Critical Essays of the Seventeenth Century*, ed. J. E. Spingarn, I, 144.) Reynolds notes three respects in which modern poets are inferior to ancient: first, they are mercenary, fawning, servile, and more interested in mere ungrounded fancy than in seeking high truths for truth's sake; again, they set such a small price on knowledge that they impart it to everyone, taking no care to wrap it in mysteries; and finally, they are ignorant of the mysteries and hidden properties of nature, such as are revealed in the fables of the ancients.

[36] John Dennis evidently was of this opinion, for in his attack on Pope's worship of classical antiquity, he declares, " 'Tis ridiculous and pedantick to imagine, that the natural Powers of the Soul were stronger or more excellent in the Ancients than they are in the Moderns. And as to Experience we have vastly the Advantage of them." In the latter connection he refers to Bacon's dictum that the moderns are the ancients, which was advanced against the theory of decay. *Reflections upon a Late Rhapsody,* 1711, in *Critical Essays of the Eighteenth Century,* ed. W. H. Durham, p. 231.

[37] Dryden, for example, in upholding the superiority of modern satire and tragedy over ancient, finds it necessary to contradict the theory: "For good sense is the same in all or most ages; and the course of time rather improves Nature, than impairs her." (*Essays,* ed. W. P. Kerr, II, 25.) Likewise, George Farquhar in his *Discourse upon Comedy,* 1702, says that since the authority of the ancients has been banished from philosophy and religion, he will attack it in criticism. "But in the first Place," he cautions the reader, "I must beg you, Sir, to lay aside your Superstitious Veneration for Antiquity, and the usual Expressions on that Score; that the present Age is illiterate, or their taste is vitiated; that we live in the decay of Time, and the Dotage of the World is fall'n to our Share . . . . 'Tis a mistake, Sir, the World was never more active or youthful, and true downright Sense was never more Universal than at this very Day." *Critical Essays of the Eighteenth Century,* ed. Durham, pp. 263-64.

Hans Baron makes clear Hakewill's indebtedness to continental writers, by virtue of which he maintains that the Renaissance humanists were on the side of the moderns rather than of the ancients. See *J. H. I.,* XX (1959), 3-22.

CHAPTER III

[1] See the preface to the *Magna Instauratio* in *Works of Francis Bacon,* ed. James Spedding, R. L. Ellis, D. D. Heath, new edition, 7 vols., London, 1879-1890, IV, 13. (All references to Bacon's works are to this edition.) In commenting on the idea entertained by some, that all sciences had reached their full perfection, Bacon exclaims, "Would it were so! But the truth is that this appropriating of the sciences has its origin . . . . in the confidence of a few persons and the sloth and indolence of the rest," a criticism which later echoes through most of the attacks on antiquity.

[2] *Works,* IV, 90, 103.

[3] *Ibid.,* III, 249; IV, 86.

[4] *Ibid.,* III, 249; IV, 13, 102. "If there be any that despond, let them look at me."

[5] *Ibid.,* II, 354; IV, 14. This contrast is employed again and again in the second and third quarters of the century.

[6] *Ibid.,* IV, 59, 69, 88, 344-45, 357.

[7] *Ibid.,* IV, 81-2, 283.

[8] *Ibid.,* IV, 14, 16, 109, 299; V, 132.

[9] *Ibid.,* IV, 82.

[10] *Ibid.,* IV, 82.

[11] *Ibid.,* III, 546; IV, 41, 52, 62.

[12] *Ibid.,* IV, 90.

[13] *Ibid.,* IV, 73, 312; V, 110.

[14] *Ibid.,* IV, 311.

15 *Ibid.*, IV, 87, 286; III, 324.

16 *Ibid.*, III, 292.

17 *Ibid.*, III, 294.

18 *Ibid.*, IV, 81, 296.

19 *Ibid.*, IV, 411.

20 For a discussion of this matter see *PMLA*, XLV (1930), 977-1009; *Jour. Eng. and Ger. Philol.*, XXX (1931), 188-217; *ibid.*, XXXI (1932), 315-31. In his undertaking Wilkins was evidently inspired by the first chapter of the sixth book of *The Advancement of Learning*, in which Bacon speaks of the "real characters" used in the Far East. (See, however, Dorothy Stimson's "Dr. Wilkins and the Royal Society," *Jour. Mod. Hist.*, III [1931], 543.) Sprat (*History of the Royal Society*, p. 113) describes the plain, simple style demanded by the Society of all the scientific communications of its members, a requirement which seems to go directly back to Bacon. The latter criticizes the natural histories of his day for being full of fables, quotations from the ancients, disputes, philology and ornaments, as if setting up a treasure-house of eloquence rather than a sound and faithful narrative of facts. (*Works*, V, 508.) So when he came to describe the proper way in which a natural history should be drawn up, it is not strange that he should wish to banish all the evils mentioned above: "And for all that concerns ornaments of speech, similitudes, treasury of eloquence, and such like emptinesses, let it be utterly dismissed." *Ibid.*, IV, 254.

21 *Works*, IV, 8, 52, 53; V, 132.

22 *Ibid.*, IV, 17, 27, 47, 50, 428. Bacon remarks that it is strange that no one has laid out a road for the human understanding direct from the senses. (*Ibid.*, IV, 80.) Of course, Bacon's method gave an important function to the understanding, but he stressed more heavily the part the senses play, because of their neglect. He defines his purpose as "a true and lawful marriage between the empirical and rational faculty, the unkind and ill-starred divorce and separation of which has thrown into confusion all the affairs of the human family." (*Works*, IV, 19.) During the Restoration this stress upon sense and distrust of reason was carried to greater lengths. In the same way in which imagination was to be controlled by reason, reason was to be controlled by the senses.

23 *Ibid.*, IV, 19.

24 *Ibid.*, IV, 26. Cf. IV, 428.

25 *Ibid.*, IV, 26, 112, 412.

26 *Ibid.*, IV, 19, 21, 31.

27 *Ibid.*, III, 293; IV, 15, 63; V, 131.

28 This characterization of the Copernican theory was amply supported by the spurious preface to Copernicus' *De Revolutionibus Orbium Coelestium*. Bacon's failure to appreciate the importance of mathematics, a deficiency lamented by his later followers, can be explained in the same way as his rejection of the Copernican theory. To mathematics he attributes the function of giving definiteness to natural philosophy, and thus as physics advances and develops new axioms, it will require fresh assistance from mathematics in many things. (*Works*, IV, 93, 371.) For mathematical systems he entertains less regard. The science, he says, has been more acutely inquired into than other matters, because it is "the nature of the human mind, certainly to the extreme prejudice of knowledge, to delight in the open plains . . . . of generalities rather than in the woods and inclosures of particulars," with the result that logic and mathematics, which should be the handmaidens of science, exercise dominion over it. (*Ibid.*, IV, 370.) The fact that mathematics, like syllogistic logic, is deductive, proceeding from general principles to particulars, was repugnant to Bacon's whole

philosophy, and so mathematics takes its place with other systems in which reason plays too great a part in proportion to observation. *Ibid.*, IV, 411.

[29] *Works*, III, 716; V, 511.

[30] *Ibid.*, IV, 183.

[31] *Ibid.*, IV, 229, 373. For instance, he says we must know the nature of spontaneous rotation, before we can decide whether the earth or the heavens move. (*Works*, IV, 123.) In short, in our astronomical theories we should seek for truth in physical nature and not in mathematics.

[32] *Ibid.*, V, 511. In another passage he says that in astronomy all theories should be suspended, because natural philosophy should embrace only "pure and separate" phenomena, as if nothing whatever had been settled by the art of astronomy. Only the observations and experiments of the latter, accurately collected, and perspicuously described, are of use.

[33] *Works*, V, 524, 515 ff., 557. Though Bacon thought the value of the microscope was distinctly limited, he did appreciate the importance of the telescope in astronomical observations, and of the discoveries made with it, which he calls noble endeavors, worthy of the human race and performed by men who should be praised highly for honesty, boldness, and clear explanations and descriptions of their proceedings. He adds that all that is needed is "constancy and great severity of judgment, to change the instruments, to encrease the number of witnesses, to try each particular experiment many times and many ways," and to consider every possible objection that can be raised. *Ibid.*, IV, 192-93; V, 512-13.

[34] *Works*, IV, 59, 185, 323.

[35] *Ibid.*, IV, 65.

[36] *Ibid.*, IV, 74, 84; V, 205. In similar manner Bacon condemned magic and astrology because they were full of superstition and contained too much of the merely imaginative and credulous, but he wished them purified rather than destroyed, for he thought there was a certain amount of physical knowledge in them which would be a contribution to natural history. In fact, he says that the experiments of magicians and astrologers have frequently proved useful. (*Works*, III, 289; IV, 349, 366-67.) Later in the century all three of the pseudo-sciences called upon Bacon for justification.

[37] *Works*, III, 247, 293; IV, 24-5, 50, 97.

[38] *Ibid.*, III, 250; IV, 357, 429. Bacon is usually careful to advance his own conclusions as tentative.

[39] *Ibid.*, IV, 32, 53. This sceptical view became a typical attitude of seventeenth-century science in England.

[40] *Ibid.*, IV, 28, 127, 251, 252; V, 211, 507.

[41] *Ibid.*, IV, 254, 298; V, 507-8.

[42] *Ibid.*, IV, 255, 296.

[43] *Ibid.*, IV, 257; V, 506. Bacon believed that "on account of the pernicious and inveterate habit of dwelling on abstractions, it is safer to begin and raise the sciences from those foundations which have relation to practice, and to let the active part itself be as the seal which prints and determines the contemplative counterpart." *Ibid.*, IV, 120-21.

[44] *Works*, IV, 257, 297.

[45] *Ibid.*, IV, 343.

[46] *Ibid.*, IV, 258, 297.

[47] "For since all Interpretation of Nature commences with the senses, and leads from the perceptions of the senses by a straight, regular, and

guarded path to the perceptions of the understanding, which are true notions and axioms, it follows of necessity that the more copious and exact the representations of the senses, the more easily and prosperously will everything proceed." *Works*, IV, 192.

⁴⁸ *Ibid.*, IV, 101, 251; III, 322.

⁴⁹ The same idea is repeated on page 328 of the third volume. This passage removes to the distant future the fulfilment of his design. It is his natural history that represents the difficulty in his method, for, as we have seen, he believed that after it had been completed, the acquiring of the knowledge of nature would be the work of only a few years. (*Works*, IV, 252.) That Bacon had in mind governmental support for his scheme is revealed not only in this passage, but also in his *New Atlantis*, in which the model of his co-operative plan, Solomon's House, is clearly a function of the government. In various works, Bacon's protégé, Thomas Bushell, makes it plain that Bacon made some headway in realizing his Solomon's House.

⁵⁰ *Works*, III, 323. As further encouragement to his age, Bacon proposed himself as an example of what one man, though busied with many things, could do to advance knowledge. What, he then asks, may not be hoped from leisure and co-operation through many ages, one man taking charge of one thing, and another of another?

⁵¹ *Works*, IV, 62-3, 109.

⁵² *Ibid.*, IV, 260-61; V, 135-36, 509.

⁵³ *Ibid.*, IV, 95, 96, 413 ff.

⁵⁴ *Ibid.*, IV, 96.

⁵⁵ *Ibid.*, IV, 50

⁵⁶ *Ibid.*, IV, 97.

⁵⁷ *Ibid.*, IV, 96, 104.

⁵⁸ *Ibid.*, IV, 29.

⁵⁹ *Ibid.*, IV, 179; V, 500.

⁶⁰ In one passage Bacon seems to repudiate the atomic philosophy: "Nor shall we thus be led to the doctrine of atoms, which implies the hypothesis of a vacuum and that of unchangeableness of matter (both false assumptions); we shall be led only to real particles, such as really exist." (*Works*, IV, 126.) He rather clearly indicates here that he considered the atoms of Democritus unreal, probably an hypothesis to explain phenomena. This interpretation is borne out by another passage in which he says, "The doctrine of Democritus concerning atoms is either true or useful for demonstration." (*Ibid.*, V, 419.) The difference between the atoms of Democritus and Bacon's particles may probably be explained by the fact that Bacon did not believe his particles were infinite or perpetually divisible. For a more complete discussion of Bacon's atomic philosophy, see C. T. Harrison, "Bacon, Hobbes, Boyle, and Ancient Atomists," *Harvard Studies and Notes in Philology and Literature*, XV (1933), 191-218.

⁶¹ There is ample evidence of Bacon's high regard for Democritus and his philosophy. See *Works*, V, 419, 421, 465, 466.

⁶² *Works*, V, 463, 464, 472, 469.

⁶³ *Ibid.*, IV, 120, 146, 150.

⁶⁴ *Ibid.*, III, 243.

⁶⁵ *Ibid.*, IV, 25.

⁶⁶ *Ibid.*, IV, 149.

⁶⁷ *Ibid.*, IV, 145.

⁶⁸ *Ibid.*, IV, 97, 146.

⁶⁹ According to Bacon's plan, there were two stages in the scientific process: the construction of a natural history, and the employment of induction upon the materials thus afforded. So scientists may be divided into two classes: theorizers, and observers and experimenters. Of the latter no unusual ability was demanded; the field was open to the joint labors of many men. But Bacon implied that the former must be men of superior intellects. He bemoaned the fact that he had to engage in the low task of laying the foundation, when he deserved to be the architect of the building. *Works*, II, 335.

⁷⁰ *Works*, III, 294; IV, 79.

⁷¹ *Ibid.*, III, 498; IV, 24, 104, 114.

⁷² *Ibid.*, IV, 29, 71.

⁷³ *Ibid.*, II, 501; IV, 95.

⁷⁴ *Ibid.*, IV, 27. See also IV, 21, 32, 91.

⁷⁵ *Ibid.*, III, 219; IV, 342; V, 112.

⁷⁶ *Ibid.*, III, 223; IV, 99, 113, 114.

⁷⁷ *Ibid.*, III, 323 ff., 326, 327, 502; IV, 89, 288, 289.

CHAPTER IV

¹ See the "Preface Magneticall" and p. 14.

² *Magneticall Animadversions. Made by* Marke Ridley, *Doctor in Physicke. Vpon certaine Magneticall Advertisements, lately published, From Maister William Barlow.* 1617.

³ *Ibid.*, pp. 7-8, 37 ff.

⁴ *A Short Treatise of Magnetical Bodies,* 1613, "To the Courteous Reader."

⁵ *Magneticall Aduertisements: Or Divers Pertinent obseruations, and approued experiments concerning the nature and properties of the Load-stone: Very pleasant for knowledge, and most needfull for practise, of trauelling, or framing of Instruments fit for Trauellers both by Sea and Land.* 1616.

⁶ *A Breife Discovery of the Idle Animadversions of Marke Ridley,* 1618, p. 12.

⁷ *Magneticall Aduertisements,* preface. In the same passage he reveals his scientific conscience: "And I propose (God willing) to tye my selfe, as strictly vnto this rule, as euer any man did that hath written of the like argument, making it euen a matter of conscience to deliuer any thing herein for certaintie, that my selfe shall not know to be sound. And so by this meanes shall I neither abuse the Reader with any vntrue assertion, nor iniurie so certaine and so excellent a knowledge with any doubtfull or vnproueable conclusions."

⁸ The preface to the first is a semi-ironical panegyric of ignorance in preference to knowledge, really a satire on disputatious, wrangling, and futile philosophical controversies, which in the second edition is supplanted by a serious defence of the liberty of scientific thought and discussion. The text of the first edition contains only twenty of the thirty "exercitationes" found in the second; the fourth, sixth, and tenth exercitations of the third decade which defend to some extent the Copernican theory, and praise Copernicus, Kepler, Brahe, and Galileo, are significantly absent. It seems that the change in place of publication made possible the expression of

much more liberal views; that, as Milton later noted with much satisfaction, England was a safer place for free-thinking than the continent. The full title throws some light on the purpose and nature of the treatise: *Philosophia Libera Duplice Exercitationum decade proposita: In qua Paradoxa quædam ad Exercenda juuenum ingenia aduersus vulgares huius temporis Philosophos suscipiuntur, validisque rationibus confirmantur.*

9 It is interesting to note that in a later work Carpenter employs one of Hakewill's arguments in combating the idea of decay, namely, that of compensation.

10 Toward this end he attempts to show that a vacuum is possible, that intelligences are not the movers of the heavenly orbs, that the circular motion of the earth is probable, and the like.

11 *Philosophia Libera,* pp. 358, 272.

12 Perhaps the most definite bit of evidence making for Baconian influence is furnished by Carpenter's interpretation of the theory of universal decay, which is exactly that of Bacon. In his *Geographie Delineated,* 1625, p. 44, he expresses one of Bacon's ideas in saying that the invention of the compass was due to chance rather than to industry.

13 *Giordano Bruno,* 1903, p. 257. McIntyre believes that Bruno influenced Bacon, and besides pointing out what he considers instances of specific indebtedness, he cites attitudes common to both, such as opposition to authority, disagreement with the idea that antiquity and general agreement are grounds for belief, and the desire to undermine Aristotle. These are also discovered in Carpenter.

14 *Geographie Delineated Forth in Two Bookes. Containing the Sphæricall and Topicall Parts Thereof.* Oxford, 1625. Unless otherwise stated, references are to the first book.

15 *Ibid.,* p. 45.

16 Elsewhere Carpenter clearly implies the importance of experiments in the promotion of science. In speaking of the disagreement between certain magnetic theories, he says, "But the decision of these differences I leaue to such as are more experimentall then my selfe, being destitute of those helpes and instruments which they enjoy." (Pp. 51-2.) "This proposition may be shewed out of euident experiment, wherein euery mans sight may be a witnes." (P. 49.) "But this hath often bin proued otherwise by diuerse experiments of Nauigations, mentioned by *Gilbert,* and *E. Wright.*" (P. 64.) "Of the Inuention of the Miridian circle, the true and exact knowledge (as we haue shewed) is endebted to heauenly obseruation, or Magneticall experiment." Pp. 130-31.

17 *Geographie Delineated,* p. 142.

18 *Ibid.,* pp. 76, 111.

19 *Ibid.,* p. 76. An interesting example of Carpenter's anti-classical spirit is revealed in a passage which ridicules the French and English for deriving their descent from the Trojans when they might have more justly done so from "the valiant nation of ancient *Gauls* and *Germans.*" (Bk. II, p. 205.) There was in England at this time an ardent worship of all things German. See the present writer's *Triumph of the English Language,* pp. 214-236.

20 *The Strange and Dangerous Voyage of Captaine Thomas Iames, in his intended Discouery of the Northwest Passage into the South Sea. Wherein the Miseries Indured Both Going, Wintering, Returning; and the Rarities obserued, both Philosophicall and Mathematicall, are related in this Iournal of it. Published by His Maiesties command. To which are added, A Plot or Card for the Sayling in those Seas. Diuers little Tables of the Author's, of the Variation of the Compasse, etc. With an Appendix concerning Longitude, by Master Henry Gellibrand Astronomy Reader of Gresham Colledge*

in London. *And an Aduise concerning the Philosophy of these late Dis-coueryes,* By W. W., 1633.

[21] *Ibid.,* p. 7.

[22] *Ibid.,* p. 15.

[23] In this connection it is interesting to note how frequently the Pillars of Hercules, with their supposed inscription *ne plus ultra,* are used to indicate the closed circle of ancient knowledge. Bacon's example and the vast increase in geographical knowledge over that of the ancients operated consciously and unconsciously in undermining the supposed superiority of the classical writers, a fact later conspicuously revealed in various geographical figures of speech used to express modern progress in knowledge.

[24] They occupy some ten pages at the end of the volume.

[25] This quotation was very popular in the seventeenth century, though it is not always found in connection with the new scientific thought. It expresses, however, the determined stand of the scientists, that truth and not authority should govern belief.

[26] *A Discours Mathematical,* p. 20. See also John Wells's *Sciographia, or the Art of Shadowes,* 1635, preface by Gellibrand, for a reference to Copernicus' *"Heroicke Hypothesis of the Earths Motion."*

[27] *The Sea-Mans Practice, Contayning A Fundamentall Probleme in Navigation, experimentally verified* . . . . By Richard Norwood, Reader of the Mathematicks. 1637.

[28] Cf. *Sea-Mans Practice,* "To the Reader" and pp. 8, 12, 41.

[29] Unless otherwise stated, all references are to Wilkins' *Mathematical and Philosophical Works.* 1708.

[30] *Ibid.,* p. 15 and preface.

[31] *Ibid.,* p. 271.

[32] *Ibid.,* p. 146.

[33] *Loc. cit.,* and p. 139.

[34] *Mercury,* 3rd ed., 1707, p. 6.

[35] *Mathematical and Philosophical Works,* pp. 152, 158. Among the causes of the opposition to the Copernican theory, Wilkins emphasizes "A servile and superstitious fear of Derogating from the Authority of the Ancients," and "that Meaning of Scripture-Phrases, wherein the supposed Infallible Church hath for a long time understood them." (P. 155.) The first, in which we are chiefly interested, clearly indicates the importance of the theory in the clash between ancients and moderns. But Wilkins himself was more concerned with answering the second, especially since, as was mentioned above, Gilbert had been severely censured for ignoring it. Most of Wilkins' arguments were inspired by the desire to defend Gilbert. These, which are beside our present purpose, introduce little that is new in the case being made out for the Copernican system, and may be found with others in Miss Dorothy Stimson's *The Gradual Acceptance of the Copernican Theory of the Universe,* 1917. It may be noted, however, that the new scientific thought was introducing a more liberal spirit into the interpretation of the Scriptures.

[36] *Philiatros,* f. 17ᵛ.

[37] *Ibid.,* f. 12ᵛ.

[38] *The Chirurgicall Lectures of Tumors and Vlcers,* 1635, p. 8.

[39] Cf. *A Short Discouerie of the Vnobserued Dangers of seuerall sorts of ignorant and vnconsiderate Practisers of Physicke in England,* 1612, which

lumps chemists with women, clergymen, witches, wizards, astrologers, etc.; and *Cotta Contra Antonium*, 1623, directed against Francis Anthony, a chemist who had written several treatises in defense of potable gold.

⁴⁰ *A Short Discouerie*, p. 10.

⁴¹ *Ibid.*, pp. 82-3.

⁴² *Ibid.*, pp. 84, 85.

⁴³ *Ibid.*, pp. 125, 127.

⁴⁴ This is another quotation popular with the scientific moderns. Cf. M. Ridley, *Magneticall Animadversions*, 1617, pp. 7-8.

⁴⁵ *Op. cit.*, p. 6. A representative example of the way in which the term "experiment" was continually gaining wider currency is furnished by an anonymous treatise entitled *A Direction to the Husbandman*, 1634. In the dedication the author speaks of "the prime *Experimentors* of the certaintie and realitie" of certain directions, of "a Demonstrative, experimented successeful truth" (Sig. B2) and of "experimentall helpes" to agriculture. (Sig. 153ᵛ) Cf. also *The Mysteryes of Nature and Art*, 1634, by J. B., in which almost every thing is called an experiment. As greater emphasis is placed upon experiments, less and less respect is paid to antiquity. Thus William Lawson considers "that Art the surest, that stands upon experimental rules .... Whereupon have I of my meere & sole experience, without respect to any former written treatise gathered these rules, & set them downe in writing ... neither is this injurious to any, though I differ from the common opinion in divers points, to make it knowne to others, what good I have found out in this faculty by long triall and experience. I confesse freely my want of curious skill in the Art of planting. And I admire and praise *Plinie, Aristotle, Virgil, Cicero*, and many others for wit and judgement in this kind, and [a sudden parting shot] leave them to their times, manner, and severall countries." (Preface to *A New Orchard and Garden*, 1638, found in G. Markham's *A Way to Get Wealth*, 1638.) Of course, the word "experiment" is used loosely, but it always includes the direct evidence of the senses. The growing practical, inventive, and utilitarian spirit of the age is manifested in John Babington's *Pyrotechnia Or, a Discourse of Artificiall Fire-Works*, 1635.

CHAPTER V

¹ Clear evidence of the growth of Bacon's prestige during the twenty years preceding the Restoration is discovered in several productions of Thomas Bushell, a mining engineer and protégé of Bacon. In *A Just and True Remonstrance of His Majesties Mines-Royall in the Principality of Wales*, 1642, Bushell is seeking renewal of his mining grant from the government, and has much to say about the difficulties of the work, expense, success, and hope for the future, but he does not once mention Bacon. In *The Case of Thomas Bushell*, 1649, which has the same end in view, he finds it worth while to state that he "had many years relation to that great Secretary of Nature, the late Lord Chancellor *Bacon;* and thereby from him learnt many secrets in discovering and extracting *minerals.*" In *Mr. Bushell's Abridgment of the Lord Chancellor Bacon's Philosophical Theory in Mineral Prosecutions*, 1659, composed of a number of letters, affidavits, testimonies, short treatises, and the like, which are followed by a reprint of *The New Atlantis*, and at least one of which, "Mr. Bushell's Minerall Overtures to the Council," had appeared in print before, he goes at great length into his interesting relationship with Bacon. In fact, he bases his claim to consideration almost solely upon his having been selected by the Chancellor to put into practice Solomon's House, a project Bacon hoped to accomplish. He repeatedly states that in his mining endeavors he is only

acting from love of, and gratitude to, Lord Verulam, whom he calls "my ever honored Master," "my heroic Master," "honored Lord and Master." In fact, Bacon is always kept in the foreground, for his name now furnished the best argument. Bushell's life as told by himself in this volume reads like a romance, throwing no little light on Sir Francis himself.

2 "Wee are upon the designe of a Publick Reformation, wherein every body is one way or other, if not engaged yet concerned, some more some lesse, some in a private; some in a Publicke way. This work hath been long in hand, many stones have been moved about it, much dust is raised in it; and to the outward appearance after all attempts, we are further from our purpose then at the beginning." John Durie, *A Seasonable Discourse*, 1649, p. 1.

3 R. H. Latham, one of the few who have noticed any relationship between Puritanism and science at Oxford, says, "The violent revolution, which had unsettled the minds of the students in each of the Universities, and in Oxford most, had produced a change in the character of the studies as well as in the discipline and government of the colleges. Nor was this change unmixed with good. There was little love of old doctrines, simply on the score of their antiquity, and there was also less tendency to acquiesce in authority than to test, observe, and criticize; for the ferment that had affected men's minds, in respect to the higher matters of theology and politics, had extended itself over the whole domain of scholarship, and philosophy. Men thought for themselves, and they did so, not because any modern method of investigation had supplanted an ancient one, nor yet because the new philosophy of Bacon was displacing the old philosophy of the schools, but because the spirit of investigation was aroused, and because there were more minds excited to a greater state of activity. I believe that at the time in question, there was, in Oxford, more activity of thought than there has been either before or since; at least, it is certain that the number of philosophical thinkers was as great as it ever has been, and that the contact or collision of the University with the world at large, along with the stimulus that such collision develops, was greater." (See "Life of Sydenham," prefixed to a translation of Sydenham's Latin works, 2 vols. 1848-50, vol. I, p. xxi. Cf. J. F. Payne, *Thomas Sydenham*, 1900, pp. 58-60, 62-3, 66.) Latham is decidedly wrong in thinking that Bacon and the new science exerted no influence on the Puritans. The time was ripe for his philosophy, and it in turn influenced the times.

4 The men most conspicuous in the movement for educational reform were Samuel Hartlib, John Durie, John Hall, and John Webster, most of whom drew their inspiration not only from Bacon, but also from Comenius. The two most influential elements in Comenius' educational reform, with the exception of his idea of pansophia were the change in language teaching to be effected through simplifying the study of grammar, and the insistence upon the study of things rather than of words. Comenius himself, however, owed much to Bacon. After stating that he received his first inspiration from Vives and Campanella, he says, "But when I chanced afterwards upon a piece of Sir *Francis Bacon Baron of Verulam, Chancellour of England,* entituled *instauratio magna* (an admirable work, and which I look upon no otherwise, then as a most bright beam of a new age of Philosophers now arising) I understood that in some particulars also of *Campanella,* such solid Demonstrations, as the truth of things requires, were wanting. Yet it grieved me again, that I saw most noble *Verulam* present us indeed with a true key of Nature, but not open the secrets of Nature, onely shewing us by a few examples, how they were to be opened; and leave the rest to depend on observations and inductions continued for severall ages. Yet I saw nevertheless, that my hopes were not quite left in suspense: in as much as I perceived my minde so enlightened by the light which it received from those severall sparks, now grown welnigh to a torch, that some great secrets of Nature, . . . were now plain, as it were of their own accord, to the

exceeding great content of my mind. For now with those, that have lighted upon a more sound way of Philosophie in this age, I saw and rested in it." (See preface to a translation with the title *Naturall Philosophie Reformed by Divine Light*, 1651.) Comenius "was invited hither by certaine Honourable and active Patrons of Learning in this present Parlament, and disappoynted of the preferment they did sincerely intend him, in regard of the great distractions which happened in the State at that juncture of time when hee came over." Epistle Dedicatory to Comenius' *A Patterne of Universall Knowledge*, translated by Jeremy Collier, 1651.

5 Another attitude, almost as general and as important as the two described, is discovered in the wide use of the term "advancement," obviously derived from the expression "advancement of learning," which indeed is employed over and over. The desire to push things forward, to open closed doors, to turn the level of civilization into an incline, in short, the desire for progress is prominent. What later becomes the well-defined idea of progress finds here a definite beginning.

6 *Motion Tending to the Publick Good of This Age, and of Posteritie. Or the Coppies of certain Letters written by Mr. John Dury, to a worthy Knight, at his earnest desire. Shewing briefly, What a Publick good is, and how by the best means of Reformation in Learning and Religion it may be advanced to some perfection. Published by Samuel Hartlib.* 1642.

7 *The Advice of W. P. to Mr. Samuel Hartlib, for the Advancement of some Particular Parts of Learning.* 1648. The treatise may now be found in the *Harleian Miscellany*, vol. VI, to which all references given below pertain. The scientific nature of Petty's outlook is revealed at the very beginning, where he says, "I have had many flying Thoughts concerning the Advancement of Real Learning in generall, but particularly of the Education of Youth, Mathematicks, Mechanicks, Physick, and concerning the History of Art and Nature." Petty made popular the use of the term "real" to denote experimental or true science. His direct indebtedness to Bacon is revealed in the following passage: "To give an exact Definition, or nice Division of Learning, or of the Advancement thereof, we shall not undertake (it being already so accurately done by the great Lord *Verulam*) intending only . . . . to point at some Pieces of Knowledge, the Improvement whereof . . . . would make much to the general Good and Comfort of all Mankind." (P. 2.) On p. 13 he refers to Bacon's "most excellent Specimen" of a history of nature, and his "exact and judicious Catalogue" of particulars.

8 *Ibid.*, pp. 1-3.

9 *Ibid.*, pp. 3 ff.

10 See an article by the present writer entitled "Science and Language in England in the Mid-Seventeenth Century," *Jour. of Eng. and Ger. Philol.*, XXXI (1932), 315-31.

11 Pp. 9-13. In his account of Petty in *Athenae Oxonienses* Wood mentions a work entitled *Advice concerning the Education of Youth*, London, 1647, which, if different from the treatise under discussion, I have not seen.

12 *A Seasonable Discourse Writen by Mr. John Dury upon the earnest requests of many, briefly shewing these Particulars.* 1. *What the Grounds and Method of our Reformation ought to be in Religion and Learning.* 2. *How even in these times of distraction, the Worke may be advanced. By the knowledge of Orientall tongues and Jewish Mysteries. By an Agency for advancement of Universall Learning. Published, by Samuel Hartlib.* 1649. This is the treatise in which Durie attributes the desire of the Puritans to reform learning to the publication of Bacon's works. He probably is thinking mainly of an edition published at Oxford in 1640, *Of the Advancement and Proficience of Learning, or the Partitions of Sciences, IX. Bookes. Interpreted by Gilbert Wats.* Whether or not this particular edition is to be

credited with stirring up the Puritans, Bacon's works became almost a second Bible with many of them.

[13] *Ibid.*, pp. 17 ff. Durie is really spreading propaganda for Hartlib and the Office of Public Address, which will be treated in the next chapter. First among the regulations which he draws up for the project is "That the Agent already design'd by the Parliament, and long ago exercised in this way, and known to the chiefe men of parts at home and abroad" shall be provided with a competence. Parliament had taken notice of Hartlib and his designs, but did not finance him sufficiently.

[14] The addition of the term "piety" or its equivalent to Bacon's "Advancement of Learning," the whole constituting a watchword of Puritanism, is nowhere more clearly attested than in "The Publisher to the Reader," written by Hartlib. The latter speaks of children being trained from infancy "to a course of Reformation, both of Virtue and Learning"; he claims he is responsible for drawing forth "from others these following Directions, towards the Reforming of Schools, and the advancement of Piety and Learning"; and he signs himself "Thy most willing Servant for the advancement of Piety and Learning." The passion for the public good, social-mindedness, an important characteristic of the times, is manifested by both Hartlib and Durie, the former stating that his usefulness to the public is the only support of his life, and the latter, that he had dedicated himself to the benefit of the public.

[15] *The Reformed School*, p. 38.

[16] *Ibid.*, pp. 40, 42, 49, 56.

[17] *Ibid.*, pp. 56-7, 68, 81, 85, 87-8.

[18] *The Reformed Library-Keeper*, p. 5. In another passage he says that only those should be made professors who, moved with "generous love of virtue and of profitable Learning," "add unto the Common stock of humane knowledg, that which others have not observed, to the end that all these degrees of Studies and Exercises of the minde of man, beeing subordinate unto the Kingdom of Jesus Christ, the happiness of Man by all Rational and Spiritual waies of improving humane Abilities, may bee advanced unto it's perfection in this life so far as may bee." (P. 7.) The advancement of learning and piety!

[19] P. 2. The expression *ne plus ultra* and the Pillars of Hercules are frequently on the lips of the progressive thinkers of the seventeenth century.

[20] *A Humble Motion to the Parliament of England Concerning the Advancement of Learning: And Reformation of the Universities.* By J. H., 1649.

[21] *Ibid.*, p. 14.

[22] *Ibid.*, pp. 15, 17, 18, 19-22.

[23] *Ibid.*, pp. 25-7.

[24] *Loc. cit.*

[25] *Ibid.*, pp. 29, 35, 39.

[26] *Ibid.*, pp. 41-2.

[27] Pp. 43-4. Another educational reformer, who dedicates his work to Durie and Hartlib as the men who urged him to publish it, expresses much the same ideas as they. He proposes as the chief problem, "how to set-up and set forward the teaching of the most useful and most needful knowledges and learnings, that may do men most good all the daies of their life; that our english youths may no longer bee taught to bee emptie Nominalists and verbalists onely, and to have no knowledg of the necessarie things and matters, that should bee taught in, and by their longsom and toilsom nouns

and verbs; but, by divine blessing, maie henceforth bee *realists* and *materialists;* to know the verie things and matters themselvs, and yet onely such matters as may best further a man for the sufficient doing of all duties and works perteining to his own profession & person; which my wel-meant project, by your ingenuous humanities, beeing then much applauded; you thereupon were verie importunate, that I would either put it in print, or let you have an exemplarie of it." Dedication to *The Right Teaching of Useful Knowledg, to fit Scholars for some honest Profession* .... 1649. Attributed to George Snell.

[28] *Matæotechnia Medicinæ Praxeos. The Vanity of the Craft of Physick. Or, A New Dispensatory. Wherein is dissected the Errors, Ignorance, Impostures and Supinities of the Schools, in their main Pillars of Purges, Blood-letting, Fontanels or Issues, and Diet, etc., and the particular Medicines of the Shops. With an humble Motion for the Reformation of the Universities, And the whole Landscape of Physick, and discovering the Terra Incognita of Chymistrie.* To the Parliament of England. By Noah Biggs, Chymiatrophilos. 1651.

[29] As, for instance, when he says that the Puritans have "upheld and stoutly maintained our liberties, worthy of praises that shall outlive time." (See section 305.) His frequent religious comments also emphasize his Puritanism.

[30] See the dedication.

[31] See section 305.

[32] *Academiarum Examen, or the Examination of Academies. Wherein is discussed and examined the Matter, Method, and Customes of Academick and Scholastick Learning, and the insufficiency thereof discovered and laid open; As also some Expedients proposed for the Reforming of Schools, and the perfecting and promoting of all kind of Science. Offered to the judgements of all those that love the proficiencie of Arts and Sciences, and the advancement of Learning. . . . In moribus et institutis Academiarum, Collegiorum, et similium conventuum, quæ ad doctorum hominum sedes, et operas mutuas destinata sunt, omnia progressui scientiarum in ulterius adversa inveniri. Franc. Bacon, de Verulamio lib. de cogitat. et vis. pag. mihi 14.* 1654. The Dedication is dated Oct. 21, 1653.

[33] *History and Antiquities of Oxford, published in English* by John Gutch, 1796, vol. 2, p. 657. This estimate discovers a touch of malice characteristic of Wood. Webster and his confrères desired a reformation, not the abolition, of learning, and he knew exactly what he wished. His treatise shows him to have been an intelligent believer in the new science and the great discoveries of the sixteenth and seventeenth centuries. It is true that some of the more fanatical Puritans wished to destroy the universities, but the more powerful faction desired only to change the curriculum and pedagogical methods. Webster makes this plain in his preface.

[34] Some idea of the relative degree of influence of the chief writers cited by Webster may be gained from the following numerical comparison: *The Advancement of Learning* is quoted or referred to ten times; the *Novum Organum,* seven times; Van Helmont, four times; Gassendi's *Exercitationes paradoxicae,* four times; Descartes's *Discours de la methode,* three times. Bacon is cited more frequently than all the others put together, but even this fact does not adequately indicate the preponderance of his influence, which is apparent on almost every page. In answer to those who may say that his book is filled with nothing but "imaginary whimseys" like Plato's *Republic,* More's *Utopia,* and Bacon's *New Atlantis,* he replies that the last "might be brought to some reasonable perfection, if the waies and means that he [Bacon] hath prescribed, were diligently observed, and persued; and if these poor lines of mine contained but any treasure comparable to any of their rich mines, I should set an higher Character of esteem upon them, than

now I ought, or they any way merit. And it is true, that supposed difficulty, and impossibility, are great causes of determent from attempting, or trying of new discoveries, and enterprises." (Epistle to the Reader.) Webster's fear of the charge of unoriginality was well founded, for Ward constantly twits him because his ideas were borrowed from Bacon, Gassendi, and Helmont, going so far as to point out parallel passages in Webster's attack on scholastic philosophy and Gassendi's *Exercitationes.* See *Vindiciæ Academiarum,* 1654, p. 32.

35 *Academiarum Examen,* Epistle to the Reader.

36 Attacks on scholastic theology find frequent expression in the treatises under discussion, in which it is characterized in a manner strikingly similar to descriptions of traditional science given by Baconians of this and the following decade. Few, however, equal Webster in intemperate language. "If we," he says, "narrowly take a survey of the whole body of their *Scholastick Theologie,* what is it else but a confused Chaos, of needless frivolous, fruitless, triviall, vain, curious, impertinent, knotty, ungodly, irreligious, thorny, and hel-hatc'ht disputes, altercations, doubts, questions and endless janglings, multiplied and spawned forth even to monstrosity and nauseousness? Like a curious spiders web cunningly interwoven with many various and subtil intertextures, and yet fit for nothing but the insnaring, manacling and intricating of rash, forward, unwary and incircumspect men." (P. 15.) In Webster's eyes the results were terrible, for "from this putrid and muddy fountain doth arise all those hellish and dark foggs and vapours that like locusts crawling from this bottomlesse pit have overspread the face of the whole earth, filling men with pride, insolency, and self-confidence, to aver and maintain that none are fit to speak, and preach the spiritual, & deep things of God, but such as are indued with this *Scholastick,* & mans *idol-made-learning.*" (P. 12.) Not only did the materialistic and utilitarian aspects of the new science recommend the latter to the Puritans, but also the revolt of science against tradition and authority, and especially the fact that in Aristotle, the corrupter both of natural philosophy and of divinity, scientist and Puritan beheld a common enemy, must have influenced favorably their attitude toward science.

37 *Academiarum Examen,* pp. 18-19.

38 *Ibid.,* p. 38.

39 *Ibid.,* p. 33.

40 *Ibid.,* pp. 33-4.

41 *Ibid.,* p. 88.

42 *Ibid.,* pp. 92-3, 98-9.

43 Webster elaborates upon the need of induction, recommending that "the chiefest and most beneficial part [of logic], which is Induction, may be improved, and that it may be serviceable and helpful for the discovering of Science, which cannot be unless some carefull, diligent, and exact means be brought into practice, for the making, trying, and observing of all sorts of experiments, both frugiferous, and luciferous, that time may not be vainly spent in needlesse altercations, disputations, speculations, and notions, but in reall, and profitable experiments, and observations; That so the end of *Logical* labour may not be to bring forth opinion, and errour, but certainty of Science, and solidity of truth." Pp. 102-3.

44 *Academiarum Examen,* p. 103.

45 *Ibid.,* p. 105.

46 *Ibid.,* p. 106.

47 *Vindiciæ Academiarum Containing, Some briefe Animadversions upon Mr. Websters Book, Stiled,* The Examination of Academies. *Together with*

an Appendix concerning what M. Hobbs, *and M.* Dell *have published on this Argument.* Oxford, 1654.

⁴⁸ *Ibid.,* p. 2. In his reply to Hobbes, Ward also denied that the universities were tied to Aristotle, claiming that they had changed since the philosopher had attended Oxford, and he might have said within the last half-dozen years. Hobbes, however, was very sceptical of this sudden transformation. "Was it so then?" he asks. "Then am I absolved, unless you can show some public act of the university made since that time to alter it. For it is not enough to name some few particular ingenuous men that usurp that liberty in their private discourses, or, with connivance, in their public disputations." (Hobbes, *English Works,* ed. Molesworth, VII, 348.) Hobbes is probably right in suspecting that the spirit of this little group of experimenters, who contributed much to the forming of the Royal Society, did not enter very deeply into the university, for certainly the latter very soon after the Restoration lost most of its scientific importance for many years, and even developed an antipathy to the Royal Society. Hobbes, who will be discussed in the next chapter, blamed the universities for their scholastic philosophy, in which only "Aristotelity" was taught, for teaching disobedience to civil authority, and for a divinity corrupted by Aristotelian philosophy. In this last respect the philosopher is in accord with the puritanic view; he makes vigorous use of St. Paul's caveat against "vain philosophy," a pillar of strength in the Puritan arguments, and he refers frequently to protestant reformers. This fact may account for the praise which Webster bestowed upon his political writings in a passage previously quoted, and for the fact that Hobbes is frequently joined with the Puritans by defenders of the universities like Ward and Wilkins. In his answer to Horne, George Kendall says, "I am ready to make it good, that *Aristotle* is opposed more then understood, by a greater Sophy then your self, and one whose *scales* are stronger then most *hornes;* the great *Leviathan,* the very *Dagon* of many young Squires or Squirrels." (*Sancti Sanciti,* 1654, Appendix X, p. 153.) In general, Wilkins and Ward agreed with Webster's and Hobbes's attack, and their answer might be summed up in the sentence: "It was so once, but we have changed it." A number of years later, Clarendon, who admitted that many errors had been introduced into philosophy through Aristotle's authority, defends the universities against Hobbes's charge, citing Wallis and Ward as examples of the new learning. *A Brief View and Survey of the Dangerous and pernicious Errors . . . . in . . . . Leviathan,* second ed., 1676 (written in 1670).

⁴⁹ See *Vindiciæ Academiarum,* pp. 25, 30, 35, 36.

⁵⁰ *Ibid.,* chap. IX and pp. 39, 45, 46.

⁵¹ *Ibid.,* p. 49.

⁵² *Histrio-Mastix,* pp. 210, 214, 215. The title springs from the fact that Hall confused his opponent with the dramatist of the same name.

⁵³ For another severe attack upon the learning taught in the universities and for a plea for experimental science see George Starkey's *Natures Explication,* 1657. (Thomason's date is Jan. 10, 1656.) Starkey gives the same picture of the traditional university training presented by Hall, Webster, Biggs, and others, a picture which, since painted by so many, must not have been far wrong. Starkey's account is perhaps the clearest and most comprehensive:

"He then who inclines to ingenuity, soon after he hath some insight into the Tongues by education at the Grammar School, he is thence transmitted to the Academy to be further cultivated. Where he learns Logically to dispute according to the Rules of *Aristotle;* and withall, exerciseth his Oratorial faculty by declamations in turnes; after that he proceedeth to read Physiology according to *Aristotles* Doctrine, which is disputed in publick *pro* and *con* with a great deal of dexterity. Ethicks are added to these studies at spare times, with the principles of Metaphysicks and the Mathematicks in general, and some things are performed in *Hebrew* and *Greek* Studies. Thus at the end of four years upon performing of publick declamations, disputations, and the like, the initiatory title of Batchelor of Arts is bestowed as a crown of their industry; from which time till the end of three years, they are to be employed in epitomizing the seven Liberal Sciences, and reducing them into Systemes and Synopses, and then with a great deal of honour they are declared Masters of the said Sciences, and have liberty granted them to apply themselves to the practise of any one or other of them, when ever called unto it.

And now the youth stands as a staffe set up, waiting which way it will fall: for this testimony concerning them hath declared them to all the world to be fit to profess any thing.

And (not to speak of the abuse of Divinity-profession, which is permitted upon these termes) if the genius of this Arts master (who hath drunk himself drunk of liberal Sciences) stand toward the profession of Physick, (which he may begin to think of when he is Batchelor in Arts) then are there publick professed Lecturers, whom he may almost daily hear declaming on that subject; which Lectures if he then attend diligently, and shall peruse Anatomists, and read *Galen* over (at least cursorily) and collect notes out of him and *Avicen*: if he turn over the Herbals, and learn to know some plants by name and sight, and in the mean time in the publick Halls shall dispute concerning the use of parts, the generation of man, the Elementall quality of some things, as namely, Camphire, Quicksilver and the like, (which two simples have, after the discharging of some hundreds, not to say thousands of *Ergoes*, first and last, out of the Canons of *Aristotle* and *Galen*, more notably puzled the Antagonists (who to this day know not on what side to determine) then ever the Devil was puzl'd to finde the meaning of *Aristotles Entelecheia*) after I say these notable performances, he shall be licensed a Practitioner, and dub'd a Doctor." *Natures Explication,* pp. 19 ff.

⁵⁴ An anonymous treatise entitled *A Vindication of Learning From unjust Aspersions,* 1646, p. 2, claims that the doctrine *"That men of learning do lesse esteeme of Christ, then the simple & ignorant"* "is growne Epidemicall among the Sectaries and some others that professe learning within the City of London."

⁵⁵ *History and Antiquities of the University of Oxford,* 1796, vol. 2, p. 657. If Wood is correct in saying that Parliament engaged certain men, among them John Webster, William Dell, and John Horne, to write against the universities, we can consider Webster's ideas as more or less representative of the Puritan government, and so the favorable attitude of the Puritans toward Baconianism appears more clearly.

⁵⁶ Besides Webster's treatise there appeared during the years 1653-54 several attacks on the universities: John Horne's *A Consideration of Infant Baptism;* and *The Stumbling-Stone; A Plain and Necessary Confutation of divers gross . . . . Errors; A Testimony from the Word Against Divinity-Degrees;* and *The Right Reformation of Learning, Schooles and Universities according to the State of the Gospel,* all by William Dell. Defences were published by, amongst others, Robert Boreman, Joseph Sedgwick, Thomas Black, George Kendall, Edward Waterhouse, and a year or two later, Edward Leigh and Edward Reynolds. In 1660 Henry Thurman came out with a belated defence.

⁵⁷ Wood, *op. cit.,* pp. 680, 683, 696-97.

⁵⁸ Edward Reynolds concedes to the Puritans that "it is well observed by a learned man that *School-men* and *Canonists* have been the fountains of that corruption which hath infected the Church of Christ . . . of which the main cause hath been the admitting of *Aristotle* and his Philosophy, *In ipsa*

adyta et penetralia Ecclesiarum, as *Hospinian* speaketh." *A Sermon Touching the Use of Humane Learning,* 1658, p. 20. (Thomason's date is Nov. 20, 1657.)

⁵⁹ That this was the position of the more important elements among the Puritans, and that it was not the abolition but the reformation of learning which they sought is revealed in William Dell's *Stumbling-Stone,* 1653, p. 27: "If the *Universities* will stand upon an *Humane* and *Civil account,* as *Schools* of good *Learning* for the instructing and educating Youth in the knowledge of the *Tongues,* and of the *liberal Arts and Sciences,* thereby to make them *usefull* and *serviceable* to the *Commonwealth,* if they will stand upon *this account,* which is the *surest* and *safest Account* they can stand on, and will be *content* to *shake hands* with their *Ecclesiastical* and *Antichristian Interest,* then let them stand, during the *good pleasure* of God." See also his "Right Reformation of Learning," contained in *The Tryal of Spirits,* 1653, in which he grants that secular learning has its place in human affairs, though not essential or helpful to the ministry. He recommends, besides the teaching of English and keeping the students busy, the erection of schools and universities throughout England, even going so far as to advocate a college in every town in England, where students might both study and work. See also the conclusion of John Horne's *Consideration of Infant Baptism,* 1654. Foster Watson says, "It may be at once said that the assumption that Cromwell's government had any ill-will against the universities and schools, or attempted to damage them, is entirely opposed to the facts." *The English Historical Review,* XV (1900), 65.

⁶⁰ Milton's essay *Of Education* should be read in the light of these early Puritan ideas on education.

⁶¹ Perhaps the most interesting example of the new valuation placed upon artisans and upon mechanical arts as objects of study is afforded by John Wilkins' *Mathematicall Magick,* 1648. The author tells a story concerning Heraclitus, who, when his followers hesitated to follow him into a tradesman's shop, remarked that the immortal gods were not absent from such places, "Intimating that a divine power and wisdome might be discerned even in those common arts, which are so much despised; And though the manuall exercise and practise of them be esteemed ignoble, yet the study of their generall causes and principles, cannot bee prejudiciall to any other (though the most sacred) profession." ("To the Reader.") Besides the delight, he remarks, which every rational mind must find in notions that carry their own evidence and demonstration, there is much "real" benefit in teaching machinery to men interested in estates. The second chapter is entirely devoted to defending his estimate of mechanical arts against the contempt of the upholders of the liberal arts: "According to ordinary signification, the word [mechanic] is used in opposition to the liberall arts: whereas in propriety of speech those employments alone may be styled *illiberal,* which require onely some bodily exercise, as manufactures, trades, etc. And on the contrary that discipline, which discovers the generall causes, effects, and properties of things, may truly be esteemed as a *species* of Philosophy." He divides mechanics into the rational, which treats of principles underlying practice, and the "Cheirurgicall or Manuall," which refers to the making of instruments and particular experiments. The first is liberal and worthy of an ingenious mind, and, being based upon mathematics and natural history, proves most useful and beneficial to navigation, husbandry, and the like. "Those other disciplines of Logick, Rhetorick, etc. doe not more protect and adorn the mind, then these Mechanicall powers doe the body. And therefore are they well worthy to be entertained with greater industry and respect, then they commonly meet with in these times; wherein there be very many that pretend to be masters in all liberall arts, who scarce understand any thing in these particulars." Pp. 8-13.

⁶² *The Relief of the Poore: And Advancement of Learning Proposed.* By Humphrey Barrow, (during the War) a Member of the Armie. 1656.

This is an appeal to Cromwell for funds in behalf of the University of Dublin and for a hospital for orphans in Ireland. In wording his title the author may very well have been influenced by the relationship under discussion, even though the pamphlet does not reveal it.

⁶³ John Wallis was appointed Savilian professor of geometry, 1649; John Wilkins, warden of Wadham College, 1648; William Petty, professor of anatomy, 1651; and Jonathan Goddard, warden of Merton College, 1651. Seth Ward was made Savilian professor of astronomy in 1649. All but the last were undeniably Puritans.

⁶⁴ C. R. Weld, *History of the Royal Society*, I, 53.

⁶⁵ *Journals of the House of Commons*, VII, 287.

⁶⁶ See the dedication to his *True and Readie Way to Learne the Latine Tongue*, 1654. Interest in scientific schemes was also carried over into Richard Cromwell's short reign. Hartlib writes Boyle, Dec. 16, 1658, "I am wondrous glad, that you have written of the present protector's intentions for countenancing and advancing of universal useful learning in due time." (Boyle, *Works*, ed. T. Birch, V, 282. Cf. Oldenburg's letter to Boyle, written a month or two earlier. *Ibid.*, V, 300.) Hartlib also has in mind Comenius' pet project of Pansophia.

⁶⁷ See in Calendar of State Papers, Domestic Series, ed. Mary A. E. Green, entries for Jan. 14, 1645 (p. 251); June 8, 1649 (p. 177); Aug. 17, Oct. 28, Nov. 7, 1650 (pp. 292, 401, 418); May 20, Oct. 6, Nov. 20, 1651 (pp. 208, 468, 27); Mar. 10, Nov. 12, 15, 22, 1652 (pp. 173, 488, 492, 502); Feb. 21, Mar. 18, Apr. 4, July 27, Oct. 18, 1653 (pp. 487, 489, 254, 53, 206); Mar. 21, Apr. 27, June 12, July 11, Aug. 1, 1654 (pp. 44, 124, 448, 249, 281); Aug. 29, 1655 (p. 303); Aug. 7, 1656 (p. 66); Apr. 20, 1660 (p. 598). Theodore Haak, who figured prominently in the first group of London experimenters, and Peter du Moulin, a sincere Baconian and later defender of the Royal Society, were granted pensions. See entry for July 16, 1649.

⁶⁸ In his excellent biography of Sydenham, one of a series entitled *Masters of Medicine*, J. F. Payne pays to the Puritans the only recognition which I have been able to discover, of their espousal of the new science. "As the movement [the new science] has been strangely called 'Anti-Puritan,' " he says, "it is necessary to draw attention to the strongly Puritan cast of the original group, and of most of the Oxford Society." He concludes in this fashion: "At all events it is quite clear that the speculative and scientific Renascence, sometimes put to the credit of the Restoration, started, both in Oxford and London, under the Commonwealth and the Protectorate." *Thomas Sydenham*, 1900, pp. 58, 60, 62-3, 66.

CHAPTER VI

¹ *A Discourse Concerning a New Planet*, 1640, in *Mathematical and Philosophical Works*, 1708, p. 146.

² The first free thinker to feel his wrath was Sir Kenelm Digby, who was enough indebted to scholastic philosophy as it was, but who in *Of Bodies*, 1644, had to some extent embraced the atomic philosophy revived by Gassendi. To Ross the "Peripatetick verities" are the solid meats and the wholesome food of the universities, "dressed in *Aristotle's kitchin*," and far superior to the frogs and mushrooms, served with French sauce (Gassendi and Descartes) which Digby provided. (*The Philosophicall Touch-Stone*, 1645, dedication.) He ridicules Digby's atoms and speaks proudly of those "who have been bred in the *peripatetick* schooles, at feet of *Aristotle*." (*Ibid.*, p. 23.) He exalts "*Aristotle* . . . . and his learned *Peripateticks*," "Though we have not sworne to defend *Aristotle* in all his Dictats." (*Ibid.*, pp. 45, 60.) He defends the Stagirite against Digby's accusation of being

a weak logician, and blames the latter for making "the *Aristotelians* speak absurdities of your owne invention." (*Ibid.*, pp. 61, 62.) Finally, he applies the stock argument of conservatism: "If you were not a Gentleman, whom, for your good parts, I honour, I could say, that . . . . you are for being too *busie* in jeering at such eminent men, and at those Maximes which have been so unanimously received by all Universities, and for so many hundred yeares constantly maintained; but your worth and my modestie enjoyne mee silence, and restrain my pen from recrimination." *Ibid.*, pp. 62-3.

³ *The New Planet, no Planet*, 1646, pp. 1-3.

⁴ *Arcana Microcosmi*, 1652, Epistle Dedicatory.

⁵ *Ibid.*, p. 207.

⁶ *Ibid.*, dedication and pp. 263 ff. The next year Ross laid out Hobbes, and in less respectful manner defended the schoolmen and scholastic philosophy against the "ignorance" of the swashbuckling philosopher on the ground that such philosophy was essential to the refutation of heretics and infidels, "for by Philosophy contradictions and absurdities are avoided, into which those ignorant souls do fall, who want it, as we finde at this time by woful experience, there being more absurd and contradictory opinions among the people of this Nation, now, in a few years since Aristotles doctrine hath been discouraged, then were all the time hitherto since Christianity was imbraced." (*Leviathan Drawn out with a Hook*, 1653, pp. 17-18.) He is pointing an accusing finger at the Puritans for the discouragement Aristotle's philosophy was receiving. Yet he bears testimony to the fact that they had not much weakened its hold upon the universities, for he says that gratitude should be felt for "the paines and industry of those brave men, which have inriched us with such monuments of learning, which the Universities of the world have received, and do to this day cherish and maintain with such applause; and not to requite them with scorn and contempt." *Ibid.*, p. 82.

⁷ See the preface to *The Anatomy of Urania Practica*, 1649.

⁸ *Ibid.*, pp. 10, 32.

⁹ A poem prefixed to Thomas Vaughan's *Magia Adamica*, 1650, reads in part as follows:

> Had *Bacon* liv'd in this *unknowing Age*
> And seen *Experience* laugh'd at on the *Stage*,
> What *Tempests* would have risen in his *Bloud*,
> To *side* an *Art*, which *Nature* hath made *Good?*
>
> . . . .
>
> Tell mee in earnest do'st thou think 'tis fit
> To *believe all* that *Aristotle* writ?
> Though he was *blinded*, yet *Experience* can
> Sever the *Clouds*, and make a *Clearer man*.

¹⁰ Elias Ashmole, *Theatrum Chemicum*, 1652, p. 445. Ashmole probably obtained his definition from Hardick Warren's *Magick and Astrology Vindicated*, 1651, p. 4. Some years later John Wagstaffe defined magic as "skill in Natural Philosophy." *The Question of Witchcraft Debated*, 1669, p. 3. See also pp. 59, 66.

¹¹ *Anthroposophia Theomagica*, 1650, preface.

¹² *Ibid.*, pp. 62-3.

¹³ *Medicina Magica Tamen Physica: Magical, but Natural Physick*. 1656, p. 97.

¹⁴ *Ibid.*, p. 179.

¹⁵ *Magick and Astrology Vindicated*, p. 9.

[16] *Anima Magica Abscondita*, p. 8.

[17] *Magia Adamica*, 1650, p. 86.

[18] *The Man-Mouse Taken in a Trap*, 1650, p. 38.

[19] The upholders of astrology claimed an experimental basis for their pseudo-science; so much so in fact that Gassendi in his attack on judicial astrology found it necessary to dispute their claim. (See English translation under the title, *The Vanity of Judiciary Astrology*, 1659.) John Allen (*Judicial Astrologers Totally Routed*, 1659, pp. 18-19) noted that the chief argument of the astrologers was based on the theory that the stars were not signs but the natural causes of events, "and indeed what they urge concerning the Sun and Moon, seems so plausible, that judicious men at first are drawn diligently to listen unto what they say concerning that, thinking they would proceed to prove the rest of their suppositions with the like evidence, not suspecting that upon such specious foundations they should so soon erect nothing but ridiculous Fables and wild absurdities." Certainly some of the astrologers clung tenaciously to the band wagon of science, and in some instances paid allegiance to new scientific ideas not generally accepted. One says, "How rashly doe the ignorant and common sort tax us for maintaining the truth and verity of the *Copernican Systeme* of the World . . . . that rare discovery of the truth now so cleerly proved by undeniable Demonstrations *Geometrical*." (George Atwel, *An Apologie*, 1660, "To the Reader.") Vaughan characterized the Ptolemaic system as "a Rumbling confused Labyrinth." *The Man-Mouse Taken in a Trap*, p. 52.

[20] When Vaughan says that Aristotle's spirit is the spirit of error and produces only notions; that his followers merely refine old notions; and that mystery lies in the terms used, for if the ideas were plainly expressed, the sense would appear empty, he is one with Bacon, Glanvill, and Sprat, to mention no others. See *Anima Magica Abscondita*, p. 6.

[21] *Magia Adamica*, pp. 3, 4.

[22] *Anima Magica Abscondita*, p. 5.

[23] *Pseuchographia Anthropomagica*, 1652, pp. 5-6.

[24] *Ibid.*, pp. 21 ff.

[25] In the dedication to the King of his *Problemata Physica*, 1662, he speaks of "meditationum mearum Physicarum partem maximam et probabilissimam," where we would in other scientists expect the word "experimentorum," instead of "meditationum." This is probably the reason that he made no contribution to scientific knowledge, but, on the other hand, frequently offered false and far-fetched explanations of pnenomena.

[26] Epistle Dedicatory, dated April 23, 1655, of *Elements of Philosophy*, 1656.

[27] *Leviathan*, ed. A. R. Waller, 1904, p. 80.

[28] See the following articles of the present writer: "Science and English Prose Style in the Third Quarter of the Seventeenth Century," *PMLA*, XLV (1930), 977-1009; "The Attack on Pulpit Eloquence in the Restoration," *Jour. of Eng. and Ger. Philol.*, XXX (1931), 188-217; and "Science and Language in England of the Mid-Seventeenth Century," *ibid.*, XXXI (1932), 315-31. These may now be found in *The Seventeenth Century*, by Richard F. Jones *et al.*, 1951.

[29] *Questions Concerning Liberty, Necessity, and Chance*, in *English Works of Thomas Hobbes*, ed. Sir William Molesworth, 11 vols., 1839-45, V, 268.

[30] *Ibid.*, p. 299.

[31] *Ibid.*, p. 359.

[32] For an important contemporary estimate of the part Hobbes played

in loosening the fetters of antiquity, see Abraham Cowley's ode to him, in which the philosopher is called the "great Columbus of the golden lands of new philosophies." Bishop Bramhall, who stood firmly on the foundation of authority and tradition, expressed great horror at the disrespect which Hobbes showed the past: "It is strange to see with what confidence now adayes particular men slight all the Schoole-men, and Philosophers, and Classick Authors of former ages, as if they were not worthy to unloose the shoe-strings of some moderne Author." (*A Defence of True Liberty*, 1655, p. 20.) He always reveals great admiration for the schoolmen and Aristotle. Evidence that the spirit of the age was making itself felt is again manifested by a passage in the preface of a work by another of Hobbes's numerous adversaries (William Pike's *Examinations, Censures, and Confutations of . . . . Leviathan*, 1656) in which the popularity of the Leviathan is attributed, "first to the *Genius* that governs this age, in which all learning, with Religion, hath suffered a change, and men are apt to entertain new opinions in any Science, although for the worse." Pike is a loyal defender of the ancients: "I think it enough for me, in most places, to confute his Doctrine, and set down the various opinions of my old acquaintance, *Aristotelean* Philosophers, which I learn'd in the contemn'd Universities." *Loc. cit.*

33 Long before this time Aristotle had found opponents aplenty on the continent. Paracelsus, Cardan, Telesio, Vesalius, Bruno, Caesalpinus, and Campanella opposed the authority of the ancients and insisted upon knowledge based upon the evidence of the senses. Pierre de la Ramèe, holding that Aristotle's logic was unsuitable for the discovery of truth, produced a rival logic which created no little stir. Gassendi perhaps exerted more influence than any other foreigner upon the anti-Aristotelian movement in England. For other anti-Aristotelians see the introduction to Thomas Fowler's edition of the *Novum Organum*, pp. 72 ff.

34 All quotations are from the preface.

35 *Mataeotechnia Medicinæ Praxeos*, 1651, section 23.

36 *Ibid.*, section 59.

37 *Ibid.*, section 18.

38 *Ibid.*, section 28.

39 *Ibid.*, section 42.

40 *Ibid.*, sections 60, 72, 80.

41 *Ibid.*, section 305.

42 The following is a typical passage: "The *Apollinian* science then, or *art* of *Physick*, is every where brought upon the Stage; and made the laughing-stock of the sick-brain'd vulgar; because *Physitians*, who have *heterodogmatiz'd*, and deviated from the ancient beaten path of clear reason and experience, put no distinction between the venerable grey-haires of *ancient* Physick, and them who weare her honourable silver livery, from the old *scurff* of *Galen* and his accomplices, benighted to the clouds of *ignorance*, and that *Tatterdemalion Linostema* of Peripatetical & Galenical *predicaments* of *qualities;* whereby, to heads of a larger size, they seem to have put out their own eyes, and willingly subject themselves, like mill-horses to grind in the Mill-house of *custome* and *Tradition*, and aforehand, to have stak'd themselves to a resolution to confine to the Custome of the Schools, and sit down to a precise Conformity, to lap up the prodigious vomits of *Aristotle*, *Galen* and other *illiterate* Ethnicks, and in effect to prescribe all the heads of the present age, as *Pupils* to the dull and doting *advisoes* of the ancient, precedent *Paper-stuffers;* and then no lesse to say, as in supernaturall things they are wont, so in naturall, to make it a kind of blasphemy, at least presumption, to step one haires breadth from the cry'd up and

vulgar receiv'd way; *So hardly in good sooth can the dotage of those who dwell upon antiquity, allow present times any share of wisdome or skill.*" Section 52.

43 *Natures Explication and* Helmont's *Vindication. Or a short and sure way to a long and sound Life: Being a necessary and full Apology for Chymical Medicaments, and a Vindication of their Excellency against those unworthy reproaches cast on the Art and its Professors (such as were* Paracelsus *and* Helmont) *by Galenists, usually called Methodists. Whose Method so adored, is examined, and their Art weighed in the ballance of sound Reason and true Philosophy, and are found too light in reference to their promises, and their Patients expectation. The Remedy of which defects is taught, and effectual Medicaments discovered for the effectual cure of all both Acute and Chronical Diseases.* By George Starkey, a Philosopher made by the fire, and a professor of that Medicine which is real and not Histrionical. 1657.

44 *Ibid.,* pp. 16 ff.

45 *Ibid.,* pp. 34 ff.

46 ". . . . for who knowes not the mighty force of education, which being once suckt in . . . is so lodged that it is with much difficulty eradicated; yea, and although an opinion to an uningaged person seem never so absurd, yet to one whom education hath ingaged, it appears not so, yea acuteness makes little to the discovering the weakness of such an opinion, but rather supplies curious and specious arguments to maintain it, and to oppose any contrary." *Natures Explication,* p. 27.

47 *Ibid.,* pp. 50-51.

48 *Loc. cit.*

49 *Ibid.,* p. 57.

50 *Ibid.,* p. 148.

51 *Ibid.,* p. 91.

52 *Ibid.,* p. 224.

53 Chap. VI, "Of Scholastick Philosophy," pp. 52-84.

54 *Academiarum Examen,* p. 82.

55 *Ibid.,* p. 67. On another page he contrasts the Baconian and traditional methods of ascertaining truth: "This *School Philosophy* is altogether void of true and infallible demonstration, observation, and experiment, the only certain means, and instruments to discover, and anatomize natures occult and central operations; which are found out by laborious tryals, manual operations, assiduous observations, and the like, and not by poring continually upon a few paper Idols, and unexperienced Authors: As though we could fathome the Universe by our shallow imaginations, or comprize the mysteries of mother nature in the narrow compass of our weak brains; or as though she would follow us into our Chambers, and there in idlenesse communicate her secrets unto us; no verily, ... ease and idlenesse are not the way to get knowledge." (*Ibid.,* p. 68.) He stresses the importance of chemistry, claiming that one year's experimenting in it would produce more real fruits than the study of Aristotelian philosophy has done in many centuries: "O that the Schools therefore would leave their idle and fruitless speculations, and not be too proud to put their hands to the coals and furnace, where they might find ocular experiments to confute their fopperies, and produce effects that would be beneficial to all posterities . . . . Neither despise this counsel because the author is no friend to *Aristotle,* nor be ashamed to cast off thy fine clothes to work in a laboratory, for without this thou mayest wax old in ignorance, and dy with guilt, because thou hast served thy generations with no better stuff than *Aristotles* dreames, and *Scholastick* fables." *Ibid.,* p. 71.

56 *An Elegant and Learned Discourse Of the Light of Nature, With several other Treatises*, 1652.

57 *Ibid.*, pp. 157-66.

58 *Ibid.*, p. 159.

59 *Morall Discourses and Essayes, Upon Severall Select Subjects*, 1655, pp. 60 ff.

60 *A Miscellany of Sundry Essayes, Paradoxes, and Problematicall Discourses, Letters and Characters*, 1659, p. 56.

61 *Ibid.*, pp. 60, 75. An interesting example of the way in which modern discoveries were undermining respect for the ancients, which will receive fuller treatment in the next chapter, is found in a treatise entitled *Diatribae duae Medico-philosophicae*, 1660, by Thomas Willis, Sedleian Professor of Natural Philosophy, who held that ancient medicine had been entirely overturned by Harvey's discovery of the circulation of the blood.

62 *Exercitationes Duæ: Altera Theologica, De Presbyteris & Episcopis: Altera Academica, De Philosophia veterum, ejusque Usu; una cum duabus Orationibus ejusdem Argumenti*, 1661.

63 See William Haller's "Before Areopagitica," *PMLA*, XLII, 875-900.

64 *The Vanity of Dogmatizing: Or Confidence in Opinions. Manifested in a Discourse Of the Shortness and Uncertainty Of Our Knowledge, And its Causes; With some Reflexions on Peripateticism; And an Apology for Philosophy*. 1661. This has been rendered accessible by the Facsimile Text Society.

65 The explanations which Glanvill gives for human ignorance (chaps. VII-XV), both in content and tone, read remarkably like those given by Bacon for the deficiency of knowledge: the unreliability of the senses (see *Novum Organum*, I, 50); the subtlety of nature (*ibid.*, I, 10); the quickness of the mind to embrace conclusions, together with the need of suspended judgment (*Advancement of Learning*, III, 293); the fallacious nature of the imagination (*ibid.*, III, 289, and *Nov. Org.*, I, 47); and the influence of the affections—custom, education, interest, etc.—on belief (*Nov. Org.*, I, 42, 49). But nowhere is Bacon's influence more clearly revealed than in chapter XV, which deals with the injury done the advancement of learning by submission to the authority of antiquity, and in which Glanvill uses the same arguments and some of the same figures of speech as his illustrious predecessor. (Bacon, *Works*, III, 289-90.) He makes use of the latter's aphorism that the antiquity of time is the youth of the world, and refers to him as "the Noble Lord Verulam," "that great man," and "The foremention'd noble *Advancer of Learning*, whose name and parts might give credit to any undertaking." *The Vanity of Dogmatizing*, p. 146.

66 It is worth noting that his joining the Society produced a decided change in his style from an imitation of the luxuriant style of Sir Thomas Browne to the plainness and simplicity demanded by the scientists. See *PMLA*, XLV, 989-98.

67 *Vanity of Dogmatizing*, pp. 146-47.

68 *Ibid.*, p. 140.

69 *Ibid.*, pp. 178, 181-83. The mathematician Thomas Baker ridiculed these prophetic visions, which do not seem so ridiculous to us today. See *Reflections upon Learning*, 1699, pp. 84-5.

70 *Vanity of Dogmatizing*, p. 152. See also pp. 166-68.

71 In arguing that all phenomena are related, Glanvill anticipates Blake and Tennyson: "And would we *know* but the most contemptible *plant* that grows, almost all things that have a being must contribute to our *knowledge*." *Vanity of Dogmatizing*, p. 214.

[72] *Ibid.*, pp. 239-40.

[73] Glanvill's book was favorably received by the liberal thinkers of the period, though his enthusiasm was not entirely appreciated (see John Worthington's *Diary*, ed. J. Crossley, April 19, 1661, vol. 1, pp. 299-301), but it drew fire from the conservatives. Two years after its publication, Thomas White, a Catholic priest much given to controversy, answered it in *Sciri Sive Sceptices et Scepticorum a Jure Disputationis Exclusio*, which the same year was englished under the title *An Exclusion of Scepticks From all Title to Dispute*. White defends Aristotle and the ancients, claiming that Glanvill's charges were pertinent only to some modern Aristotelians. What is of more interest is his attack on the new science for being manual and mechanical, a charge frequently asserted at this time. "His [Glanvill's] nineteenth Chapter inveighs against *Aristotle's* Doctrine as unfruitful and barren; but, weakly and falsely. Weakly, because all the inventions he speaks of belong to Artificers and Handy-craft-Men; not Philosophers, whose office 'tis to *make use of* Experiments for Science, not to make them." (*Ibid.*, p. 73.) Science, he claims, can only be learned through metaphysics, for which we must go to Aristotle; and he employs metaphysics to show that Boyle's discoveries as regards the possibility of a vacuum and the pressure of gases are wrong. *Ibid.*, p. 77.

[74] See T. Baker's *Reflections upon Learning*, 1699, pp. 76-78.

[75] H. Rigault, *Histoire de la Querelle des Anciens et des Modernes*, 1856.

CHAPTER VII

[1] Evelyn was under the very definite impression that Hartlib was a Lithuanian who came to England to escape persecution. (See *Diary*, ed. Bray, Sept. 27, 1653, III, 391, and letter to William Wotton, Sept. 12, 1703, III, 391.) He also states that Hartlib recommended himself to many, including Robert Boyle, by his industry and by his communicating many secrets in chemistry, improvements in agriculture, and "other useful novelties," as well as by his general correspondence abroad; and that he was readily taken up by the Puritans. As early as 1639 he was searched as a Puritan for seditious papers. See G. H. Turnbull, *Samuel Hartlib*, 1920, p. 16.

[2] *The Dictionary of National Biography* gives 1628. Professor G. H. Turnbull believes Hartlib was at Cambridge University as early as 1621. *Hartlib, Dury and Comenius*, 1947, p. 15.

[3] Robert Boyle, *Works*, ed. T. Birch, 1744, I, 21-2. All references to Boyle's works are to this edition.

[4] Henry Dircks, *A Biographical Memoir of Samuel Hartlib*, 1865, pp. 10-12. To show that an attempt was made by the Puritans to secure Hartlib a position at Oxford, Dircks quotes from the Journals of the House of Commons, Mar. 31, 1647, as follows: "*Ordered*—That it be referred to the Committee for the University of Oxon, to take into their especial consideration the deserts of Mr. Hartlib, both from this Parliament, and from all that are well-wishers to the advancement of learning: and to recommend him to some place of benefit in the University of Oxon; where he may have an encouragement, and future supply." This is additional evidence, to what has been said about the appointment of Wilkins, Wallis, Petty, and Goddard to positions at Oxford, of Puritan interest in the new science. Theodore Haak, another sedulous advancer of experimental science, also received a pension from the Puritan Parliament "for the Advancement of Arts and Learning." (Turnbull, *Samuel Hartlib*, p. 49.) Both Hartlib and Haak naturally lost their pensions at the Restoration.

[5] Some idea of Hartlib's multifarious and untiring activities and interests may be gained from his correspondence with Robert Boyle. See especially

Boyle's *Works*, V, 256, 257, 261, 264, 268, 271, 272, 278, 279, 281, 282, 283, 286. Professor Turnbull gives much valuable information regarding Hartlib's activities in the two books cited above.

6 G. H. Turnbull, *Samuel Hartlib*, p. 62.

7 Cf. the preface to John Worthington's *Diary*, ed. J. Crossley (Chetham Society, vols. XIII and XXXVI). All references to this diary are to this edition.

8 After stating that he wrote the treatise at Hartlib's earnest entreaty, Milton says, "Nor should the laws of any private friendship have prevail'd with me to divide thus, or transpose my former thoughts but that I see those aims, those actions which have won you with me the esteem of a person sent hither by some good providence from a far country to be the occasion and incitement of great good to this Island. And, as I hear, you have obtain'd the same repute with men of most approved wisdom, and some of highest authority among us. Not to mention the learned correspondence which you hold in forreign parts, and the extraordinary pains and diligence which you have us'd in this matter both here, and beyond the Seas; either by the definite will of God so ruling, or the peculiar sway of nature, which also is Gods working." The utilitarianism in Milton's tractate, as well as the defence of philosophic freedom in *Areopagitica*, is another manifestation of the relationship between Puritanism and Baconian science, upon which we have been insisting. Though Milton's ideas on education are loftier and broader than those of his contemporaries, his spirit is not far removed from that of Durie, Hartlib, Petty, and others.

9 Boyle, *Works*, V, 265.

10 *Ibid.*, V, 258, 263.

11 *Ibid.*, V, 270, Jan. 7, 1658; 272, Feb. 2, 1658.

12 *Ibid.*, V, 278-79, June 1, 1658; 281, Sept. 14, 1658; 282, Dec. 16, 1658.

13 *Considerations Tending to the Happy Accomplishment of Englands Reformation in Church and State*, 1647, pp. 47, 50-51.

14 Worthington, *Diary*, Sept. 5, 1661.

15 *The True and Readie Way to Learne the Latine Tongue*, 1654, p. 1.

16 See Cressy Dymock's *A Discoverie for Division . . . . of Land*, 1653, p. 16, and *An Invention of Engines of Motion*, 1651. The latter is headed by a two-page translation from Bacon's *Novum Organum* extolling the virtues of invention. Both treatises are generally listed under Hartlib.

17 A full description is given in Hartlib's *Considerations Tending to the Happy Accomplishment*. Though, as Professor Turnbull has shown (*Hartlib, Dury and Comenius*, p. 80) Hartlib got the idea of his project from Theophraste Renaudot's *Bureau D'Addresse*, he made it his own.

18 *A further Discoverie of The Office of Publick Addresse For Accommodations*, 1648. This would seem to be by Sir William Petty, with whom Hartlib was very closely associated; in fact, it is possible that Hartlib had him in mind for some position in the Office. There is, however, a passage in one of Boyle's letters to Hartlib, May 8, 1647, that points to Durie as the author. (Boyle, *Works*, I, 24.) This letter reveals Boyle's interest in, and close association with, Durie and Hartlib. See Dircks, *A Biographical Memoir of Samuel Hartlib*, pp. 19-20.

19 The humanitarian spirit, one of the important justifications of the scientific program, always burns brightly in Hartlib. On Sept. 14, 1658, he writes Boyle, "I do not remember to have told you of the like charitable design to be founded at *Durham* [in 1656 Hartlib had been appointed to the Committee of the Puritan college at Durham. Turnbull, *Hartlib*, p. 63]; I

mean of a charitable physician, or laboratory, for the poor." Boyle, *Works*, V, 281.

[20] *A further Discoverie*, pp. 24-5.

[21] *Considerations Tending to the Happy Accomplishment*, p. 45.

[22] *Ibid.*, p. 47.

[23] *Ibid.*, pp. 50-51.

[24] It seems that ten years later some headway was made toward realizing this ambitious scheme, for in the latter part of May, 1657, appeared a sheet entitled *The Publick Adviser, Weekly Communicating unto the whole Nation the several Occasions of all persons that are any way concerned in matter of Buying and Selling, or in any kind of Imployment, or dealings whatsoever; according to the intent of the Office of Publick Advice newly set up in several places, in and about* London *and* Westminster. *For the better Accommodation and Ease of the People, and the Universal Benefit of the Commonwealth in point of Publick Intercourse.* The preface contains the information that "the Undertakers of the *Office of Public Advice* have already declared at large their whole Design in the management of this new way by them found out for the Advantage of the Commonwealth and did to that purpose, on the 14 instant, set forth a Printed Sheet (which is to be had at the several Book sellers shops in *London*) containing all the particulars of the said Design, wherein they engaged themselves to appear this day thus in Print, and so weekly from this day forward, upon every Tuesday." The paper contains information concerning the departure and sale of ships, houses to be let or sold, persons wishing houses, physicians with advertised cures, lessons in hair-dressing and shorthand, persons wishing employment, wet and dry nurses, board and lodging, places and times of departure of stage-coaches and carriers, and stolen property. At the end are given the addresses of eight branch offices, which fact, together with the notice contained in the title, indicates that Hartlib's scheme had actually become operative. The last number of the Publick Adviser, no. 8, appeared July 13, but it was followed the next week by *The Weekly Information From the Office of Intelligence Established in severall places in and about the Cities of* London *and* Westminster, *by Authority granted under the Great Seal of England, and conferred upon* Oliver Williams. *Communicating Intelligence to all Merchants, Tradesmen, and all other Persons concerned in buying, selling, or any ways disposing of any Lands, Houses, Ships, or Goods, and of any other Imployment, Businesse or Occasion whatsoever relating to this Office; For the Propagation of Trade, the general Benefit of the Nation, and the Accommodation of every particular person.* The preface states that Oliver Williams issued on June 29 a sheet describing the method of the "Office of Intelligence, as it may be most advantagious to the Common-Wealth according to the best experience gained from forraign Parts." At the end are given six addresses of offices of intelligence. Though the information given by the *Publick Adviser* shows it to be an outgrowth of Hartlib's Office of Public Address, one cannot state certainly that he had anything to do with it. It seems clear, however, that the first venture was cut short by Williams' stealing the idea and getting governmental sanction for his activities.

[25] *A further Discoverie*, p. 2.

[26] The Puritan spirit is as evident in Hartlib as the Baconian; in fact, they are so blended as almost to appear one and the same. "The Advancement of Learning and Piety," with which we have become acquainted in an earlier chapter, is also a slogan in the treatises discussed in this chapter. In the way described below Hartlib's project is to fulfill a Puritan purpose. After speaking of the duty of a magistrate to make all work who can, he continues, "And to fulfill this part of his [magistrate's] Duty, all Vulgar Trades belonging to Husbandry, to Manufactures, to Merchandize

and Commerce by Sea or Land in the severall kindes of Usefull Commodities; or employments about Commodities and all Honorable Offices and Charges belonging to the Common-wealth in Generall, or to any part of it in Country and City in Particular; are to bee ranked in their proper places: and all the Unlawfull and Unprofitable Ways whereby Men or Women get a livelihood, or spend their time in Idleness, in riot and vanity, are to be taken notice of; that such Employments as foment naughty Superfluities causing Pride and Sin to abound in a Nation, or such persons as live disorderly, and cannot be reduced to any certain Employment, may be banished the Common-wealth, even as weeds are to be rooted up and cast out of a fruitfull garden." (*Considerations*, pp. 22-3.) This passage is suggestive of the relationship between Puritan asceticism and the spirit of capitalism (see Max Weber, *The Protestant Ethic and the Spirit of Capitalism*, tr. Talcott Parsons, 1930, chap. V), and indicative of the place science assumed in this relationship.

[27] Hartlib was conscious of his true function. In the preface to Cressy Dymock's *An Essay For Advancement of Husbandry-Learning*, 1651, he says that though he can contribute little to the advancement of agricultural knowledge, "yet being carried forth to watch for the *Opportunities* of provoking others, who can do more, to improve their Talents, I have found experimentally that my *Endeavours* have not been without *effect* as to this *undertaking*."

[28] *Samuel Hartlib His Legacie: Or an Enlargement of the Discourse of Husbandry Used in Brabant and Flanders; Wherein are bequeathed to the* Common-Wealth of England, *more Outlandish and Domestick Experiments and Secrets in reference to Universal Husbandry*. 1651. A second edition appeared in 1652 under the title given above, and in 1659 a third was published called *The Compleat Husbandman*.

[29] Another work sponsored by Hartlib was *The Reformed Common-Wealth of Bees. Presented in severall Letters and Observations to Samuel Hartlib Esq. With the Reformed Virginian Silk-Worm. Containing Many Excellent and Choice Secrets, Experiments, and Discoveries for attaining of National and Private Profits and Riches*, 1655, which consists of sundry letters from various correspondents, involving both Wilkins and Wren, and in which great stress is placed upon experiments, and the utilitarian and humanitarian elements are pronounced, the poor coming in for frequent notice. Another work for which Hartlib may have been responsible was Ad. Speed's *Adam out of Eden, Or, An abstract of divers excellent Experiments touching the advancement of Husbandry. Shewing, among very many other things, an Aprovement of Ground by* Rabbiss *from 200L. annual Rent, to 2000L. yearly profit, all charges deducted, 1659*. The "To the Reader" states that publication was due to "a *Publick-spirited* Gentleman, (to whose industry in severall other things our *Age* is obliged)," probably Hartlib. (See also p. 139.) Speed bases the validity of his ideas squarely upon experiments.

[30] *The Reformed Husband-Man; Or a Brief Treatise Of the Errors, Defects, and Inconveniences of our English Husbandry, in ploughing and sowing for Corn; With the Reasons and general Remedies; and a large, yet fruitful Offer or Undertaking for the benefit of them that will joyn in this good and publick Work: Imparted some years ago to Mr. Samuel Hartlib; And now by him re-imparted to all ingenuous* English-men, *that are willing to advance the Prosperity, Wealth and Plenty of their Native Countrey*. 1651.

[31] *A Discoverie for Division or Setting out of Land, as to the best Form. Published by Samuel Hartlib Esquire, for Direction and more Advantage and Profit of the Adventurers and Planters in the* Fens *and other Waste and undisposed Places in* England *and* Ireland. *Whereunto are added some other Choice Secrets or Experiments of Husbandry. With a Philosophical Quere concerning the Cause of Fruitfulness. And An Essay to shew How*

all Lands may be improved in a New Way to become the ground of the increase of Trading and Revenue to this Common-Wealth. 1653.

[32] *An Essay For Advancement of Husbandry-Learning: Or Propositions For the Erecting a Colledge of Husbandry: And in order thereunto, for the taking in of Pupills or Apprentices. And also Friends or Fellowes of the same Colledge or Society.* 1651.

[33] In *A Discoverie for Division* we find direct evidence of Bacon's influence, as is true of most of these treatises. In speaking of the difficulty of meeting some problems connected with agriculture, one of the writers remarks, "The Lord *Bacon* hath gathered stubble (as he ingenuously and truly affirms) for the bricks of this foundation; but as yet I have not seen so much as a solid foundation plainly laid by any, on which an ingenious Man might venture to raise a noble Fabrick." P. 16.

[34] *An Essay for Advancement of Husbandry-Learning,* p. 1.

[35] *Ibid.,* p. 4. Turnbull attributes to Hartlib (*Samuel Hartlib,* p. 68) a slight treatise entitled *Cornu Copia; A Miscellanium of lucriferous and most fructiferous Experiments, Observations, and Discoveries, immethodically distributed; to be really demonstrated and communicated in all Sincerity.* (It can now be found in *An Harleian Miscellany,* VI, 26-34.) Dymock was probably the author, for it contains most extravagant proposals regarding enormous yields from any sort of ground, and such matters as grapes grown on oak trees, proposals obviously false but claiming experiments as their basis: "And, by a Way likewise newly experimented, to make a barren Field quite out of Heart . . . . to yield an extraordinary good Crop of Corn, without Muck, and with no Charge, and but little Trouble." *Ibid.,* p. 30.

In *A Catalogue of Chymicall Books . . . . Collected by Will Cooper, Bookseller,* 1675, one item reads, "Sam. Hartlib *his Chymical, Medicinal* and *Chirurgical* Addresses, In nine Treatises . . . . 1655." I have not seen this, but the description of the treatises given in the title indicates that Hartlib was doing for chemistry what he had done for agriculture.

[36] *An Invention of Engines of Motion Lately Brought to perfection. Whereby May be dispatched any work now done in England or elsewhere, (especially Works that require strength and swiftness) either by Wind, Water, Cattel or Men. And that with better accomodation, and more profit than by any thing hitherto known and used.* 1651.

[37] This may be the reason that Hartlib lost interest in the matter.

[38] Technological unemployment. Hartlib was entirely taken in by the "Honest Mr. Dymock," who, he says, "is blamed almost by every body [for his *Invention of Engines*], though the fault, in my judgment, be as much, if not more, in them that blame him." When he heard that the Marquis of Worcester had purchased Vauxhall to make it a college of artisans, he had great hopes of securing Dymock a position in it, though he preferred that the latter should head a college of agriculture. Boyle, *Works,* V, 264, May 8, 1654.

[39] Cf. John Evelyn's *Diary,* Nov. 27, 1653.

[40] Cf. Boyle, *Works,* V, 271, Jan. 7, 1658; 278; May 25, 1658; 283, Apr. 5, 1659; 286, Apr. 19, 1659. The reference to sunbeams reminds us of Laputa.

[41] For a discussion of this movement and its relation to science see *Jour. of Eng. and Ger. Philol.,* XXXI (1932), 315-31. The ascription of *A Common Writing* to Hartlib is far from certain.

[42] Boyle, *Works,* V, 264, May 8, 1654.

[43] G. H. Turnbull, *Samuel Hartlib,* pp. 10-13.

[44] Thomas Powell, *Humane Industry: Or, A History Of Most Manual*

Arts, Deducing the Original, Progress, and Improvement of them. Furnished with variety of Instances and Examples, shewing forth the excellency of Humane Wit. 1660.

45 Boyle, *Works*, V, 264. Elsewhere Hartlib writes that "Fauxhall is to be sett apart for publick uses, by which is meant making it a place of resort for artists, mechanics, etc., and a dépôt for models and philosophicall apparatus" where "Experiments and trials of profitable inventions should be carried on . . . . which will be of great use to the Commonwealth." By an act of the House of Commons, July 17, 1649, Vauxhall was exempted from crown property to be sold, as well as "the grounds, houses, buildings, *models*, utensils, or other necessaries for *practical inventions therein contained;* but that they should remain for the use of the Commonwealth, to be employed and disposed of by the Parliament, as they shall think fit." Upon this and other evidence Dircks holds that Worcester probably "had established a laboratory or workshop there, years before the Civil War broke out; that from its extent it was proposed to retain it for the benefit of the State; and that on his own release from the Tower he sought to regain possession of the premises, but possibly, for political reasons, in the name of his faithful workman, Caspar Kaltoff." (Henry Dircks, *The Life, Times, and Scientific Labours of the second Marquis of Worcester. To which is added, a Reprint of his Century of Inventions.* 1865, pp. 266-67.) According to Hartlib, however, Worcester was in the Tower when negotiating for the place.

46 Dircks, *op. cit.*, pp. 217, 223.

47 Cf. Dircks, *op. cit.*, pp. 268, 291-92. In 1663 appeared his only published work: *A Century of the Names and Scantlings of such Inventions, As at present I can call to mind to have tried and perfected, which (my former Notes being lost) I have, at the instance of a powerful Friend, endeavoured now in the year 1655, to set these down in such a way as may sufficiently instruct me to put any of them in practise.* 1663.

48 Another example of the passion of the age for inventions and practical discoveries is furnished by Thomas Bushell's *Abridgment of the Lord Chancellor Bacon's Philosophical Theory in Mineral Prosecutions,* 1659. Bushell states that Bacon made definite efforts to realize his Solomon's House, having been promised assistance by King James in the way of funds raised by working abandoned mines with convict labor, and that he passed on the work to Bushell. The latter seems never to have lost sight of this scheme, but tried every means of achieving his goal, even securing promises of assistance from Cromwell. It appears that he made some little headway in his plan, for Hartlib writes Boyle, May 8, 1654, "Yesterday I was invited by the famous *Thomas Bushel* (for I suppose you have seen his *Mineral Overtures* in print) to *Lambeth-Marsh*, to see part of that foundation or building, which is designed for the execution of my lord *Verulam's New-Atlantis.*" (Boyle, *Works*, V, 264.) His project survived for at least five years longer, for in 1659 he says that if he is given governmental assistance, he "doth hereby declare, that if it shall please the Almighty to prosper him in this his undertaking, that he shall then provide a certain convenient Place, and skilfull Artists, for the tryal and perfecting of all such Natural Experiments, as have been imparted to him by his said honoured Lord and Master [Bacon]; and in the mean time, he hath prepared a Place very convenient, in *Lambeth-Marsh;* to which house, all persons concerned may direct their discoveries, to be assayed and tryed at the charge of me." (*Abridgment,* p. 8.) Bushell constantly speaks of the public good, to which he was, like Hartlib, a martyr, for he was imprisoned for debt incurred by carrying on his ambitious scheme. The enterprises at Vauxhall and Lambeth-Marsh show how secure a hold the Baconian spirit had secured on the age.

49 J. F. Payne, *Harvey and Galen*, 1897, p. 13. Payne considers this the reason for Harvey's lack of sympathy for Bacon, and he defends Galen, as well as Aristotle, because he relied on observations and experiments. He

calls Galen "a modern experimental physiologist," and intimates that the
anatomical revival, represented by Vesalius, Harvey, and others, was due
to Galen's insistence upon the importance of anatomy. "Thus modern
natural science," he says, "grew out of the ancient science of the Greeks."
(*Op. cit.*, pp. 23, 25, 26, 30, 41, 47.) We must keep in mind, however, that
the great problem which confronted early modern science was the changing
of the attitude which regarded ancient science as the depository of all
knowledge and the ancients as absolute authority, and which prevented
scientific progress by discouraging investigation and by inciting men to read
books rather than to observe and experiment. To undermine this authority
it was necessary to attack the ancients on the score of their mistakes and
lack of knowledge, and to encourage the moderns by pointing out the superiority of modern discoveries and, in a less degree, by attacking ancient
methods. Recognition of the true scientific spirit of Aristotle, Galen, and
others is not lacking in the general attack on antiquity, but greater emphasis, for the purpose stated above, is laid upon the inadequacy and in-
accuracies of their science.

⁵⁰ See Payne, *op. cit.*, p. 4, and Sir William Hale-White, *Bacon, Gilbert
and Harvey*, 1927, pp. 29-30.

⁵¹ *Diatribæ duæ Medico-philosophicæ, Quarum prior agit De Fermen-
tatione Sive De motu intestino particularum in quovis corpore. Altera De
Febribus, Sive De motu earundem in sanguine Animalium. . . . Studio
Thomae Willis Medici Oxoniensis.* 2d ed., 1660.

⁵² Hale-White holds that Harvey was great, not so much because of his
discovery as because "he was among the earlier scientists to show the value
of properly conducted experiments." (*Op. cit.*, p. 1.) He also claims, in an
interesting though not entirely convincing manner, that Harvey in this
respect was influenced by Bacon.

⁵³ This is an English version of the *Tractatus de Rachitide* which had
appeared in the preceding year. It was begun in collaboration with several
other doctors, but the latter, perceiving Glisson's superior ability, relinquished the task to him. It is among the very earliest of English medical
monographs. Glisson was a member of the group of London experimenters,
who first met in 1645, and Willis was a member of the Oxford group, which
had much to do with the founding of the Royal Society.

⁵⁴ See the preface and p. 9.

⁵⁵ *Observations upon some part of Sr Francis Bacon's Naturall History
as it concernes, Fruit-trees, Fruits, and Flowers: especially the Fifth, Sixth,
and Seaventh Centuries, Improving the Experiments mentioned, to the best
advantage.* Oxford, 1658. See the dedication to Robert Boyle.

⁵⁶ *Ibid.*, "To the Reader."

⁵⁷ Not satisfied with his own apology, he secured the services of a friend,
R. Sharrock, to write a preface to absolve Austen from presumption in con-
tradicting some of Bacon's views. Sharrock claims that Austen "has more
mind to communicate to the World (for publique profit) what he has found
by triall certaine, than to make a book; and indeed am Witnesse my selfe to
the truth of most of his *Experiments*, the subjects of which no man dares
call too low for the pen, that Remembers the Author whose writing first
gave occasion to these *Animadversions*." He points out the great benefit
that would accrue to science from the lowest mechanic's divulging the
mysteries of his art, which hitherto had been thwarted by selfish interest.
"But," he proudly asserts, "tis *Heroicall* and *Noble Charity* when theres
nothing but *selfe Interest* hindring, to deny that for the *publique good*."
Certainly the "public good" was a potent catchword and one of the dominant
values of the period. The preceding quotations reveal Bacon's influence in
dignifying the lowest manual contact with nature and in creating the

democratic attitude which raised to a high plane of importance the farmer, artisan, and mechanic.

58 Gilbert Clerke lays all stress upon experiments, making them the test of theories, in a treatise which, though written in defence of Cartesianism, praises Bacon highly. It was his belief that certain divine geniuses by joining reason, mathematics, and experiments had demonstrated a new method of philosophizing, and it was the insistence which his age was placing on this procedure in preference to the disputatious method of Peripateticism which gave him hope for the future of science. This method, which he attributes to Bacon, he considers the foundation of natural philosophy, though he agrees with his great predecessor that such philosophy is not the work of one man nor of one age. *De Plenitudine Mundi Brevis & Philosophica dissertatio,* 1660, Praefatio.

59 See the account of Childrey in Anthony à Wood's *Athenae Oxonienses.*

60 *Britannia Baconia: Or, The Natural Rarities of England, Scotland, & Wales. According as they are to be found in every Shire. Historically related, according to the Precepts of Lord* Bacon; *Methodically digested; and the Causes of many of them Philosophically attempted. With Observations upon them, and Deductions from them, whereby divers Secrets in Nature are discovered, and some things hitherto reckoned Prodigies, are fain to confess the cause whence they proceed. Useful for all ingenious men of what Profession or Quality soever.* 1660. All quotations are from the preface.

61 See *Works,* ed. T. Birch, 1744, I, 20-21.

62 *New Experiments,* "To the Reader."

63 *Ibid.,* p. 122.

64 Another indication of Boyle's missionary spirit is revealed in Robert Sharrock's preface entitled "Friendly Reader," which explains Boyle's writing in English on the ground that he wished to reach men who "may be suppos'd to be able to make a better account, by employing their Studies and Time on Matter then Words, and so are justly impeded from learning Languages"—further evidence of the constant antithesis between words and material things, characteristic of this period. It also speaks of Boyle as a great "Patron and Friend to Experimental Learning, and all true Wisdom," who possesses "a publick Spirit for the Advancement of Learning."

65 "In the next place, these Experiments may teach us, what to judge of the vulgar Axiom receiv'd for so many Ages as an undoubted Truth in the Peripatetick Schools; That Nature abhorres and flys a *Vacuum.*" *Ibid.,* p. 246.

66 See *New Experiments,* pp. 37, 111-12, 268, 333, 360-61.

67 *Certain Physiological Essays, Written at distant Times, and on several Occasions.* 1661, pp. 9-10.

68 *Ibid.,* p. 25-26.

69 "And truly, *Pyro:* if men could be perswaded to mind more the Advancement of Natural Philosophy than that of their own reputations, 'twere not me-thinks very uneasie to make them sensible, that one of the considerablest services that they could do Mankind were to set themselves diligently and industriously to make Experiments and collect Observations, without being over-forward to establish Principles and Axioms, believing it uneasie to erect such Theories as are capable to explicate all the Phænomena of Nature, before they have been able to take notice of the tenth part of those Phænomena that are to be explicated." *Ibid.,* p. 8.

70 *Ibid.,* p. 17. Relative to this attitude of Boyle's, Birch says, "Though he wanted no capacity or abilities to have worked up a glorious system, and

erected a more pompous, ostentatious, and perhaps a more durable structure of natural and chemical philosophy, than had ever appeared in the world before; he nobly despised this poor satisfaction and mean gratification, telling us plainly and expressly, that, notwithstanding all he had done, all the labour, pains, and expense bestowed in a life of natural inquiries, notwithstanding the vastly numerous and important observations and discoveries he made, he saw nothing but the first dawnings of science; has drawn only the rudiments of natural knowledge, and leaves it in charge to posterity, for their own sakes to consider him but as a beginner, and to pursue philosophical enquiries in general without stopping to raise petty systems by the way." Boyle, *Works*, I, 94.

[71] Boyle believed that the errors in Aristotle and Campanella arose from their insufficient experiments, and that a system so erected was sure to be proved false by later experiments. (*Certain Physiological Essays*, p. 7.) He himself was so afraid of forming ideas before they were coerced by experiments that he refused to read systematically Gassendi, Descartes, and even the *Novum Organum*. The danger of forming hasty theories from insufficient data continued to haunt him. In the preface to *New Experiments and Observations touching Cold*, 1665, he asserts his unwillingness to produce any hypotheses until "the historical part" having been finished, he can survey the data to determine what theory will best fit them. In his eyes framing a general hypothesis was a work of great difficulty, "And whatever applause is wont in this age to attend a forwardness to assert hypotheses, yet, though fame were less to be sought than truth, this will not much move me, whilst I observe, that hypotheses hastily pitched upon do seldom keep their reputation long; and divers of them, that are highly applauded at the first, come, after a while, to be forsaken, even by those, that devised them." (*Works*, II, 239.) This emphasis upon experiment rather than explanation is clearly seen in the Latin motto taken from Bacon, which appears on the title page of this treatise as well as on that of a treatise on colors published the preceding year: "Non fingendum, aut excogitandum, sed inveniendum, quid Natura faciat, aut ferat." Boyle uses the term "history" in the Baconian sense to designate a collection of data from which theories may ultimately be drawn.

[72] *Certain Physiological Essays*, pp. 18-24.

[73] Boyle is glad "to acknowledge the great service that the Labours of Chymists have done the Lovers of useful Learning; nor even on this occasion shall their Arrogance hinder my Gratitude," but "when I acknowledge the usefulness of the Labours of *Spagyrists* to Natural Philosophy, I do it upon the score of their experiments, not upon that of Their Speculations." *The Sceptical Chymist*, 1680, p. 311. The first edition appeared in 1661.

[74] *Ibid.*, preface.

[75] Preface to *Some Specimens of an Attempt to make Chymical Experiments Usefull to Illustrate the Notions of the Corpuscular Philosophy*, 1661.

[76] *Sceptical Chymist*, pp. 42, 289.

[77] "If the Chymists . . . . had been so modest, or so Discreet, as to propose their Opinion of *Tria Prima*, but as a Notion useful among Others, to increase Humane knowledge, they had deserv'd more of our thanks, and less of our Opposition; but since the Thing, that they pretend, is not so much to contribute a Notion toward the Improvement of Philosophy, as to make this Notion (attended by a few less considerable ones) pass for a New Philosophy it self," the case is different. *Sceptical Chymist*, pp. 308-9.

[78] The term Boyle generally uses to express this philosophy is "Corpuscular" or "Corpuscularian." He sees as a common element in atomism, revived and elaborated upon by Gassendi, and in Cartesianism the belief that all natural phenomena are deducible from matter and local motion. As

points of difference between the two philosophies he notes the possibility of a vacuum, the origin of motion, and the indefinite divisibility of matter. (Preface to *Some Specimens of an Attempt to make Chymical Experiments Usefull to Illustrate the Notions of the Corpuscular Philosophy*. 1661.) In his earlier works Boyle reveals greater indebtedness to Gassendi than to Descartes, but in his later productions the situation is reversed.

[79] Years later, in a letter dated Oct. 16, 1684, Ralph Cudworth tells Boyle that "The writers of hypotheses in natural philosophy will be confuting one another a long time, before the world will ever agree, if ever it do. But your pieces of natural history are unconfutable, and will afford the best grounds to build hypotheses upon. You have much outdone Sir *Francis Bacon* in your natural experiments. . . ." Boyle, *Works*, I, 76.

[80] The seeds of this suspicion of reason are to be discovered in Bacon's insistence upon experiment and direct observation of material things and in his excessive caution against the use of reason in devising theories. This attitude continued strong throughout the period in which modern science was establishing itself. In fact, the anti-rationalistic implications in the propaganda for the new science led one of the important members of the newly-established Royal Society to say, in opposition to the rising confidence of the day in the advancement of science, "I rather fear our tumbling into the greatest barbarity and the most profound ignorance, the way to solid knowledge by cultivating of our reasons, and inuring them to compare, compute, and estimate well, begins now to be wholly despised," and ignorant young men who can "rally ancient learning, and talk of experiments" can silence the most learned. (Timothy Clerke, *Some Papers Writ in the year 1664*, 1670, p. 2. See also pp. 16-17.) The Duchess of Newcastle engaged in a vigorous defence of reason against the experimental philosophy, complaining that the experimentalists employ their senses more than their reason, and insisting that reason is the best guide to truth, and that speculative philosophy is nobler than the experimental. The mechanic should be only a servant to the student. "Neither ought Artists," she says, "in my opinion, to condemn Contemplative Philosophy, nay, not to prefer the Experimental part before her; for all that Artists have, they are beholden for it to the conceptions of the ingenious Student, except some few Arts which ascribe their original to chance; and therefore speculation must needs go before practice; for how shall a man practise, if he does not know what or which way to practise? Reason must direct first how sense ought to work, and so much as the Rational knowledg is more noble then the Sensitive, so much is the Speculative part of Philosophy more noble then the Mechanical. But our age being more for deluding Experiments then rational arguments, which some cal a *tedious babble*, doth prefer Sense before Reason, and trusts more to the deceiving sight of their eyes, and deluding glasses [telescopes and microscopes], then to the perception of clear and regular Reason; nay, many will not admit of rational arguments, but the bare authority of an Experimental Philosopher is sufficient to them to decide all Controversies, & to pronounce the Truth without any appeal to Reason." (*Further Observations*, pp. 1-4, found in *Observations upon Experimental Philosophy*, 1666. See also pp. 79-80, and *Observations*, preface and pp. 7, 10-12.) When the Duchess later visited the Royal Society she seemed, according to Pepys, full of admiration at the experiments performed in her honor. (See the *Diary*, May 30, 1667, and T. Birch, *History of the Royal Society*, II, 177.) As we have seen in an earlier chapter, Hobbes depreciated the Royal Society because its members stressed experiments and observation more than reason. Much of the scorn levelled at the mechanical activities of the Baconians can be explained by their apparent depreciation of reason. In the preface to his proposal to establish a philosophical college (1661), Cowley subjects reason to the restraint of the evidence of the senses in much the same way in which critics were beginning to insist upon the restraint of the imagination by reason: "And certainly the solitary and unactive Contemplation of Nature,

by the most ingenious Persons living, in their own private Studies, can never effect it [improvement of knowledge]. Our Reasoning Faculty as well as Fancy, does but Dream, when it is not guided by sensible Objects. We shall compound where Nature has divided, and divide where Nature has compounded, and create nothing but either Deformed Monsters, or at best pretty but impossible Mermaids." Cf. Hobbes's treatment of the imagination in the second chapter of the *Leviathan*, 1651.

For another aspect of the sceptical attitude toward reason, see Louis Bredvold's "Dryden, Hobbes, and the Royal Society," *Modern Philology*, XXV (1928), 425.

[81] In his correspondence with Worthington, Hartlib quotes a letter of John Beale, an early member of the Royal Society, which reveals clearly the difference in attitude of a typical scientist of the day toward the mechanical and other theories and toward the experimental philosophy, as well as the part Bacon played in this difference. The letter also discloses the sanguine hopes for science aroused by Boyle's works. "For the Corpuscularian Philosophy, I had long ago complained to you, that Sir K.[enelm] Digby had said enough of it to make me giddy in their Pro et Contra, & my refuge was Ld Bacon's 'Circa Ultimates rerum frustranea est Inquisitio?' But now I see a stay for the light of reason and experience. Neither had I anything in chase, then the cause of firmness, wch in Ld Bacon's language (who first awaken'd my attention to it) is frequently called consistency. . . . 'Tis my great joy that Mr. B.[oyle] is so far engaged to give us the rest of his notes and following experiments. In these he hath obliged all the intelligent inhabitants of this world, and hath given us hope, that we shall shortly complete humane sciences. Some families amongst us have answered all Ld Bacon's votes for advancement of learning. . . . To those that have been so tired and wearied, as I have been, in the several ways of Teles.[ius] Flood, Gassendus, the Cartesian, & Atomical or Corpuscularian Philosophers, and all others that I could hear of, ever since I was concerned in the victory of school sophisters. To those, I say, that have condescended to take any deep notice of the insufficiency of language and conjectures & ungrounded ratiocinations, and have submitted their patience to the severity of Ld Bacon's Inquisitions, & (amongst these) to myself, here are offered such pleasing refreshments, as gain us the relish of . . . . Virgilian simplicity. . . . For philosophical satisfactions, I did chiefly address [myself] to philosophical experiments, in wch I seemed to have the best overtures of aid from Ld Bacon, but of this I complained, that in the progress of late years we had not brought his experiments or added our own, to any degree of ripeness. And this was indeed my discouragement. Now I confess I am surprised with wonder at the present advancement, & I dare promise our posterity, that knowledge shall in this following age abound in very great perfection, & to the best of noble operations." Worthington, *Diary*, I, 369-73. Cf. Boyle's *Works*, I, 38-9.

For other evidence of the relative importance of Bacon and Descartes in the English scientific movement see the dedication of G. Havers' *A General Collection of Discourses of the Virtuosi of France*, 1664, and Sprat's *Observations on Monsieur de Sorbière's Voyage into England*, 1665, p. 165.

[82] *Certain Physiological Essays*, 1661, pp. 115, 98; *New Experiments and Observations Touching Cold*, 1665, preface; *Some Considerations touching the Usefulnesse of Experimental Naturall Philosophy*, 1663, p. 398; *The Excellency of Theology*, 1674, p. 170; *Works*, V, 2; *The Christian Virtuoso*, 1690, p. 8.

[83] Cf. *Certain Physiological Essays*, pp. 14, 70, 71, 98, 115. *Works*, V. 77, 106. Boyle says, "no writer, that I know of has so early and so well, both urged the necessity of natural history, and promoted divers parts of it by precepts and specimens, as the illustrious lord *Verulam*." *Works*, V, 77.

[84] Cf. *The Excellency of Theology*, pp. 170 ff.; *Works*, III, 518; IV, 48, 163; *New Experiments*, pp. 293, 296.

[85] Birch quotes Herman Boerhaave to the effect that Bacon was the father of experimental philosophy, and that "Mr. Boyle, the ornament of his age and country, succeeded to the genius and inquiries of the great chancellor *Verulam* . . . . from his works may be deduced the whole system of natural knowledge." Birch also quotes from one of John Hughes's contributions to the *Spectator*, no. 554, to the effect that Boyle, born in the year in which Bacon died, "was the person, who seems to have been designed by nature, to succeed to the labours and inquiries of that extraordinary genius . . . . By innumerable experiments, he in a great measure filled up those plans and out-lines of science, which his predecessors had sketched out." Boyle, *Works*, I, 91.

[86] In "Democritus to the Reader" prefixed to the *Anatomy*, Robert Burton laughs at the *New Atlantis*, together with More's *Utopia*, and as late as 1658 Bishop Bramhall lists Bacon's works with Plato's and More's creations as being beyond the domain of mortal achievement (*Castigations of Mr. Hobbes*, p. 182); but Peter Heylin considered the *New Atlantis* far superior to More's *Utopia* in the "Excellency and feasibility of his [Bacon's] invention." (*Cosmographie*, 1652, Appendix, p. 196.) In the dedication of *Instructions Concerning Erecting of a Library*, 1661, translated from the French of Gabriel Naudé, Evelyn tells Clarendon, "There is certainly nothing more expedient, than in pursuite of that stupendous *Idea* of your Illustrious *Predecessor* [Bacon], to set upon a Design no way beneath that of his *Solomons House;* which, however lofty, and to appearance *Romantic*, has yet in it nothing of Impossible to be effected."

[87] Boyle, *Works*, V, 278-79. In this letter Hartlib quotes from a letter from Worsley which complains that the government wished to take over money which Hartlib and others had raised for the scheme, and which Worsley thinks should be managed entirely by Boyle, Durie, Sadler, Hartlib, Beale, and himself. "If the state," he continues, "have a mind to set up such an institution, let them do it of their own, and dispose it to their own ministers. If other men are willing to lay a foundation of so much good, let not the state hinder them," and he speaks of the possibility of Durie's having a conference with Cromwell, Fleetwood, and others about it. The letter bears evidence of the fact that Hartlib was pushing the matter with the government.

[88] The plan called for the establishment of a wide scientific correspondence through a learned secretary, the collecting, abstracting, and methodizing of books, the purchasing and making of experiments, mechanical models, and works, and the collecting of "rarities and representations of all common natural and artificial things." A subscription form for raising the money reads as follows: "Whereas Mr. John Dury, library keeper of St. James's, and Samuel Hartlib Esq. have accepted the trust of receiving and disposing such summes of moneys, as well affected persons and lovers of the publick good shall be pleased to give, and to bequeath, towards a most useful agencie and councell for advancement of universal learning and arts; as is more fully specified and explained, in the memorial hereunto annexed; as likewise in the lord *Verulam's* workes [my italics]; in the advice for the advancement of some particular parts of learning; in the Reformed library keeper, in the Seasonable discourse; and in the discourses, which describe the offices of address for accommodation and communications;

Wee, whose names are here underwritten, highly approving these pious and noble aimes, and earnestly desiring that towards the accomplishment thereof, the designed agencie and councell may bee erected and supported;

Being also very well satisfyed with the approved faithfulness and abilities of the afore named trustees to manage and direct the foresaid undertaking; doe to the said ends and purposes (besides the conscionable engaging ourselves to draw in others to the cheerfull assistance of so commendable an enterprise) voluntarily subscribe these several summes of money yearly to be continued during pleasure; and to be paid in whole at once, or by equal portions every quarter, to the aforesaid trustees, or to such collectors, as the said trustees shall employ." (See Turnbull, *Samuel Hartlib*, pp. 57 ff., and Dircks, *Hartlib*, p. 18). Comenius' pansophic ideas exerted much influence on the project.

[89] Boyle, *Works*, I, 28.

[90] Boyle, *Works*, V, 280-81. Petty's design was the founding of a college of twenty good Latinists to teach Latin "(as other vulgar languages are learnt) merely by use and custom," which, together with a history of trades, he considered "the great pillars of reformation of the world." The Baconians viewed the emphasis upon linguistic study as one of the great impediments to the advancement of science, and they especially deprecated the amount of time spent upon the classical languages.

[91] Worthington, *Diary*, I, 163. Hartlib's hopes had been considerably brightened by the seemingly favorable attitude of Richard Cromwell toward his design. On Dec. 16, 1658, he writes Boyle, "I am wondrous glad, that you have written of the present protector's intentions for countenancing and advancing of universal useful learning in due time," and he expresses the hope that Wilkins will be "president of the forementioned standing council of universal learning." Boyle, *Works*, V, 282.

[92] Worthington, *Diary*, I, 180-81. In the middle of November he thought his scheme was about to be realized: "The truth is, I design all such and the like works or tracts be printed upon the charges of *Macaria*, whose scope it is most professedly to propagate religion, and to endeavour the reformation of the whole world. But it is scarce one day (or hour in the day) or night, being brim full with all manner of objects of that publick and most universal nature, but my soul is crying out,

> *Phosphore! redde diem, quid gaudia nostra moraris?*
> *Phosphore, redde diem!"* Boyle, *Works*, V, 293.

[93] Nov. 29, 1659, Boyle, *Works*, V, 296. Hartlib's sympathies had been too strongly puritanic and his commerce with the previous government too close for him to be countenanced by the royalist government. It also appears, though one hates to believe it, that Boyle, for the same reason, discontinued his association with Hartlib.

[94] Crossley, the editor of Worthington's *Diary*, believes that the two societies were one and the same. He was influenced probably by what Hartlib says in a letter of Oct. 15, 1660: "We were wont to call the desirable Society by the name of Antilia, and sometimes by the name of Macaria." (*Diary*, I, 211, 228.) But subsequent references show conclusively that they represented two different societies, and that Hartlib merely applied the name of his long cherished dream to the other. Worthington says, "Macaria . . . . is too good a word for this late pretending company." *Diary*, I, 244.

[95] *Ibid.*, I, 228, 239-41, 255.

[96] *Ibid.*, I, 238, 244-45.

[97] On July 20, 1659, Hartlib writes Worthington, "The adjoined printed Discourse I believe you have not yet seen. The subject of it is very lovely and desirable. I like the exhortation better than the model itself, wch methinks is not comprehensive enough. Capt. Shane, an Irish gentleman, has given a thousand acres of land in Ireland toward it. And my Lord Newport is like to give his college in Northamptonshire for the same purpose." (*Diary*, I, 149.) Crossley thinks that this project, which I have been unable to identify, was a philosophical college, and certainly Hartlib's words

strongly suggest it. In Oldenburg's letter to Boyle, Oct. (?) 1658, there may be a reference to another such project in which Boyle was interested. See Boyle's *Works*, V, 300.

98 Worthington, *Diary*, I, 245-46.

99 *Ibid.*, I, 257, 295-96.

100 *Ibid.*, I, 342.

101 Only a few years later Cowley, in urging that agriculture be put on a solid and experimental basis, refers in a complimentary way to Hartlib, "if the Gentleman be yet alive." (See his essay *Of Agriculture.*)

102 John Evelyn, *Diary and Correspondence*, ed. William Bray, new edition, 4 vols., 1902, III, 116-19.

103 *New Atlantis. Begun by the Lord Verulam, Viscount St. Albans: And Continued by R. H. Esquire. Wherein is set forth a Platform of Monarchical Government. With a pleasant intermixture of divers rare Inventions, and wholsom Customs, fit to be introduced into all Kingdoms, States, and Common-Wealths.* 1660. Bacon comes in for exaggerated praise in the preface and in a prefatory poem by G. Herbert.

104 "We have in the three Universities, Colledges (besides those for Divinity, Law and Philosophy) for Mathematicians, Historians, Poets, Musicians, Stage-players, Alchymists, Florists, Herbarists, Chirurgians, Anatomists, and Physitians also. Unto the last are adjoyning large Physic Gardens, Theatres and Schools. In all of them all the Students are to apply themselves particularly unto that study and art they first undertake, and none other, that thus all experiments may be sooner reduced unto perfection, all Arts and Sciences soon learnt; all these severall Colledges being very fairly endowed." *New Atlantis*, p. 42.

105 *New Atlantis*, pp. 54-71.

106 Cowley may have been incited to his undertaking by the society of virtuosi which had been organized at Gresham College on November 28, 1660. In February, 1661, he was proposed as a member and in March was elected. Furthermore, Sprat says that Cowley's *Proposition* hastened "very much" the contrivance of the Royal Society. (*History of the Royal Society*, 1667, p. 59.) As we have seen, schemes for some kind of scientific organization were plentiful about this time, and must have been considered by almost every member of the society meeting at Gresham College. There is some evidence that Cowley was familiar with Evelyn's plan for a philosophical college. They both propose as the model of their buildings the Carthusian monastery, and both are interested in analyzing to the smallest detail the amount of money necessary for their schemes. Their doubtful attitudes toward the possibility of realizing Bacon's Solomon's House, which nevertheless inspired their projects, are the same. Evelyn says, "we are not to hope for a mathematical college, much less, a Solomon's House" (*Diary*, ed. Bray, III, 116), and Cowley remarks, "we do not design this after the Model of *Solomons* House in my Lord *Bacon* (which is a Project for Experiments that can never be experimented) but propose it within such bounds of Expence as have often been exceeded by the Buildings of private Citizens." (*Essays*, ed. A. B. Gough, 1915, p. 33.) News of Cowley's *Proposition* had got abroad in the news-books by April 1, and by the latter part of June the treatise had appeared. See Worthington, *Diary*, I, 284, 339.

107 Sprat found it necessary to discuss this aspect of Cowley's proposal, and to explain why the Society did not adopt it. *History of the Royal Society*, pp. 59-60.

108 Cowley's pedagogical ideas are closer to Petty's than to those of any other of the Puritans, although in the main they are common to most of the educational treatises of the period of the rebellion. It is possible that Petty and Cowley had discussed the matter.

[109] Cowley even proposed scientific materials for poetry: and "because we want good Poets (I mean we have but few) who have purposely treated of solid and learned, that is, Natural Matters (the most part indulging to the weakness of the world, and feeding it either with the follies of Love, or with the Fables of gods and Heroes) we conceive that one Book ought to be compiled of all the scattered little parcels among the ancient Poets that might serve for the advancement of Natural Science, and which would make no small or unuseful or unpleasant Volume." The abandonment of love and the restricting of poetry to "solid and learned" matters represent greater inroads upon the poetic spirit than the usefulness of experimental philosophy to poetry which Sprat points out, namely, the furnishing of "*beautiful Conceptions,* and inimitable *Similitudes." History of the Royal Society,* p. 417.

[110] One in *A Defence of the Royal Society,* 1678, pp. 7-9; the other in his autobiography contained in "The Publisher's Appendix to his Preface" (p. 127) prefixed to Thomas Hearne's edition of Peter of Langtoft's *Chronicle,* 2 vols., Oxford, 1725. This latter, dated Jan. 29, 1697, was evidently copied from the first account, though some minor changes were introduced. I have combined the two in my account.

[111] John Wilkins, Dr. Jonathan Goddard, Dr. George Ent, Dr. Francis Glisson, Dr. Charles Scarborough, Dr. Christopher Merrett, John Wallis, Samuel Foster, professor of astronomy at Gresham College, Theodore Haak, a German of the Palatinate who "gave the first occasion, and first suggested those meetings," and "some others." It is to be noted that over half of these men were doctors.

[112] Anthony à Wood, whose account is borrowed from Wallis, adds that the group had no idea of founding a society, but operated for diversion's sake only in an innocent and virtuous manner. The fact, however, that they called themselves a society, and their enthusiastic faith in the new method, as well as their seriousness in carrying on investigations, would belie Wood's statement. See *History and Antiquities of Oxford,* Bk. I, pp. 632-34.

[113] Wallis enumerates them as follows: the circulation of the blood, the valves in the veins, the venae lacteae, the lymphatic vessels, the Copernican hypothesis, the nature of comets and new stars, the satellites of Jupiter, the oval shape of Saturn, the spots in the sun and its turning on its axis, the inequalities and selenography of the moon, the several phases of Venus and Mercury, the improvement of telescopes and grinding of glasses for that purpose, the weight of the air, the possibility or impossibility of a vacuum and nature's abhorrence of it, the Torricellian experiment in quicksilver, the descent of heavy bodies and the degrees of acceleration therein, "and divers others of like nature."

[114] *Works,* I, 20.

[115] *Loc. cit.*

[116] See Boyle's letter to Hartlib, May 8, 1647, *Works,* I, 24.

[117] Sprat says, "The *University* had, at that time, many Members of its own, who had begun a *free way* of reasoning; and was also frequented by some *Gentlemen,* of Philosophical Minds, whom the misfortunes of the Kingdom, and the security and ease of a retirement amongst Gown-men, had drawn thither." (*Op. cit.,* p. 53.) Of this group Wood remarks, "These persons gave themselves the name of Vertuosi, and pretended to go beyond all others in the University for knowledge, which causing envy in many, some, especially those of the old stamp, that had been eminent for School and Polemical Divinity, and other polite parts of Learning, look upon them very inconsiderably, and their Experiments as much below their profound Learning and the Professors of them. Undervalued by the Aristotelians, Galenists, profound disputants and schoolmen, who looked upon their

operation as much below their philosophy, and rather to be embraced by the Quack Salvers and Apothecaries boys than by them. . . ." (*Hist. and Ant. of Oxford*, pp. 632-34.) The original minutes of this philosophical society are in the Ashmolean Museum, and a MS letter addressed to the heads of the colleges shows that the new philosophers were trying to placate those described by Wood and to enlist them in the service of science: "We would by no means be thought to slight or undervalue the philosophy of Aristotle, which hath for many ages obtained in the schools. But have (as we ought), a great esteem for him, and judge him to have been a very great man, and think those who do most slight him, to be such as are less acquainted with him. He was a great enquirer into the history of nature, but we do not think (nor did he think), that he had so exhausted the stock of knowledge of that kind as that there would be nothing left for the enquiry of after-times, as neither can we of this age hope to find out so much, but that there will be much left for those that come after us." The letter earnestly asks aid toward the great work of advancing scientific knowledge. See C. R. Weld, *A History of the Royal Society*, I, 34-5.

[118] Sprat ascribes the source of the Royal Society to these meetings. (*Op. cit.*, pp. 52 ff.) The discrepancy between his and Wallis' accounts has been frequently noted. Had Sprat alone been responsible for his *History*, the inconsistency would offer little difficulty, for unlike Wallis, he had not come in contact with the earlier group, but since the history was in a way an official pronouncement of the Society, in which Sprat had been assisted by other members, especially Wilkins, and the progress of which had been closely followed by the Society as a whole, the completed book being heartily approved by the virtuosi, it is strange that this discrepancy was not noticed by at least one member who had also belonged to the early London group. Wood insinuates that Sprat gave an inaccurate account of the origin of the Royal Society in order that his alma mater, Wadham College, might gain the honor of being the home of the beginnings of the Society. (*Hist. and Ant. of Oxford*, Bk. II, p. 633.) There may be more truth than malice in his statement, for in his *History* (p. 53) Sprat admits the personal equation in mentioning "that, which will be for the honor of that place, where I receiv'd a great part of my Education." Wilkins, who was closely associated with Sprat in the composition of the *History*, may have connived at the inaccuracy because of the honor done him and the college of which he was warden. The seventeenth century in general accepted Sprat's story. (See William Holder, *Supplement To the Philosophical Transactions of July, 1670, 1678*, p. 4; Walter Pope's *The Life of the Right Reverend Father in God Seth, Lord Bishop of Salisbury*, 1697, pp. 29, 110.) In inquiring why Glanvill should profess love for the universities, Henry Stubbe asks, "Is it because the first New Philosophy was so much promoted, and the R. S. as it were embryonated there?" *An Answer to the Letter of Dr. Henry More*, 1671, p. 45. Also see Robert Hooke's *Philosophical Experiments and Observations*, ed. W. Derham, 1726, pp. 388-91.

[119] Boyle, *Works*, V, 298.

[120] Sprat, *History of the Royal Society*, pp. 57-8. Wallis makes no mention of this interruption. Sprat claims that it was the coming of the Oxonians to London in 1658 that established the London group, which was joined by several other distinguished persons. From this time on, experimental activities languished at Oxford. During the plague enough fellows of the Royal Society fled to Oxford to constitute a quorum of the Society, according to Oldenburg, who urges them, through Boyle, "so to insinuate the designs of the Royal Society into the Oxonians, that they may relish them as much, as most of them have been reported to disgust them, and give them cause to prefer that solidity of knowledge, the said Society aims at, before scholastical intentions." (Letter to Boyle, Oct. 5, 1665, in Boyle's *Works*, V, 337.) Two months later he asks Boyle about the attitude at

Oxford toward the new philosophy. (*Ibid.*, V, 342.) But Oxford was more backward than Cambridge in showing favor to the new science. Perhaps the reaction against the espousal of the Baconian philosophy by Puritan appointees and intruded fellows had something to do with the matter. It is not strange that at the end of the century the Christ Church crowd lined up solidly against Bentley, Wotton, and the moderns in the controversy which led to the *Battle of the Books.*

121 For the account given in the Society's *Journal Book*, see Weld, *History of the Royal Society*, pp. 64 ff.

122 See Worthington's *Diary*, I, 215, 247, 249.

123 See the dedication of *Instructions Concerning Erecting of a Library,* . . . *by Gabriel Naudeus . . . and now interpreted by Jo. Evelyn, Esquire,* 1661. In his *Diary*, Nov. 16, 1661, Evelyn speaks of his presenting his translation to Clarendon, and some days later, on Dec. 3, mentions a vote of thanks from "our philosophic assembly" because of his mention of them as the Royal Society. That the name was being used with some frequency is revealed in the second dedication of John Graunt's *Natural and Political Observations*, which speaks of Moray as the "President of the Royal Society." The first dedication is dated Jan. 25, 1662.

CHAPTER VIII

1 See Joseph Glanvill's "To the Royal Society" prefixed to *Scepsis Scientifica*, 1665, and More's letter to Glanvill, contained in the latter's *A Præfatory Answer to Mr. Henry Stubbe*, 1671, pp. 154-58.

2 See the preface to *Enchiridion Metaphysicum*, 1671, and Glanvill's *Præfatory Answer to Mr. Henry Stubbe*, p. 155. Of course, when the atheistic implications in Cartesian materialism became more fully revealed in Hobbes's philosophy, the scientists found it increasingly necessary to draw a clear line of distinction between their philosophy and Descartes's, but such a distinction was real. The experimental philosophy involved a method to which all scientists without question subscribed; the mechanical philosophy was a theory with which they might or might not agree. The exuberant hope of the period lay in the former, not in the latter. The mechanical philosophy was merely one explanation, widely accepted, to be sure, and destined to revolutionize opinions of nature, of the data furnished by the experimental philosophy; and as it has been pointed out in the preceding chapter, the collecting of authentic data, which show how nature acts, and which thus enable men to command nature for the satisfaction of their own needs, was more important than the devising of hypotheses as to why nature acts so. Moreover, as we have seen and shall see, all theories continue to be somewhat suspect, owing to the emphasis which Bacon placed upon the necessity of the most comprehensive data for the establishment of any hypothesis, and to the horrible example of traditional reasonings and theories unsupported by natural facts. The experimental method was employed to test the truth of the mechanical theory, and we find complaints that Descartes had based his philosophy upon insufficient experiments.

It is not difficult to find evidence of the critical attitude of various Baconians toward the mechanical philosophy and of their realization of the difference between it and the experimental learning. Richard Baxter furnishes some. He says that "that eminently learned and industrious man, Dr. *Willis*," rejected the mechanical theory, and he refers to the avowed scepticism of Glanvill's *Vanity of Dogmatizing*. He quotes Samuel Parker, whom we shall consider soon, as follows: "I am lately grown such a despairing Sceptick in all Physiological Theories, that I cannot concern myself in the truth or falsehood of any hypotheses. For although I prefer the Mechanical hypotheses before any other, yet me-thinks their contexture is too slight and brittle to have any stress laid upon them: and I can resemble them to nothing better than your glass drops, from which if the least portion be broken, the whole compages immediately dissolves and shatters into dust and atoms: for their parts . . . . rather lie than hang together, being supported only by the thin film of a brittle conjecture, (not annealed by experience and observation)." (See *The Reasons for the Christian Religion*, 1667, p. 496-97. The quotation is from Parker's *A Free and Impartial Censure of the Platonick Philosophie*, 1666.) Here is clearly disclosed the Bacon-inspired distrust of theory, together with emphasis upon actual observation. Baxter himself furnishes an interesting example of the difference between the experimental and mechanical philosophies. He argues strenuously against the latter, but is very favorably disposed toward the former. He tells Boyle that he takes pleasure in the latter's experimental learning, and in reply the scientist says that Baxter is "none of those narrow-souled divines, that, by too much suspecting natural philosophy, tempt many of its votaries to suspect theology." (See Baxter, *op. cit.*, and Boyle, *Works*, V, 553, 556. For other evidence of the sceptical attitude toward Cartesianism, see the preceding chapter, especially note 81.) How important, however, some of the virtuosi considered the mechanical philosophy in making ideas of nature intelligible is revealed by John Wallis in *Philosophical Transactions*, No. 16, p. 264.

One interesting result of the materialism of the mechanical philosophy, given wide currency by Hobbes's insistence that there is no such thing as an incorporeal substance, or, for that matter, any spiritual beings at all, was the strengthening of the belief in witchcraft, adherence to which by some scientists has puzzled scholars. In his *Reasons for the Christian Religion*, Baxter frequently cites witches as evidence of spirits and incorporeal substances, and in *A Saint or a Brute*, 1662, (p. 135) he says that those who do not believe that there are spiritual inhabitants of other worlds go against experience, "because that many a hundred *Witches*, and many *Apparitions* and *haunted houses* have put the matter out of question." In the preface to *Of the Immortality of Mans Soul*, 1682, he recommends to infidels the reading of Glanvill's and More's accounts of witches, which prove "the existence and Individuation of Spirits, and the future life of separated Souls." Baxter also wrote a book, *The Certainty of the Worlds of Spirits*, composed entirely of witch- and ghost-stories, to prove the existence of spirits. In a letter to Glanvill, Sept. 18, 1677, Boyle emphatically states his belief in the usefulness to religion of witch-stories and approves of Glanvill's proposed work on witches, cautioning him, however, to be sure of the truth of his narratives, since most stories of this kind are false, and thus tend to strengthen rather than refute atheism springing out of a materialistic conception of the universe. Boyle, *Works*, V, 244. Cf. Moody E. Prior, "Joseph Glanvill, Witchcraft, and Seventeenth-Century Science," *Modern Philosophy*, XXX (1932), 167-93.

[3] Patrick's treatise purports to be an answer to a letter, prefixed to it, which asks information about the "Latitude-Men," who, the author says, are the chief topic of conversation wherever he goes, and who, as far as he can gather, had their rise at Cambridge and for the most part embraced the new philosophy. Patrick also bears witness to the wide notice the Latitudinarians were securing, and states that they received their title by virtue of their opposition to the strait-laced spirit which prevailed during the Puritan domination. Patrick, though believing that too much reliance should not be placed on reason, is one with the Cambridge Platonists in trying to reconcile philisophy and religion, in denying the separation of them advocated by Bacon, and in holding that truth in divinity is truth in philosophy.

[4] *A Brief Account of the new Sect of Latitude-Men*, p. 19.

[5] He believed that because of the discovery of America men should not

restrict themselves to Strabo's geography, and that in view of other discoveries they should not use ancient maps only.

⁶ *Ibid.*, pp. 19-21.

⁷ *Ibid.*, pp. 22-4.

⁸ *A Free and Impartial Censure of the Platonick Philosophie Being a
Letter Written to his much Honoured Friend Mr. N. B.*, Oxford, 1666. Before
he had been converted to experimental science by Bathurst, Parker says that
he looked upon Platonism "as the loftiest and sublimest knowledge in the
world, but when I came to survey it more closely, I soon found that it was
nothing else but words." *Ibid.*, p. 77.

⁹ *Ibid.*, p. 53.

¹⁰ *Ibid.*, pp. 56-7.

¹¹ *Ibid.*, pp. 63-4, 78. From this attack on Platonism, Parker passes
naturally to an onslaught on scholasticism, in which the charges which are
by now familiar to us are repeated: empty and insignificant words, frivolous
and confused distinctions, useless and imaginary notions, precarious and
uncertain suppositions, senseless and unintelligible discourses, and the like.
He bemoans the time which he, contrary to Bathurst's advice, had wasted
on these "senceless authors." He also condemns the incorporation of Aristotle's philosophy in scholastic divinity, with its "subtle and nice Hypotheses
. . . . out of which are of necessity generated an infinite number of idle
or unprofitable Altercations, or . . . . (as my *Lord Bacon* prettily stiles
them) *Vermiculate Questions*, because they are generated from the putrefaction of true and solid knowledge, like worms from putrified substances;
though perhaps they may ere long deserve that Epithete upon another
score." *Ibid.*, p. 89.

¹² The influence which this concrete, materialistic spirit exerted upon
stylistic views is clearly apparent in Parker, who, distrusting the imagination, inveighs sharply against the use of figurative language in prose. See
The Seventeenth Century, by Richard Foster Jones *et al.*, pp. 75-160.

¹³ Other evidence of Parker's view of the distinction between experiments
and theories, data and explanations is to be found in his attitude toward
the chemists, who, he says, "darken by their smoke" "what they discover
by their fire." He is like Bacon and Boyle in thinking that chemical experiments are valuable, though the theories offered in explanation of them are
untenable. Parker's favorable opinion of the mechanical hypothesis, however, is manifested in his *Tentamina Physico-Theologica De Deo*, 1665, in
which he attempts to reconcile that hypothesis with Christian doctrine.

¹⁴ Parker's work met with the approval of the scientists, and though
Glanvill was identified with the Platonists, Oldenburg expresses the opinion
that Glanvill "will rather yield and fly the field, than handle a weapon
unhandsomely." Boyle, *Works*, V, 357.

¹⁵ *A Free and Impartial Censure*, p. 98. Bacon's influence appears
throughout Parker's treatise. He frequently quotes him, and in upholding
the need of experiments instead of theories, adduces the *Advancement of
Learning* as authority. *Ibid.*, p. 57.

¹⁶ The treatise was evidently composed some time before its publication,
for the Imprimatur bears the date of Aug. 5, 1663, and the preface, Aug. 1,
1661. This fact probably accounts for the absence of any reference to the
Royal Society as such, but the conclusion, which is headed "To the generous
Virtuosi, and Lovers of Experimental Philosophy," seems certainly to be
addressed to the society at Gresham College. In fact, there is on p. 149 a
reference to "our Noble Society of *Gresham-Colledge*."

¹⁷ See preface and p. 82. Other evidence of his indebtedness to Descartes

is seen in his appreciation of the new idea of the immensity of space, his belief that the world was not created primarily for man, and his support of Descartes's "supposition" that fluidity consists in motions of parts of bodies, which, he says, Boyle demonstrated.

18 *Experimental Philosophy*, pp. 90, 107, 183 ff., 190. The adjectives applied to the new thought of this period are of some interest. The terms "new" and "free" are applied indiscriminately to both the experimental and mechanical philosophies, and connote, of course, freedom from authority. The words "solid" and "real" are used chiefly of the experimental philosophy. The term "mechanical" when used independently of "philosophy" does not necessarily signify Descartes's theory. It may have reference only to the dealing with material things (as seen in the word "mechanic"), or to the manual nature of experimental operations, or to the employment of instruments in experiments. Power says that without "mechanical assistance [i.e., microscopes, telescopes, and the like] our best philosophers will but prove empty Conjecturalists, and their profoundest Speculations herein, but gloss'd outside Fallacies." *Ibid.*, Preface.

19 *Ibid.*, pp. 90, 107, 121, 191.

20 *Ibid.*, pp. 149, 191.

21 *Ibid.*, pp. 183 ff.

22 *Loc. cit.*

23 *Loc. cit.*

24 The last part of this passage is similar to Patrick's description of the function of science quoted on page 186 *supra.* In fact, the idea was the chief support of those who attempted to reconcile science and religion.

25 *Micrographia: or Some Physiological Descriptions of Minute Bodies Made by Magnifying Glasses. With Observations and Inquiries thereupon.* By R. Hooke, Fellow of the Royal Society. 1665. The book was printed by order of the Royal Society, the order being dated Nov. 23, 1664.

26 This idea is similar to what Bacon said in support of his method; namely, that its operation did not require men of genius. Boyle, though to a certain extent agreeing with Bacon, thought fit to introduce certain reservations into this hardly complimentary estimate of the intellectual qualifications required of a scientist.

27 In another passage of the preface Hooke declares his intention of describing a method "of compiling a Natural and Artificial History, and of so ranging and registring its Particulars into Philosophical Tables, as may make them most useful for the raising of *Axioms* and *Theories*," in which he must have had in mind the comprehensive method which Bacon partly elucidated in the *Novum Organum*, as may be seen in his treatise entitled "The Present State of *Natural Philosophy*, and wherein it is deficient," published in *Posthumous Works*, 1705, and based on the *Novum Organum*. Hooke was also very much interested in the proper method of experimenting, outlining a plan of procedure and drawing up rules for the Royal Society.

28 Though Hooke entertained no unfavorable opinion of Descartes, he does not hesitate to take him to task for failure to experiment sufficiently: "For if the most Acute *Des Cartes* had applied himself experimentally to have examined what substance it was that caused that shining of the falling Sparks struck from a Flint and a Steel, he would certainly have a little altered his *Hypothesis*." *Micrographia*, p. 46.

29 Unless otherwise stated, all quotations are from the preface. In view of Hooke's clear announcement of the utilitarian purpose of the Royal Society, and his recognition of the relationship of applied science and industrial and commercial forces, it is worth noting that he praises Sir John

Cutler for endowing a lecture for the promotion of the mechanic arts, to be governed by the Society, in words that are prophetic of the affinity of humanitarianism and public-spiritedness with the utilitarian scientific spirit. "This Gentleman has well observ'd, that the *Arts* of life have been too long *imprison'd* in the dark shops of Mechanicks themselves, there *hindred from growth,* either by ignorance, or self-interest [Bacon's idea] . . . . We have already seen many other great signs of Liberality and a large mind, from the same hand: For by his diligence about the *Corporation for the Poor;* by his honorable *Subscriptions* for the rebuilding of St. *Paul's;* by his chearful *Disbursment* for the replanting of *Ireland,* and by many other such *publick works,* he has shewn by what means he indeavours to *establish* his Memory; and now by this last gift he has done that, which became one of the *wisest Citizens* of our Nation to accomplish, seeing one of the *wisest of our Statesmen,* the *Lord Verulam,* first propounded it."

³⁰ Hooke does not fail to notice the ignorance of the ancients due to lack of magnifying instruments and to the falsity of Peripatetic doctrines. His indebtedness to Bacon is revealed in citations from him, (pp. 93, 204), in his analysis of the mistakes due to the imperfections of human faculties, and in the stress he places upon experiments and a natural history.

³¹ The "Address" was read to the Society on Dec. 10, 1664, and won for Glanvill admission into the organization. In a letter to Boyle, Oldenburg speaks of it is a "pretty long dedication," and while expressing pleasure that the design of the Society was beginning to be so well understood, he voices a fear that the high expectations aroused by Glanvill's enthusiastic praise might not be realized. Boyle, *Works,* V, 328.

³² Glanvill says that in attacking Aristotelianism he seeks protection under the Royal Society, the members of which, he feels sure, are sympathetic with his position, since they have deserted the old verbose and disputatious philosophy for mathematics and the *"experimental* way of *Enquiry,* and *Mechanical Attempts* for solving the *Phænomena."* Yet he says only some of them have embraced "the *Cartesian,* and *Atomical* Hyphotheseis." Elsewhere he asserts, however, that "the ingenious World" has grown weary of qualities and forms, and has declared in favor of the mechanical hypothesis.

³³ *Some Considerations touching the Usefulnesse of Experimental Naturall Philosophy, Propos'd in Familiar Discourses to a Friend, by way of Invitation to the Study of it.* Oxford, 1663. This volume consists of "The First Part. Of its Usefulnesse in reference to *the Minde of Man,"* and the first section of "The Second Part. Of its Usefulnesse to promote the *Empire of Man* over things *Corporeal."* The second section of "The Second Part. *Of its Usefulnesse* to the *Empire of Men over inferior Creatures,"* did not appear until 1671, though the "Preamble" states that it was written in 1658. Next to Sprat's *History of the Royal Society,* this was the most important document in the scientific propaganda of the period, and also the most popular. Two years after its appearance, Oldenburg says it was much desired by all sorts of men. See letter to Robert Boyle, Sept. 18, 1665, in Boyle's *Works,* V, 336.

³⁴ As we have already seen, this was a popular argument; Hooke, in the preface to *Micrographia,* and Glanvill, in his "Address to the Royal Society," had employed it.

³⁵ See Glanvill's "Address to the Royal Society"; Richard Baxter's *Reasons of the Christian Religion,* 1667, pp. 519-21, and dedication to *More Reasons of the Christian Religion,* 1672; and Beale's letter to Boyle, June 26, 1682, in Boyle's *Works,* V, 505.

³⁶ Boyle's confession of faith represents fairly well the typical attitude of the religious scientists of the seventeenth century: "That there are some Actions so peculiar to Man, upon the account of his Intellect and Will, that

they cannot be satisfactorily explicated after the mann'r of the Actings of meer corporal Agents, I am very much inclin'd to believe: And whether or no there may be some Actions of some other Animals, which cannot well be Mechanically explicated, I have not here leisure or opportunity to examine. But for (most of) the other *Phænomena* of Nature, methinks we may, without absurdity, conceive, that God . . . . having resolved, before the Creation, to make such a World as this of Ours, did divide . . . that matter which he had provided into an innumerable multitude of variously figur'd Corpuscles, and both connected those Particles into such Textures or particular Bodies, and plac'd them in such Scituations, and put them into such Motions, that by the assistance of his ordinary preserving Concourse, the *Phænomena*, which he intended should appear in the Universe, must as orderly follow, and be exhibited by the Bodies necessarily acting according to those Impressions or Laws, though they understand them not at all, as if each of those Creatures had a Design of Self-preservation, and were furnish'd with Knowledge and Industry to prosecute it; and as if there were diffus'd through the Universe an intelligent Being, watchful over the publick Good of it, and careful to Administer all things wisely for the good of the particular Parts of it, but so far forth as is consistent with the Good of the whole, and the preservation of the Primitive and Catholick Laws established by the Supreme Cause. As in the formerly mention'd Clock of *Strasburg*, the several Pieces making up that curious Engine, are so fram'd and adapted, and are put into such a motion, that though the numerous Wheels, and other parts of it, move several ways, and that without any thing either of Knowledge or Design; yet each performs its part in order to the various Ends for which it was contriv'd, as regularly and uniformally as if it knew and were concern'd to do its Duty; and the various Motions of the Wheels, and other parts concur to exhibit the *Phænomena* design'd by the Artificer in the Engine, as exactly as if they were animated by a common Principle, which makes them knowingly conspire to do so, and might, to a rude Indian, seem to be more intelligent than *Cunradus Dasypodius* himself, that published a Description of it." *Some Considerations, The First Part*, pp. 70-2.

³⁷ *Ibid.*, Second Part, pp. 3-5.

³⁸ This firm conviction of the great contribution which science could make to the material welfare of mankind did much to inspire that humanitarian spirit which denounced war and exalted the triumphs of a peace blessed with science. It is a greater credit, says Glanvill, "to know the wayes of *captivating Nature*, and making her *subserve* our *purposes* and *designments;* then to have *learnt* all the *intrigues* of *Policy*, and the *Cabals* of *States* and *Kingdoms:* yea, then to *triumph* in the *head* of *victorious Troops* over *conquer'd Empires.* Those *successes* being more *glorious* which bring *benefit* to the *World;* then such *ruinous ones* as are dyed in *humane blood*, and *cloathed* in the *livery* of *Cruelty* and *Slaughter.*" ("Address to the Royal Society," *Scepsis Scientifica.*) Oldenburg writes Boyle, Mar. 24, 1666, "Let princes and states make war and shed blood; let us cultivate virtue and philosophy, and study to do good to mankind." (Boyle, *Works*, V, 355.) Other evidence of this new and high evaluation of science is revealed in those whom the age selected to make heroes of, for hero worship is a sure index to the standards of a period. Parker says that scientific curiosity is a "heroical Quality" (*Censure*, p. 78), and Power, that to solve scientific difficulties is a "Heroick attempt." The latter speaks enthusiastically of "the winged Souls of our modern Hero's," who had given to the world so many useful inventions. *Experimental Philosophy*, pp. 190, 191. See also Glanvill's "Address to the Royal Society" and Sprat's *History of the Royal Society*, p. 392.

For even clearer evidence of the association of the humanitarian spirit with science see Boyle's *Some Considerations, The First Part*, pp. 301-4, in which the author reveals his conception of the nobility of man, hatred of war, sympathy with human suffering, and firm faith in the efficacy of science in promoting man's happiness. Evelyn likewise expresses himself in the boldest terms on the matter: "I had rather be the Author of one good and beneficial *Invention*, than to have been *Julius Cæsar*, or the great *Alexander* himself; and do range the Names of a *Gilbert*, a *Bacon*, a *Harvey*, a *Guttemberge*, *Columbus*, *Goia*, *Metius*, *Janellius*, *Thyco*, *Galileo* . . . . who gave us the *Use of the Load-stone*, *Taught us the Art of Printing*; found out the *Circulation of the Blood*, *detected new Worlds*, *invented the Telescope*, and other *opticall Glasses*, *Engines* and *Automates*, amongst the *Heroes*, whom they *Deifi'd*, and placed above the Stars; because they were the Authors of ten thousand more worthy Things, than those who had never been named but for their blood-shed and cruelty, pride and prodigious lusts . . . . but for the Pens of such great *Genius's* and learned men, of whom some of them did the least deserve." (Dedication of *Instructions Concerning Erecting a Library*.)

[39] *Some Considerations touching the Usefulnesse of Experimental Naturall Philosophy, Second Part*, pp. 5, 7.

[40] *Ibid.*, p. 46.

[41] "Chymists, (even as matters now stand with them) may considerably adde to the Pharmaceutical part of Physick. But if the Operations of Chymistry were seriously enquir'd into, and throughly understood, I make little doubt, but by a skilfull application of them, and especially by a *series* of them, in rationall and orderly way, succeeding one another, there may be found out a great many preparations of Remedies, both very differing from the common ones, and far more noble then they." (*Ibid.*, p. 196.) Boyle called Van Helmont "a benefactor to experimental learning" in spite of some extravagant ideas. *Works*, II, 333.

[42] *Experiments and Considerations Touching Colours*, 1664, and *New Experiments and Observations Touching Cold, or An Experimental History of Cold Begun*, 1665. Both are headed by a quotation from Bacon.

[43] *Experiments and Considerations Touching Colours*, p. 88. In the "Publisher to the Reader" Henry Oldenburg expresses confidence that the Royal Society will "exhort our Author to the prosecution of this Argument, considering how much it is their design and business to accumulate a good stock of such accurate Observations and Experiments, as may afford them and their Offspring genuine Matter to raise a Masculine Philosophy upon, whereby the Mind of Man may be enobled with the Knowledge of the solid Truths, and the Life of Man benefited with ampler accommodations, than it hath been hitherto." Bacon's natural history again.

[44] *Medela Medicinæ. A Plea for the Free Profession, and a Renovation of the Art of Physick, Out of the Noblest and most Authentic Writers. Shewing the Public Advantage of its Liberty. The Disadvantage that comes to the Publick by any sort of Physicians, imposing upon the Studies and Practise of others. The Alteration of Diseases from their old State and Condition. The Causes of that Alteration. The Insufficiency and Uselessness of meer Scholastick Methods and Medicines, with a necessity of new. Tending to the Rescue of Mankind from the Tyranny of Diseases; and of Physicians themselves, from the Pedantism of old Authors and present Dictators*, 1665. The book was licensed Sept. 2, and the dedication is dated Nov. 26, 1664.

One reason for the intensification of the conflict between chemists and Galenists is to be found in the former's petition to the King to found a royal chemical society in emulation of the Royal Society and probably in opposition to the Royal College of Physicians. The royal favor which the experimental philosophy had received in the establishment of the Royal Society inspired the chemical doctors, who had for long maintained a noisy allegiance to the experimental method, with the hope that they too might reap a similar advantage.

[45] *Medela Medicinæ*, p. 19.

[46] *Ibid.*, pp. 97, 193, 211.

47 "And if so, then certainly the *Scholastick Method and Medicins* erected upon such Scholastic Foundations, cannot in reason stand any longer, but ought to be turned up as Insufficient, and of little use, in an Age wherein better things are known." (*Ibid.*, pp. 256-57.) Nedham is very severe in his strictures upon the ancients, for which in one place he half-way apologizes on the ground that " 'tis necessary to be a little brisk in expression, because the world is apt to dote upon old Authors, especially when they hear of honors little less than Divine given to these Erroneous Heathens" (*ibid.*, p. 238), the result of which dotage is that "the whole Fabrick of Natural Science hath hitherto stood upon mere opinion obtruded upon us by gross Heathens, upon whom it is a shame to hear what high Elogies are bestowed by some of their Christian Commentators." (*Ibid.*, p. 249.) Nedham seems to echo the Puritans in his attitude toward "heathenish philosophers." *Ibid.*, p. 238.

48 *Ibid.*, pp. 437-38, 464, 255.

49 His reference to Hakewill (p. 10) is significant in showing how important the *Apologie* was in supporting the break with antiquity which accompanied the rise of the Royal Society.

50 Cf. *ibid.*, pp. 10-18, 27-28, 339.

51 Cf. *ibid.*, pp. 6, 8, 213, 227-28, 234-35, 244, 245, 361, 451-52.

52 *Ibid.*, pp. 451-52.

53 " 'Tis not vast reading and learning of other men's speculations that makes a Physician, but a near Approach unto Nature (as my Lord *Bacon* calls it;) a strict and constant observing of her Motions and manner of Operations, is that which gives a man light how to trace her in the darkness of obscure Causes, when she is out of order, and to reduce her into order; and an observing of the various effects, or non-effects, and operations of all sorts of Medicaments, both old and new, together with a man's own manual operation in making them, will inlighten him and conduct him so far, as to perceive Causes more evidently, and take off a great part of that Scandal hitherto imputed unto our Profession, that it is but . . . . a meer Conjectural Art." *Ibid.*, pp. 227-28.

54 Nedham urges that all encouragement be given investigators, "whose industry and experiments may contribute toward the compleating of our Art; the Harvest is great, and the Labourers but few; what vast crops are appearing in the spatious Field of Nature, to find work for Ages to fetch them forth!" (*Op. cit.*, p. 216.) He frequently repeats the idea that scientific discovery will continue ad infinitum: "could I live 1000 years, I might always find work in a Laboratory and a Study, for the advancement of Physick both in Theory and Practise" (p. 205); and again: "he who will study, could he live . . . . a thousand Ages, would even then find matter and occasion for new Enquiry and Discovery." *Ibid.*, p. 10. Cf. also pp. 4-6.

55 *Medela Ignorantiæ: Or a Just and plain Vindication of Hippocrates and Galen from the groundless Imputations of M. N. Wherein the whole Substance of his Illiterate Plea, intituled Medela Medicinæ is occasionally considered.* 1665.

56 Elsewhere Sprackling goes more thoroughly into the matter by claiming that the doctrine of elements and humours explains the appearance of bodies, while chemistry, the advancement of which is to be applauded, goes deeper and divides bodies into their constituent parts; yet it must not be forgotten, he says, that the medicine based upon the humours and temperaments has cured diseases for many hundred years, and that future discoveries, while explaining more clearly the nature of diseases, will not change the remedies. *Op. cit.*, pp. 45-6.

57 The next year this opposition becomes apparent in the *Medicina*

Veterum Vindicata of John Twysden, who speaks of finding Nedham's "hand subscribed under an engagement entred into by some of your company, to endeavour the instituting an Incorporation of Professors of Physick, *Onely by Hermetick or Chymical Physick, and in those their endeavours to be assistant to one another, and never to relinquish that their engagement for any temporal respects whatsoever*." (*Op. cit.*, pp. 208-9.) Although Twysden scoffs at this college as lying "in *Utopia*, or buried in the middle of the Atlantic Sea" (a thrust at the *New Atlantis*), he was visibly disturbed at the prospect. (See *op. cit.*, preface.) "Neither have they rested here," he complains, after having called attention to the chemists' attack upon classical medicine, "but endeavoured to draw His most Gracious Majesty's concurrence to their undertakings, and by that dangerous way of innovation through subscriptions of Hands, laboured to erect a new Society of Chymical Physitians in *London* in opposition to that Body already setled"; and he expresses the fear that the liberty of practising will be granted to any pretender in the profession. (See the Epistle Dedicatory.) The conservative physicians were beginning to be apprehensive of the inroads upon their domains made by the chemical doctors. One chemist, George Thomson, expressed the hope that the King, out of his love for chemical truth, "will be pleased to Erect a Learned Chymical Society, which may be a President to Foreign Nations to imitate; for which, present and Future Ages will be bound to bless him." Such a society would not only benefit medicine, but "if such a Philosophical Chymical Company be Authorized and Countenanced, Mechanicks will be much promoted, their Manual faculties will be more dextrously carried on: The Art of Gilding, Painting, Writing, Artillery may be wound up to a higher pitch. Metals might be more intimately purged, refined, volatile made more fixed, the terrestrial parts more sublimed, their colour and sound exalted, those friable made Ductile, the Ignobler provected, and Generous augmented." (*Syrma Ortho Chymicum*, [n.d.], pp. 25 ff., 197. Chemistry meant more to Thomson than improved medicines.) Later Thomson says that the King professed no small kindness for the design, but was finally influenced against it by "some malevolent, ill-disposed persons." *Galeno-pale*, 1665, pp. 103 ff.

⁵⁸ *Op. cit.*, p. 134.

⁵⁹ *Medicina Veterum Vindicata; Or An Answer to a Book, entituled* Medela Medicinæ; *In which the ancient Method and Rules are defended, and farther shewed, that there is no such change in the Diseases of this Age, or their Nature in general, that we should be obliged to an alteration of them. Against the Calumnies and bitter Invectives of an Author who calls himself M. N. Med. Londinens. but in his Epistle before a Book, put out by Mr.* Bolnest, *gives himself the name of* Mar. Nedham. 1666.

⁶⁰ *Op. cit.*, pp. 1, 44.

⁶¹ Twysden ignores entirely Bacon's insistence upon the inadequacy of ancient learning and upon the evils of subserviency to its authority. He really tries to dodge the issue of the relative merits of modernity and antiquity, involved in the propaganda in behalf of the experimental method, by granting much to the upholders of the latter and at the same time saving the face of the ancients. "My Lord *Bacon*," he says, "and those other Worthy persons that have encouraged men to make further search into the things of Nature, and those Noble persons that have written and still labour in Experimental Philosophy, do not do it to disparage the Ancients, but search into the Reasons of the works of Nature, and discover new Truths, and establish the old by new Confirmations." (*Op. cit.*, p. 102.) The sanction which the new philosophy had received from royal and noble quarters did much to soften criticism of it. Twysden tries to remove Boyle from the side of his opponents by insisting that Nedham had twisted the meaning of many of the passages from Boyle cited by him.

⁶² Thomson had already opened the attack in a treatise entitled *Syrma Ortho Chymicum*, which, though bearing no date, must have been published a short while before *Galeno-pale*, and which emphasizes experimentation in contrast to dogmatical rules derived from antiquity, pleads for the establishment of experiments as a criterion of scientific truth and for the introduction of experimental philosophy into the schools, speaks of the author's relationship with Willis, Merrett, Goddard, and "other of our London College," and, as we have seen, expresses a fervent wish that the King would found a chemical society.

⁶³ *Op. cit.*, p. 68.

⁶⁴ πλανο-πνιτμος, *Or a Gag For Johnson*, 1665, pp. 38, 39, 44, 45.

⁶⁵ It seems that Christopher Merrett in *A Letter Concerning The Present State of Physick*, 1665, p. 58, was the first to make this contrast. Merrett had great admiration for, and interest in, chemistry, but he looked with suspicion upon most of the theories of the spagyrists.

⁶⁶ *Op. cit.*, pp. 43 ff. Other treatises of Thomson's which continue the controversy are *Loimologia*, 1665, which contains a running attack on the Galenists, and *Loimotomia*, 1666, in which he contrasts his own ideas, "found visibly and experimentally true," with the "dogmatical fancies" of his opponents. He also indignantly denies that surgery is manual labor and so beneath the dignity of a doctor, and earnestly seeks to free himself from being associated with quacks, whom he upbraids for their attack on learning.

⁶⁷ *A Letter Concerning the Present State of Physick*, pp. 59, 33, 18-19.

⁶⁸ A few years later Merrett again bears witness for the experimental philosophy: "I say that within these few last experimental years, the practical part of Physick hath been much improved (as well as Anatomy) especially by such as have put their hands to work; and therefore till such improvement, this [the Pharmacopoeia] could not be well amended." *A Short View of the* Frauds, *and* Abuses *Committed by* Apothecaries, 1669, p. 34.

⁶⁹ *A Letter*, pp. 44, 55.

⁷⁰ *Ibid.*, pp. 6-8.

⁷¹ *Some Papers Writ in the year 1664. In Answer to a Letter, Concerning the Practice of Physick in England.* By Dr. C. T., 1670, p. 16. Clerke is obviously mistaken in the year 1664, which should be 1665. The treatise is more of an approval of, than an answer to, Merrett's book.

⁷² *Op. cit.*, pp. 16-17. Clerke pays a very high compliment to the members of the Royal Society for their ability, innocence, and unselfishness, "*yet they cannot escape the malicious* Scommas *and detractions of covetous Persons who either hope from them the Phylosophers stone, or some wonderful projects for the getting of money, or of ignorant and pretending Curiosos, who would presently see a perpetual motion, or* the squaring of the Circle. *Beside the liberty some men take, and think it the only witt too, of turning every thing, into* Ridicule, *neither* the Scripture, *nor the great author of it* God *himself excepted.*" [*Loc. cit.*] The wits had already begun their ridicule of the Royal Society, to which fact a year or two later Sprat bears witness, and they continued for a long time to make merry with the men of Gresham.

⁷³ *Op. cit.*, p. 29.

⁷⁴ *Ibid.*, pp. 2-3. To Clerke's account of the ignorant adherents to the new science should be added the picture John Eachard gives of the Cartesian novice and amateur Baconian. (*Some Observations upon the Answer to an Enquiry*, 1671, pp. 162-69.) There were many camp followers in the Baconian army, who must have been a thorn in the flesh to the reputable scientists, yet "to rally ancient learning and talk of experiments" sums up

tersely the extensive propaganda which great and small, respectable and disreputable were spreading in behalf of the new science.

[75] *A Discourse*, pp. 12-13.

[76] *Ibid.*, pp. 6, 35.

[77] A few years later one who signs himself "a Lover of Truth and the Good of Mankind" issued a plea to those physicians who were members both of the Society and the College, that they should pursue their scientific investigations under the auspices of the latter, in order that the College might have the honor of their discoveries, and also because it offered more facilities for their work. He praises the work of the fellows of the College as being "grounded upon reiterated experiments, . . . new and elaborate, disengaged from party and faction, industriously aiming at verity, and profitable knowledg." Though he fully approves of the Royal Society, he wishes the Royal College to gain some of the honors which the new method was bringing to the former. See *An Essay For the Regulation of the Practice of Physick*, 1673, pp. 16-17, 25.

[78] J. F. Payne notes that during the civil wars and Commonwealth, disregard of traditional opinion put a premium on unorthodoxy and gave quacks a great vogue; the laxity of professional restrictions degenerated into unbridled license vexing to orthodox physicians. Conditions, he says, were no better after the Restoration, because the growth of the scientific movement as represented in the establishment of the Royal Society fostered scepticism of traditional doctrines and tolerance of novel theories and systems of medicines. Charles II encouraged quackery by giving patronage to quacks of every kind. (*Thomas Sydenham*, 1900, pp. 159, 161.) Actually the quacks were clinging to the skirts of the experimental method, and as that became more and more prominent, they secured more of a hearing. The trouble, as Timothy Clerke pointed out, lay in the fact that the true method of experimentation had only partially evolved (Bacon insisted as one of the strong arguments for his philosophy that his method did not require men of unusual ability), and anyone who could "rally the ancients" and call Experiment! Experiment! was likely to be treated with more seriousness than he deserved.

[79] Sir George Newman, *Thomas Sydenham, Reformer of English Medicine*, 1924, p. 32.

[80] J. F. Payne, *Thomas Sydenham*, pp. 230-31.

[81] Payne, *op. cit.*, pp. 177, 146.

[82] "He had read in Bacon that the systematic and wide examination of facts was the first thing to be done in science, and that until this had been done faithfully and impartially theories and generalizations must be adjourned." Newman, *op. cit.*, pp. 30-1.

[83] See *Medical Observations*, 3rd ed., 1676, found in *The Works of Thomas Sydenham* M.D., tr. R. G. Letham, vol. I, Dedication and pp. 12, 15, 17-18, 24. The first edition, 1666, was dedicated to Robert Boyle.

[84] One interesting, though hardly important, critic of the experimental philosophy was the Duchess of Newcastle, who was incited by the propaganda in behalf of it to express earnest and rather delightful views on the matter in "Observations upon Experimental Philosophy," "Further Observations upon Experimental Philosophy," and "The Description of a New World," all published in the same volume, 1666. She states the position of the Baconians in words which strongly suggest Lord Verulam: "It is no wonder that our power over natural Causes and Effects is so slowly improved, seeing we are not onely to contend with the obscurity and difficulty of the things whereon we work and think, but even the forces of our minds conspire to betray us: And these being the dangers in the process of Humane Reason, the remedies can only proceed from the Real, the Mechanical, the

Experimental Philosophy, which hath this advantage over the Philosophy of discourse and disputation, That whereas that chiefly aims at the subtilty of its deductions and conclusions, without much regard to the first ground-work, which ought to be well laid on the sense and memory, so this intends the right ordering of them all, and making them serviceable to each other." ("Observations upon Experimental Philosophy," p. 5.) The Duchess objects to the experimental philosophy because it lays emphasis upon the senses, which are deceptive, rather than upon reason. Worse still, she turns the tables upon the Baconians by insisting that their activities, especially those connected with the telescope and microscope, possess no utility. This last is, in fact, the argument which she seeks to drive home the hardest. Against the two inventions just mentioned she is unusually severe, because, she says, they have so intoxicated men's minds that the latter eschew better arts and studies, and even consider them unprofitable to the Commonwealth. The discoveries made in the heavens by the telescope possess no practical value, and the microscope cannot reveal the inner nature of things, and falsifies the exterior by making a louse look like a lobster. Thus they serve no useful end; gunpowder, printing, and the compass owe nothing to them. The good Duchess believed that "regular" reason with a minimum of observation was the more reliable avenue to truth. She also defends the ancients against the rising tide; she expresses great admiration for Aristotle and other classical philosophers, who, she thinks, are often nearer the truth than the moderns. They possessed as good minds, and perhaps even more profitable arts and inventions, which have been lost in the course of time (an idea Sir William Temple put forth many years later). She subscribes to Hakewill's idea of circular progression, and if she escapes the idea of nature's decay, she equally misses the thrill of the idea of progress.

85 Thomas Birch, *History of the Royal Society,* 1756, II, 3, 47, 51, 138, 161, 163, 197.

86 See Oldenburg's letter to Boyle, Nov. 24, 1664, in Boyle's *Works,* V, 325, in which he says that the Society also wished Boyle to read it. In the same letter Oldenburg states that Sprat intended to begin printing the *History* the following week; so it must have been largely completed before the end of 1664. On p. 120 of the *History* Sprat says that the plague and great fire interrupted the printing at that point. Thus the *History* must have been in printable shape before the Society was three years old.

87 *History of the Royal Society,* p. 94.

88 *Ibid.,* pp. 6, 7, 9, 11, 12, 20-1, 25, 46-51, 80 1, 105, 117-18, 152, 154.

89 *Ibid.,* p. 83.

90 *Ibid.,* p. 312.

91 "This, perhaps, is more allowable in matters of *Contemplation,* and in a *Gentleman,* whose chief aim was his own delight; and so it was in his own choice, whether or no, he would go farther to seek it, than his own mind: But it can by no means stand with a practical and universal *Inquiry.*" (*Ibid.,* p. 96.) Here again the utilitarian experimental philosophy is contrasted with the Cartesian.

92 *Ibid.,* pp. 82-3, 95, 117, 222, 311-12. The criticism of Descartes, that his theory depended more upon speculation than upon experiments, to which we have several times called attention, continues to the end of the century: "Monsieur *Des Cartes* .... did not perfectly tread in his [Bacon's] Steps, since he was for doing most of his Work in his Closet, concluding too soon, before he had made Experiments enough; but then to a vast Genius he joined exquisite Skill in Geometry, and working upon intelligible Principles in an intelligible Manner, though he very often failed of one Part of his End, namely, a right Explication of the Phænomena of Nature, yet by marrying Geometry and Physicks together, he put the World in Hopes of a

Masculine Off-spring in process of Time, though the first Productions should prove abortive." At this point in the development of science, Wotton says, the Royal Society was founded. William Wotton, *Reflections upon Ancient and Modern Learning*, 1694, p. 30.

[93] *History of the Royal Society*, pp. 38-9.

[94] As has been noted on a previous page, Sprat blames antiquity, not for failing to discover natural laws, but for not collecting "a vast pile of experiments."

[95] "Their purpose is, in short, to make faithful *Records*, of all the Works of *Nature*, or *Art*, which can come within their reach: that so the present Age, and posterity, may be able to put a mark on the Errors, which have been strengthened by long prescription: to restore the Truths, that have lain neglected: to push on those, which are already known, to more various uses: and to make the way more passable, to what remains unreveal'd. This is the compass of their Design." *Ibid.*, p. 61.

[96] *Ibid.*, p. 86. "I shall lay it down, as their *Fundamental Law*, that whenever they could possibly get to *handle* the subject, the *Experiment* was still perform'd by some of the *Members* themselves. The want of this *exactness*, has very much diminish'd the credit of former *Naturalists*." *Ibid.*, p. 83.

[97] "To this I shall add, that they have never affirm'd any thing, concerning the cause, till the trial was past: whereas, to do it before, is a most venemous thing in the making of Sciences: for whoever has fix'd on his *Cause*, before he has experimented; can hardly avoid fitting his *Experiment*, and his Observations, to his own *Cause*, which he had before imagin'd; rather than the *Cause* to the truth of the *Experiment* it self." (*Ibid.*, p. 108. See also p. 312.) This is exactly Bacon's idea, which ignores the necessity of some kind of theory to serve as a principle of selection in determining what experiments are to be made.

[98] *Ibid.*, p. 115.

[99] *Ibid.*, pp. 100, 101, 362.

[100] *Ibid.*, pp. 83, 86, 89, 91-2, 101-14, 108, 115, 119, 312.

[101] *Ibid.*, p. 119.

[102] "They have come as furiously to the purging of *Philosophy*, as our *Modern Zealots* did to the *reformation* of *Religion* . . . Nothing will suffice either of them, but an utter *Destruction*, *Root*, and *Branch*, of whatever has the face of *Antiquity*." (*Ibid.*, pp. 328-29.) Sprat really reveals here the underlying affinity between Puritanism and science.

[103] Ward, Boyle, Wilkins, Petty, Wren, Wallis, Goddard, Willis, Bathurst, and Rooke. We have already seen to what shifts Ward and Wilkins were put in defending their institution from the charges contained in Webster's *Academiarum Examen*. See *Vindiciae Academiarum*.

[104] Sprat probably had in mind Wilkins' and Ward's answer to Webster's book mentioned in the preceding note. The Restoration and the return to power of the conservative elements in Oxford rendered imperative some attempt to placate the latter.

[105] *Ibid.*, p. 327.

[106] *Ibid.*, p. 329.

[107] *Ibid.*, p. 152.

[108] *Ibid.*, p. 345. Sprat speaks of "the Arguments of some devout men against Knowledge and chiefly that of *Experiments*." *Ibid.*, p. 347.

[109] *Ibid.*, p. 366.

[110] *Ibid.*, p. 372.

111 *Ibid.*, pp. 352-53, 366-67.

112 **Cf.** Wordsworth's lines:
   One impulse from a vernal wood
   May teach you more of man,
   Of moral evil and of good
   Than all the sages can.

113 *History of the Royal Society*, pp. 82-3, 341-45.

114 The interest of the scientists in the English language is revealed in Sprat's proposal to establish an academy for the improvement of the mother tongue. (*Ibid.*, pp. 40 ff.) He also points out how important the development of science is to the spread of English abroad. What is, perhaps, more noteworthy is the stylistic platform which the Royal Society adopted, and which is fully described by Sprat. (*Ibid.*, pp. 111-13.) Early science considered a reformation in style only a little less important than a reformation in learning. Sprat clearly states that to achieve their scientific ends, the virtuosi found it necessary "to separate the knowledge of *Nature*, from the colours of *Rhetorick*, the devices of *Fancy*, or the delightful deceit of *Fables*." (*Ibid.*, p. 62.) For the important influence which science exerted upon the simplification of English prose, see *supra*, Chapter VI, note 28.

115 *History of the Royal Society*, pp. 2, 67, 109-10, 118-19, 190, 245, 437-38.

116 *Ibid.*, pp. 118, 121, 129.

117 **Cf.** *ibid.*, pp. 72, 76, 398, 408.

118 *Ibid.*, pp. 35-6, 144.

119 Other references to Bacon may be found on pages 75, 89, 98, 245.

120 *Ibid.*, pp. 67-8, 110, 123-24.

121 *Ibid.*, pp. 73, 344, 435.

122 *Ibid.*, pp. 39, 56-7, 62-4, 86, 102, 333.

123 *Ibid.*, pp. 1, 71, 77.

124 *Ibid.*, pp. 4, 71, 79, 153, 321, 397, 417. There is sufficient evidence to show that the satirists, especially Samuel Butler, were beginning to make merry over the virtuosi. (See C. S. Duncan, *The New Science and English Literature*, chap. IV.) Ballads, too, were being composed on the Society. (See *Ballad of Gresham College*, edited by Dean Dorothy Stimson in *Isis*, XVIII [1932], 103-17.) In a letter to Robert Boyle, Oct. 31, 1665, Oldenburg quotes five lines from a ballad on the Royal Society, which, he understood, contained twenty-six stanzas. See Boyle, *Works*, V, 341.

125 Oldenburg speaks of those "who ask continually, what have they [virtuosi] done." Boyle, *op. cit.*, V, 358-59.

CHAPTER IX

1 See *Propositions For the Carrying on A Philosophical Correspondence, Already begun in the County of Sommerset, Upon incouragement given from the Royal Society*. The purpose of the organization was to further Bacon's natural history by a collection of the rarities which the district afforded. Glanvill was appointed secretary.

2 Glanvill says that he sent the content of Crosse's book to a friend in London, who published it as the *Chue Gazett*, but that less than a hundred copies were printed, and these remained in private hands in order that the author's shame might not be made public. I have discovered no copy of it. See Glanvill's *A Præfatory Answer to Mr. Henry Stubbe*, 1671, p. 187.

[3] See Anthony à Wood's account of Crosse in *Athenae Oxonienses*. Glanvill says Crosse "travelled up and down to tell his Stories of the *Royal Society*, and to vent his spite against that Honourable Assembly. He took care to inform every *Tapster* of the Danger of their Designs; and would scarce take his Horse out of an *Hostler's* hands, till he had first let him know how he had *confuted* the *Virtuosi*. He set his *everlasting* Tongue at work in every Coffee-House, and drew the Apron-men about him, as *Ballad-singers* do the Rout in *Fairs* and *Markets;* They admir'd the man, and wondred what the *strange thing* call'd the *Royal Society* should be." *A Præfatory Answer to Mr. Henry Stubbe,* p. 3.

[4] *Plus Ultra: Or, the Progress and Advancement of Knowledge Since the Days of* Aristotle. *In an Account of some of the most Remarkable Late Improvements of* Practical, Useful *Learning: To encourage Philosophical Endeavours. Occasioned By a Conference with one of the Notional Way.* 1668. Cf. Thomas Birch's *History of the Royal Society,* II, 197 n.

[5] *Plus Ultra,* preface and p. 108.

[6] *Ibid.,* preface.

[7] Glanvill took his account of ancient anatomy from Hakewill, to whom he refers as "a Learned Man of our own." Cf. Hakewill's *Apologie,* 3rd ed., 1635, p. 271, and *Plus Ultra,* p. 13.

[8] *Plus Ultra,* pp. 51, 75.

[9] *Ibid.,* p. 87.

[10] *Ibid.,* pp. 89, 91.

[11] *Ibid.,* p. 109.

[12] *Ibid.,* p. 91.

[13] *A Letter of Meric Casaubon D. D. etc. To Peter du Moulin D. D. and Prebendarie of the same Church: Concerning Natural experimental Philosophie, and some books lately set out about it.* Cambridge, 1669.

[14] He objects to Sprat's calling miracles divine experiments and to his considering men of honesty, trade, and experiments better testifiers to religion than philosophers, which he properly interprets as an attack on the ancient fathers, schoolmen, and controversial writers. He also denies a strong talking point of the scientists, that one versed in the experimental science could better see God's wisdom revealed in the universe than an ordinary man.

[15] The year before his Letter appeared, Casaubon had published *Of Credulity and Incredulity, In things* Natural, Civil, *and* Divine. *Wherein, Among other things, the* Sadducism *of these times, in denying* Spirits, Witches, *and* Supernatural Operations, *by pregnant instances, and evidences, is fully confuted:* Epicurus *his cause, discussed, and the* jugling *and* false dealing, *lately used, to bring* Him *and* Atheism, *into credit, clearly discovered: the use and necessity of* Ancient Learning, *against the Innovating humor, all along proved, and asserted.* 1668. In this work the author upholds the ancients against the moderns, attacks the mechanical philosophy as leading to a purely materialistic view of life, and employs the evidence of witch stories as proof of the existence of immaterial substances and as disproof of atheism. Scientific scepticism, he says, "hath bereav'd it [man's nature] of its more noble function, the contemplation of *things spiritual, and eternal;* not discernable with bodily eyes, but by the light of *faith . . . .* upon *sound reason* and certain *experience* also." (*Op. cit.,* p. 2.) Mathematics, he asserts, has established such dominion over men's minds that religion itself is tested by it. In a work published after the *Letter* he continues his attack on the new science for its atheistic materialism. In speaking of the deception of the senses and the mutability of

visible things, he says, "How well this will agree with the Philosophy of these dayes, I know not, by which all profitable knowledge is reduced to *experiments*, and *natural Philosophy."* *Of Credulity and Incredulity; in Things Divine and Spiritual,* 1670, p. 198.

Glanvill wrote an answer to the *Letter,* but did not publish it, because, he says, of the fame and learning of Casaubon and of the civility with which he himself had been treated. It is also possible that Glanvill recognized the weight of the very real objections which the aged ecclesiastic advanced. See *A Further Discovery of M. Stubbe,* 1671, p. 12, and the preface to *Essays on Several Important Subjects in Philosophy and Religion,* 1676.

16 *The Plus Ultra reduced to a Non Plus: Or, A Specimen of some* Animadversions *upon the* Plus Ultra *of Mr. Glanvill, wherein sundry* Errors *of some* Virtuosi *are discovered, the Credit of the* Aristotelians *in part* Re-advanced; *and Enquiries made about the Advantages of the* Ancient Education *in* England *above the* Novel *and* Mechanical. *The* old Peripatetick *notion of the* Gravity *of the* Air, *and the* Pressure of the aereal Columne *or* Cylinder. *The* Deceitfulness of Telescopes. *The* World in the Moon, *and* a Voyage *thither. The* Original *and* Progress of Chymistry. *The* Use of Chymical Medicaments. *The* Usefulness of the Peripatetick Philosophy *in reference to the* Practice of Physick. *The* Original *and* Progress of Anatomy. *The* First Inventor of the Circulation of the Blood. *The* Transfusion of Blood, the first Proposers *and* Inventers thereof; *and* its Usefulness. *The different* Nature of the Blood, *and the* variety of Phænomena *appearing upon the* burning thereof, *and* mixing *of it with* several liquors. *Some* Trials *in order to a discovery of the* Nature of English Baths. By Henry Stubbe, Physician at Warwick. Printed for the Author, 1670. The preface gives the reasons for his engaging in the controversy. The idea that Stubbe was hired to attack the Society (Harcourt Brown, *Scientific Organizations in Seventeenth Century France,* pp. 256-57) is hardly credible. No bribe could have inspired the intensity of his hatred.

17 See *Campanella Revived,* 1670, p. 14.

18 *Plus Ultra reduced to a Non Plus,* pp. 138-56.

19 *Ibid.,* p. 12.

20 *Ibid.,* p. 175.

21 *Ibid.,* pp. 20-1.

22 *Ibid.,* p. 14.

23 Stubbe established a definite relationship between Paracelsus and the new scientists. He finds in the former all the qualities which characterize the latter: condemnation of the study of languages, vilification of syllogistic logic because it leads to endless disputes; opposition to the reading of ancient authors and contempt for them; hostility to established methods; a sceptical and tentative mind; and allegiance to experiments, even those of old women, mountebanks, and farmers.

24 *Plus Ultra reduced to a Non Plus,* p. 14.

25 *Ibid.,* p. 159.

26 *Ibid.,* p. 73.

27 *Ibid.,* p. 75.

28 *Ibid.,* p. 13.

29 *Ibid.,* p. 17.

30 *Ibid.,* p. 1. Stubbe clearly perceived Sprat's materialistic and fallacious argument in upholding the existence of the soul on the basis of the numberless invisible corpuscles in the blood, and he insisted that as long as one

sought proofs of such a belief in the material world, he would never arrive at spiritual truth. Cf. *Plus Ultra reduced to a Non Plus*, pp. 173-74.

[31] *Ibid.*, pp. 3-10.

[32] *Ibid.*, pp. 69, 160 ff.

[33] *Ibid.*, pp. 3-10, 24-9, 50-70. Stubbe frequently indicates that he is opposing the moderns as well as the Royal Society. "Mr. *Glanvill* doth not so much as know who writ well upon the several *subjects*, in which he pretends that the *Moderns* have out-done the *Ancients*." (*Plus Ultra reduced. to a Non Plus*, p. 21.) And again, in seeking to carry the argument into fields in which ancient superiority would appear obvious and in which Glanvill later flatly denied modern superiority, he says, "I am surprized to finde, that Mr. *Glanvill* doth not make the *Moderns* to surpass the *Ancients*, in *Architecture, Sculpture, Picture*, and several other Arts of *ingenious Luxury.*" *Ibid.*, p. 24.

[34] *Ibid.*, pp. 65-6, 113, 116-56.

[35] *Ibid.*, pp. 98 ff.

[36] *Ibid.*, preface.

[37] See the dedication to *Legends no Histories*.

[38] *Ibid.*, preface.

[39] *A Censure upon Certaine Passages Contained in the History of the* Royal Society, *As being Destructive to the* Established Religion *and* Church *of England.* Oxford, 1670.

[40] *A Censure*, p. 38. Cf. Sprat's *History of the Royal Society*, p. 349.

[41] *Ibid.*, p. 63. Cf. Sprat, *op. cit.*, pp. 414-15.

[42] *Ibid.*, p. 40. Cf. Sprat, *op. cit.*, p. 355.

[43] *Ibid.*, p. 1. Cf. Sprat, *op. cit.*, p. 47.

[44] *A Præfatory Answer to Mr. Henry Stubbe, The Doctor of Warwick. Wherein the Malignitie, Hypocrisie, Falshood of his Temper, Pretences, Reports, and the Impertinency of his Arguings & Quotations In his Animadversions on Plus Ultra, Are discovered.* 1671. Much of the book is devoted to revealing Stubbe's former Puritan affiliations and writings, in which Stubbe himself was sympathetic with the new science. The preface contains an interesting allusion to "old Alexander Ross" who was the staunchest defender of antiquity when Baconianism was getting a start in the Puritan period.

[45] *A Præfatory Answer*, pp. 69-74, 83.

[46] *Ibid.*, p. 144.

[47] *Ibid.*, pp. 72-3, 94-5.

[48] *Ibid.*, pp. 116, 184. Glanvill refused, however, to widen the controversy to include the arts, as Stubbe had suggested. "I do not make," he says, "the *Moderns to surpass the Ancients in Architecture, Sculpture, Picture, and several other Arts of ingenious Luxury.*" (P. 164.) The English controversy between the ancients and moderns was in the seventeenth century almost exclusively confined to science and philosophy.

[49] *Plus Ultra reduced to a Non Plus*, p. 173.

[50] Glanvill printed the letter in *A Præfatory Answer to Mr. Henry Stubbe*, pp. 154-58.

[51] *A Præfatory Answer*, p. 155.

[52] *A Brief Vindication of the Royal Society: From the late Invectives and*

Mis-representations of Mr. Henry Stubbe: By a Well-Wisher to that Noble Foundation. 1670.

[53] *Ibid.*, pp. 3-4.

[54] This is found among several pieces written by Stubbe and appended to the second edition of his *Censure*. The complete title of the latter describes the contents of the volume: *A Censure upon Certain Passages Contained in the History of the* Royall Society, *As being destructive to the* Established Religion *and* Church *of* England. *The second Edition corrected and enlarged.* Whereunto is added *The Letter of a* Virtuoso *in Opposition to the* Censure, *A Reply unto the Letter Aforesaid, and A Reply unto the Præfatory Answer of* Ecebolius Glanvill, *Chaplain to* Mr. Rouse *of* Eaton (late member of the Rump Parlament) *Rectour of* Bath & *Fellow of the* Royal Society. *Also an Answer to the Letter of* Dr. Henry More, *relating unto* Henry Stubbe, *Physician at* Warwick. Oxford, 1671.

[55] See the preceding note.

[56] *A Reply unto the Letter*, p. 20.

[57] See letter to Boyle, June 4, 1670, in Boyle's *Works*, I, 59.

[58] *A Reply unto the Letter*, p. 16.

[59] *Ibid.*, pp. 29, 31, and *A Reply to a Letter of Dr. Henry More*, p. 34.

[60] *A Reply to a Letter*, p. 47. Stubbe ridicules More's rather feeble attempt to distinguish between atheistic, mechanical philosophy and his own brand, and he tries to cast upon the Royal Society the odium attached to the former. At times he shifts his attack and interprets Sprat's tolerant attitude as a desire to destroy Protestantism and introduce Catholicism.

[61] Boyle, *Works*, I, 59.

[62] The next year, however, during Gunning's absence, du Moulin succeeded in publishing the poem in another volume of verses. (See his Παρεργα, 1670, and Παρεργων Incrementum, 1671; and Boyle's *Works*, I, 60-1; V, 594.) The poem, entitled "Pro Regia Societate Londinensi," is a mediocre performance of two or three pages. Du Moulin attributed to Glanvill's books much of the hostility to the new science, which he hoped Boyle's moderation might soften. Boyle, *Works*, V, 594.

[63] Letter dated July 17, 1669, Boyle's *Works*, V, 514. See also Evelyn's *Diary*, July 9, 1669.

[64] *A Further Discovery of M. Stubbe, In a Brief Reply to His Last Pamphlet Against Jos. Glanvill.* 1671.

[65] *Ibid.*, pp. 15, 17.

[66] *Ibid.*, pp. 8-17.

[67] *Ibid.*, pp. 10-11.

[68] *The Lord Bacons Relation of the Sweating-sickness Examined, in a Reply to George Thomson, Pretender to Physick and Chymistry. Together with a Defence of Phlebotomy . . . . In Opposition to the same Author, and the Author of* Medela Medicinae, Doctor *Whitaker, and* Doctor *Sydenham . . . . And a Reply, by way of Preface to the Calumnies of* Eccebolius Glanvile. 1671.

[69] See the dedication and p. 2.

[70] *The Lord Bacons Relation*, pp. 21, 27.

[71] "An Epistolary Discourse Concerning Phlebotomy. In Opposition to G. *Thomson* Pseudo-Chymist, a pretended Disciple of the Lord *Verulam*," separately paginated.

⁷² *Ibid.*, pp. 9, 49, 73, 151, 168.

⁷³ *Ibid.*, p. 222. Cf. pp. 175, 178. The epithets which the angry doctor applies to his opponents eloquently bespeak the position which Bacon held in the scientific movement: Baconical virtuosi, Baconical experimentators, Baconical philosophers, Baconists, disciples of Lord Bacon, Bacon-faced pyrotechnists, Verulamians, etc., etc. For other evidence of the part which Bacon played in the development of the experimental philosophy see Oldenburg's dedications to the *Philosophical Transactions* for 1670, 1672, 1676.

⁷⁴ " 'Tis true, there is a Professur of *Natural Philosophy* in *Oxford;* and that the *Scholars* in the *Course of their Studies,* are *obliged* to employ a part of their time in *Logick, Physicks,* and *Metaphysicks:* And the reason, is because that *the interest of our Monarchy* is an *Interest of Religion,* and the *support of the Religion established* by *Law* is complicated with, and depends upon *those Studies:* 'Tis no less then impossible for any man to *understand* or *manage* the controversies with the *Papists* (and our *Church* is framed *principally* in Opposition to *them;* as appears by our *Articles,* and *Homilies:* and the *Monarchy* subsists only by that *opposition*) without a *deep knowledge* of *those Sciences:* As any man mustknow who hath inspected no more than the Controversy of the *Eucharist,* wherein the Doctrine of *Substance & accidents,* of *Quantity* distinct from *Matter* of *Ubication,* etc. is so requisite to be *understood,* that the *protestants* [cannot] sustain the dispute without them: For if we change our *Notions* in *Natural Philosophy,* we then differ in the *principals* of *discourse;* and where men differ *therein* 'tis impossible for them to *proceed* . . . . By the change agitated, and now *pursued,* we make our selves *incapable* of convincing a *papist:* and considering the prejudices of *long Education,* and the *Authority* of the *Catholic Church,* we must render our selves in their *judgment* as *Perfect Fooles; and not to be able to proceed,* is in this case all one as to be *baffled.*" "A Preface to the Reader," pp. 29-30.

⁷⁵ "Preface to the Reader," p. 30. In his defence of the universities Stubbe attacks Nedham and Evelyn, whose criticism of the schools has been discussed in previous chapters.

⁷⁶ *Μισοχυμίας "Ελεγχθος*: Or, a Check given to the insolent Garrulity of Henry Stubbe: in Vindication of My Lord Bacon, and the Author; With an Assertion of Experimental Philosophy: Also Some Practical Observations exhibited for the Credit of the true Chymical Science . . . . 1671.

⁷⁷ *Ibid.*, p. 28. Cf. pp. 15, 16, 28.

⁷⁸ *Ibid.*, pp. 9, 40.

⁷⁹ *Ibid.*, p. 57. The relations between the two organizations seem to have been friendly. In his answer to Stubbe's *Campanella Revived,* Christopher Merrett explained Stubbe's statement that the College refused to incorporate the Society with themselves, by saying that when a noble benefactor proposed to build a common meeting place for the two bodies, the Society refused the offer from fear of being obscured. He also contradicts Stubbe's claim that only one or two members of the College were members of the Society, stating that thirty-five were members and some were on the council. Boyle also insisted that harmonious relations existed between the two institutions. See Merrett's *A Short Reply to the Postscript, etc. of H. S.,* 1670, pp. 2-3, and Boyle's *Works,* I, 58.

⁸⁰ *Hydrologia Chymica: Or, The Chymical Anatomy of the Scarbrough, And other Spaws in York-Shire. Wherein are Interspersed Some Animadversions upon Dr. Wittie's lately Published Treatise of the Scarbrough-Spaw* . . . . *Also, a Vindication of Chymical Physick; where a probable way is propounded for the Improvement of Experimental Philosophy; With a Digression concerning an Universal Character.* 1669.

81 See p. 207 and the section entitled "A probable way propounded for the improvement of Experimental Philosophy." (Pp. 213-18.) Simpson had faith in his age and confidence in the feasibility of Bacon's ambitious program, "For an active plodding *Genius* is now at work, which is bringing all former received opinions to the Test, and examines them by matter of Fact, in *Experiments;* and what is found consonant to *Truth*, made forth by collateral *Observations*, is approved, the rest (as frivolous and uncertain) is rejected. And as to what I have propounded, in order to the raising a Structure of *Natural Philosophy*, by the Collection of *Experiments* of all sorts, etc. though it may seem to be but an *Utopian Product*, better discoursed of then performed; yet may it probably, and not unaptly, be looked upon as a direct, even Road, toward a real *Science* of *Experimental Physiology:* to the promoting and perfecting of which, the *Pyrotechnical Art* is of no small use . . . ." ("To the Reader.") Bacon's program had been taken over by the chemists, and thus two great forces against antiquity were united. In contrast to Simpson's faith in the experimental philosophy is his opinion that the mechanical philosophy is a slight hypothesis. *Ibid.*, pp. 195-96.

82 *Pyrologia Mimica, Or, an Answer To Hydrologia Chymica of William Sympson Phylo-Chymico-Medicus; In Defence of Scarbrough-Spaw . . . . Also a Vindication of the Rational Method and Practice of Physick called Galenical, and a Reconciliation betwixt that and the Chymical . . . .* 1669.

83 "I would not here be understood to condemn those Systems, and Methods of Arts and Sciences, that are necessary to the instructing of Youth therein, but that they are laudable and allowable, and have their proper and peculiar uses and benefits, but not so far as to set a stop to farther Discoveries by Experiment and Observation, especially in Natural Philosophy; for all the knowledge (doubtless) that is yet had, is but a small part of that which may be known, and lies yet undiscovered, or found out." *Preface.* Cf. his *Displaying of Supposed Witchcraft*, 1677, pp. 19-20.

84 *Some Considerations Touching the Usefulnesse of Experimental Naturall Philosophy, Propos'd in a familiar discourse to a friend by way of Invitation to the study of it. The Second Tome. Containing the later Section Of the Second Part.* Oxford, 1671.

85 *Some Considerations*, Essay IV, pp. 1-5.

86 John Beale speaks of "Hobbians and Stubbians, atheists and scoffers" who have done more dishonor to the King than the dissenters themselves. See letter to Boyle, June 26, 1682, in Boyle's *Works*, V, 505-6.

87 See C. S. Duncan, *The New Science and English Literature in the Classical Period*, chaps. III, IV, V. In 1678 appeared a translation of Rapin's *Reflexions upon Ancient and Modern Philosophy*, which in general exalts ancient science and philosophy over modern.

88 See that weird book of Nathaniel Fairfax, *A Treatise of the Bulk and Selvedge of the World*, 1674, the dedication and preface of which praise extravagantly the Royal Society and the new philosophy; William Simpson's *Philosophical Dialogues concerning the Principles Of Natural Bodies: Wherein The Principles of the Old and New Philosophy are stated, and the New demonstrated, more agreeable to Reason, from Mechanical Experiments and its usefulness to the benefit of Man-kind*, 1677, which had been composed during the Stubbe controversy, and the title of which declares its contents; John Webster's continued support of Baconianism in *The Displaying of Supposed Witchcraft*, 1677, which was directed against Casaubon's and Glanvill's belief in witches, and was sanctioned by the Royal Society, much to Glanvill's concern, and in which the author refers to Stubbe and the wide-spread hostility to the virtuosi, and admits that philosophical liberty is now prevalent "to compleat a perfect History of Nature" (p. 20); Theophilus Gale's *The Court of the Gentiles*, the fourth part of which, con-

taining an attack on Aristotle and scholastic philosophy, was published in 1677; Robert Wittie's *A Survey of the Heavens*, 1685, which is dedicated to the members of the Royal Society, and which contrasts ancient and modern knowledge in astronomy; and William Molyneux's *Sciothericum Telescopium*, 1686, the dedication of which attempts to show that experimental science is worthy of Clarendon's favor by pointing out in some detail its superiority to traditional learning.

[89] On April 27, 1681, the work was discussed and approved by the Royal Society. (See T. Birch, *History of the Royal Society*, IV, 83.) The dedication to Charles II praises the King for his patronage of true science, which is worth more than "all the dead Learning of the Schools." In the text he continues to praise the new philosophy at the expense of the old: "Besides, the Nature of the Subject does direct us sufficiently; for when we contemplate or treat of Bodies, and the material World, we must proceed by the Modes of Bodies, and their real Properties, such as can be represented either to Sense or Imagination, for These Faculties are made for corporeal Things; but Logical Notions, when applied to particular Bodies, are meer shadows of them, without Light or Substance. No Man can raise a Theory upon such Grounds . . . . nor render any Service, or invent any Thing useful in human Life. And accordingly we see, that for these many Ages, that this dry Philosophy hath govern'd Christendom, it hath brought forth no Fruit, produc'd nothing good to God or Man, to Religion or human Society." (See *The Sacred Theory of the Earth*, 7th ed., 2 vols., 1759, I, 404. Cf. pp. 405-6.) This last sentence represents essentially Glanvill's attitude.

[90] With Burnet's book Temple couples Fontenelle's *Digression sur les anciens et les modernes* as being responsible for his essay, but he dismisses the subject of poetry, which was the chief concern of Fontenelle's digression, with the remark that it would require separate treatment. Two or three pages of his treatise attempt to show ancient superiority in prose literature, but with this slight exception he is concerned only with learning or science. The controversy between the ancients and moderns in seventeenth-century England was concerned almost exclusively with science and philosophy.

[91] See T. R., *An Essay Concerning Critical and Curious Learning*, 1698, p. 47.

[92] *Reflections Upon* Ancient *and* Modern *Learning. To which is now added A Defense Thereof, In Answer to the Objections of Sir W. Temple, and Others. With Observations upon the Tale of a Tub* . . . . Third Edition. Corrected. 1705, p. 475. Page 393 makes it plain that Wotton associated Temple's attack on modern learning with those of Stubbe and others upon the new science, and that he looked upon it as merely carrying the warfare on.

[93] For a more detailed account of this last battle between ancients and moderns see the present writer's "The Background of the Battle of the Books," *Washington University Studies*, Humanistic Series, vol. VII, no. 2, 1920, pp. 142-62; and *The Seventeenth Century* by Richard F. Jones *et al.*, pp. 21-40.

INDEX

INDEX

In the text synonyms are frequently used to express the more important conceptions, but reference may be found only under the term most frequently employed to express the idea. For example, the references listed under "experimental philosophy" may refer to pages on which the idea is expressed by such terms as "experimental method," "experimental science," and "new science." Numerical order is maintaind only in the segments of the entries, which are separated by semicolons, but for the most part the segments are introduced as they occur in the text. Names and titles contained in quotations have not been indexed.

A CATALOGUE OF
SELECTED DOVER BOOKS
IN ALL FIELDS OF INTEREST

A CATALOGUE OF SELECTED DOVER
BOOKS IN ALL FIELDS OF INTEREST

CELESTIAL OBJECTS FOR COMMON TELESCOPES, T. W. Webb. The most used book in amateur astronomy: inestimable aid for locating and identifying nearly 4,000 celestial objects. Edited, updated by Margaret W. Mayall. 77 illustrations. Total of 645pp. 5⅜ x 8½.
20917-2, 20918-0 Pa., Two-vol. set $9.00

HISTORICAL STUDIES IN THE LANGUAGE OF CHEMISTRY, M. P. Crosland. The important part language has played in the development of chemistry from the symbolism of alchemy to the adoption of systematic nomenclature in 1892. ". . . wholeheartedly recommended,"—Science. 15 illustrations. 416pp. of text. 5⅝ x 8¼.   63702-6 Pa. $6.00

BURNHAM'S CELESTIAL HANDBOOK, Robert Burnham, Jr. Thorough, readable guide to the stars beyond our solar system. Exhaustive treatment, fully illustrated. Breakdown is alphabetical by constellation: Andromeda to Cetus in Vol. 1; Chamaeleon to Orion in Vol. 2; and Pavo to Vulpecula in Vol. 3. Hundreds of illustrations. Total of about 2000pp. 6⅛ x 9¼.
23567-X, 23568-8, 23673-0 Pa., Three-vol. set $27.85

THEORY OF WING SECTIONS: INCLUDING A SUMMARY OF AIR-FOIL DATA, Ira H. Abbott and A. E. von Doenhoff. Concise compilation of subatomic aerodynamic characteristics of modern NASA wing sections, plus description of theory. 350pp. of tables. 693pp. 5⅜ x 8½.
60586-8 Pa. $8.50

DE RE METALLICA, Georgius Agricola. Translated by Herbert C. Hoover and Lou H. Hoover. The famous Hoover translation of greatest treatise on technological chemistry, engineering, geology, mining of early modern times (1556). All 289 original woodcuts. 638pp. 6¾ x 11.
60006-8 Clothbd. $17.95

THE ORIGIN OF CONTINENTS AND OCEANS, Alfred Wegener. One of the most influential, most controversial books in science, the classic statement for continental drift. Full 1966 translation of Wegener's final (1929) version. 64 illustrations. 246pp. 5⅜ x 8½.   61708-4 Pa. $4.50

THE PRINCIPLES OF PSYCHOLOGY, William James. Famous long course complete, unabridged. Stream of thought, time perception, memory, experimental methods; great work decades ahead of its time. Still valid, useful; read in many classes. 94 figures. Total of 1391pp. 5⅜ x 8½.
20381-6, 20382-4 Pa., Two-vol. set $13.00

DRAWINGS OF WILLIAM BLAKE, William Blake. 92 plates from Book of Job, *Divine Comedy, Paradise Lost,* visionary heads, mythological figures, Laocoon, etc. Selection, introduction, commentary by Sir Geoffrey Keynes. 178pp. 8⅛ x 11. 22303-5 Pa. $4.00

ENGRAVINGS OF HOGARTH, William Hogarth. 101 of Hogarth's greatest works: *Rake's Progress, Harlot's Progress, Illustrations for Hudibras, Before and After, Beer Street and Gin Lane,* many more. Full commentary. 256pp. 11 x 13¾. 22479-1 Pa. $12.95

DAUMIER: 120 GREAT LITHOGRAPHS, Honore Daumier. Wide-ranging collection of lithographs by the greatest caricaturist of the 19th century. Concentrates on eternally popular series on lawyers, on married life, on liberated women, etc. Selection, introduction, and notes on plates by Charles F. Ramus. Total of 158pp. 9⅜ x 12¼. 23512-2 Pa. $6.00

DRAWINGS OF MUCHA, Alphonse Maria Mucha. Work reveals drafts-man of highest caliber: studies for famous posters and paintings, render-ings for book illustrations and ads, etc. 70 works, 9 in color; including 6 items not drawings. Introduction. List of illustrations. 72pp. 9⅜ x 12¼. (Available in U.S. only) 23672-2 Pa. $4.00

GIOVANNI BATTISTA PIRANESI: DRAWINGS IN THE PIERPONT MORGAN LIBRARY, Giovanni Battista Piranesi. For first time ever all of Morgan Library's collection, world's largest. 167 illustrations of rare Piranesi drawings—archeological, architectural, decorative and visionary. Essay, detailed list of drawings, chronology, captions. Edited by Felice Stampfle. 144pp. 9⅜ x 12¼. 23714-1 Pa. $7.50

NEW YORK ETCHINGS (1905-1949), John Sloan. All of important American artist's N.Y. life etchings. 67 works include some of his best art; also lively historical record—Greenwich Village, tenement scenes. Edited by Sloan's widow. Introduction and captions. 79pp. 8⅜ x 11¼. 23651-X Pa. $4.00

CHINESE PAINTING AND CALLIGRAPHY: A PICTORIAL SURVEY, Wan-go Weng. 69 fine examples from John M. Crawford's matchless private collection: landscapes, birds, flowers, human figures, etc., plus calligraphy. Every basic form included: hanging scrolls, handscrolls, album leaves, fans, etc. 109 illustrations. Introduction. Captions. 192pp. 8⅞ x 11¾. 23707-9 Pa. $7.95

DRAWINGS OF REMBRANDT, edited by Seymour Slive. Updated Lipp-mann, Hofstede de Groot edition, with definitive scholarly apparatus. All portraits, biblical sketches, landscapes, nudes, Oriental figures, classical studies, together with selection of work by followers. 550 illustrations. Total of 630pp. 9⅛ x 12¼. 21485-0, 21486-9 Pa., Two-vol. set $15.00

THE DISASTERS OF WAR, Francisco Goya. 83 etchings record horrors of Napoleonic wars in Spain and war in general. Reprint of 1st edition, plus 3 additional plates. Introduction by Philip Hofer. 97pp. 9⅜ x 8¼. 21872-4 Pa. $4.00

THE SENSE OF BEAUTY, George Santayana. Masterfully written discussion of nature of beauty, materials of beauty, form, expression; art, literature, social sciences all involved. 168pp. 5⅜ x 8½.   20238-0 Pa. $3.00

ON THE IMPROVEMENT OF THE UNDERSTANDING, Benedict Spinoza. Also contains *Ethics, Correspondence,* all in excellent R. Elwes translation. Basic works on entry to philosophy, pantheism, exchange of ideas with great contemporaries. 402pp. 5⅜ x 8½.   20250-X Pa. $4.50

THE TRAGIC SENSE OF LIFE, Miguel de Unamuno. Acknowledged masterpiece of existential literature, one of most important books of 20th century. Introduction by Madariaga. 367pp. 5⅜ x 8½.
                                                           20257-7 Pa. $4.50

THE GUIDE FOR THE PERPLEXED, Moses Maimonides. Great classic of medieval Judaism attempts to reconcile revealed religion (Pentateuch, commentaries) with Aristotelian philosophy. Important historically, still relevant in problems. Unabridged Friedlander translation. Total of 473pp. 5⅜ x 8½.   20351-4 Pa. $6.00

THE I CHING (THE BOOK OF CHANGES), translated by James Legge. Complete translation of basic text plus appendices by Confucius, and Chinese commentary of most penetrating divination manual ever prepared. Indispensable to study of early Oriental civilizations, to modern inquiring reader. 448pp. 5⅜ x 8½.   21062-6 Pa. $5.00

THE EGYPTIAN BOOK OF THE DEAD, E. A. Wallis Budge. Complete reproduction of Ani's papyrus, finest ever found. Full hieroglyphic text, interlinear transliteration, word for word translation, smooth translation. Basic work, for Egyptology, for modern study of psychic matters. Total of 533pp. 6½ x 9¼. (Available in U.S. only)   21866-X Pa. $5.95

THE GODS OF THE EGYPTIANS, E. A. Wallis Budge. Never excelled for richness, fullness: all gods, goddesses, demons, mythical figures of Ancient Egypt; their legends, rites, incarnations, variations, powers, etc. Many hieroglyphic texts cited. Over 225 illustrations, plus 6 color plates. Total of 988pp. 6⅛ x 9¼. (Available in U.S. only)
                                    22055-9, 22056-7 Pa., Two-vol. set $16.00

THE STANDARD BOOK OF QUILT MAKING AND COLLECTING, Marguerite Ickis. Full information, full-sized patterns for making 46 traditional quilts, also 150 other patterns. Quilted cloths, lame, satin quilts, etc. 483 illustrations. 273pp. 6⅞ x 9⅝.   20582-7 Pa. $4.95

CORAL GARDENS AND THEIR MAGIC, Bronsilaw Malinowski. Classic study of the methods of tilling the soil and of agricultural rites in the Trobriand Islands of Melanesia. Author is one of the most important figures in the field of modern social anthropology. 143 illustrations. Indexes. Total of 911pp. of text. 5⅝ x 8¼. (Available in U.S. only)
                                                           23597-1 Pa. $12.95

THE PHILOSOPHY OF HISTORY, Georg W. Hegel. Great classic of Western thought develops concept that history is not chance but a rational process, the evolution of freedom. 457pp. 5⅜ x 8½.   20112-0 Pa. $4.50

LANGUAGE, TRUTH AND LOGIC, Alfred J. Ayer. Famous, clear introduction to Vienna, Cambridge schools of Logical Positivism. Role of philosophy, elimination of metaphysics, nature of analysis, etc. 160pp. 5⅜ x 8½. (Available in U.S. only)   20010-8 Pa. $2.00

A PREFACE TO LOGIC, Morris R. Cohen. Great City College teacher in renowned, easily followed exposition of formal logic, probability, values, logic and world order and similar topics; no previous background needed. 209pp. 5⅜ x 8½.   23517-3 Pa. $3.50

REASON AND NATURE, Morris R. Cohen. Brilliant analysis of reason and its multitudinous ramifications by charismatic teacher. Interdisciplinary, synthesizing work widely praised when it first appeared in 1931. Second (1953) edition. Indexes. 496pp. 5⅜ x 8½.   23633-1 Pa. $6.50

AN ESSAY CONCERNING HUMAN UNDERSTANDING, John Locke. The only complete edition of enormously important classic, with authoritative editorial material by A. C. Fraser. Total of 1176pp. 5⅜ x 8½.
20530-4, 20531-2 Pa., Two-vol. set $16.00

HANDBOOK OF MATHEMATICAL FUNCTIONS WITH FORMULAS, GRAPHS, AND MATHEMATICAL TABLES, edited by Milton Abramowitz and Irene A. Stegun. Vast compendium: 29 sets of tables, some to as high as 20 places. 1,046pp. 8 x 10½.   61272-4 Pa. $14.95

MATHEMATICS FOR THE PHYSICAL SCIENCES, Herbert S. Wilf. Highly acclaimed work offers clear presentations of vector spaces and matrices, orthogonal functions, roots of polynomial equations, conformal mapping, calculus of variations, etc. Knowledge of theory of functions of real and complex variables is assumed. Exercises and solutions. Index. 284pp. 5⅝ x 8¼.   63635-6 Pa. $5.00

THE PRINCIPLE OF RELATIVITY, Albert Einstein et al. Eleven most important original papers on special and general theories. Seven by Einstein, two by Lorentz, one each by Minkowski and Weyl. All translated, unabridged. 216pp. 5⅜ x 8½.   60081-5 Pa. $3.50

THERMODYNAMICS, Enrico Fermi. A classic of modern science. Clear, organized treatment of systems, first and second laws, entropy, thermodynamic potentials, gaseous reactions, dilute solutions, entropy constant. No math beyond calculus required. Problems. 160pp. 5⅜ x 8½.
60361-X Pa. $3.00

ELEMENTARY MECHANICS OF FLUIDS, Hunter Rouse. Classic undergraduate text widely considered to be far better than many later books. Ranges from fluid velocity and acceleration to role of compressibility in fluid motion. Numerous examples, questions, problems. 224 illustrations. 376pp. 5⅝ x 8¼.   63699-2 Pa. $5.00

THE AMERICAN SENATOR, Anthony Trollope. Little known, long un-available Trollope novel on a grand scale. Here are humorous comment on American vs. English culture, and stunning portrayal of a heroine/villainess. Superb evocation of Victorian village life. 561pp. 5⅜ x 8½.
23801-6 Pa. $6.00

WAS IT MURDER? James Hilton. The author of *Lost Horizon* and *Good-bye, Mr. Chips* wrote one detective novel (under a pen-name) which was quickly forgotten and virtually lost, even at the height of Hilton's fame. This edition brings it back—a finely crafted public school puzzle resplendent with Hilton's stylish atmosphere. A thoroughly English thriller by the creator of Shangri-la. 252pp. 5⅜ x 8. (Available in U.S. only)
23774-5 Pa. $3.00

CENTRAL PARK: A PHOTOGRAPHIC GUIDE, Victor Laredo and Henry Hope Reed. 121 superb photographs show dramatic views of Central Park: Bethesda Fountain, Cleopatra's Needle, Sheep Meadow, the Blockhouse, plus people engaged in many park activities: ice skating, bike riding, etc. Captions by former Curator of Central Park, Henry Hope Reed, provide historical view, changes, etc. Also photos of N.Y. landmarks on park's periphery. 96pp. 8½ x 11. 23750-8 Pa. $4.50

NANTUCKET IN THE NINETEENTH CENTURY, Clay Lancaster. 180 rare photographs, stereographs, maps, drawings and floor plans recreate unique American island society. Authentic scenes of shipwreck, lighthouses, streets, homes are arranged in geographic sequence to provide walking-tour guide to old Nantucket existing today. Introduction, captions. 160pp. 8⅞ x 11¾. 23747-8 Pa. $6.95

STONE AND MAN: A PHOTOGRAPHIC EXPLORATION, Andreas Feininger. 106 photographs by *Life* photographer Feininger portray man's deep passion for stone through the ages. Stonehenge-like megaliths, fortified towns, sculpted marble and crumbling tenements show textures, beauties, fascination. 128pp. 9¼ x 10¾. 23756-7 Pa. $5.95

CIRCLES, A MATHEMATICAL VIEW, D. Pedoe. Fundamental aspects of college geometry, non-Euclidean geometry, and other branches of mathematics: representing circle by point. Poincare model, isoperimetric property, etc. Stimulating recreational reading. 66 figures. 96pp. 5⅝ x 8¼.
63698-4 Pa. $2.75

THE DISCOVERY OF NEPTUNE, Morton Grosser. Dramatic scientific history of the investigations leading up to the actual discovery of the eighth planet of our solar system. Lucid, well-researched book by well-known historian of science. 172pp. 5⅜ x 8½. 23726-5 Pa. $3.50

SECOND PIATIGORSKY CUP, edited by Isaac Kashdan. One of the greatest tournament books ever produced in the English language. All 90 games of the 1966 tournament, annotated by players, most annotated by both players. Features Petrosian, Spassky, Fischer, Larsen, six others. 228pp. 5⅜ x 8½. 23572-6 Pa. $3.50

ENCYCLOPEDIA OF CARD TRICKS, revised and edited by Jean Hugard. How to perform over 600 card tricks, devised by the world's greatest magicians: impromptus, spelling tricks, key cards, using special packs, much, much more. Additional chapter on card technique. 66 illustrations. 402pp. 5⅜ x 8½. (Available in U.S. only) 21252-1 Pa. $4.95

MAGIC: STAGE ILLUSIONS, SPECIAL EFFECTS AND TRICK PHO-TOGRAPHY, Albert A. Hopkins, Henry R. Evans. One of the great classics; fullest, most authorative explanation of vanishing lady, levitations, scores of other great stage effects. Also small magic, automata, stunts. 446 illus-trations. 556pp. 5⅜ x 8½. 23344-8 Pa. $6.95

THE SECRETS OF HOUDINI, J. C. Cannell. Classic study of Houdini's incredible magic, exposing closely-kept professional secrets and revealing, in general terms, the whole art of stage magic. 67 illustrations. 279pp. 5⅜ x 8½. 22913-0 Pa. $4.00

HOFFMANN'S MODERN MAGIC, Professor Hoffmann. One of the best, and best-known, magicians' manuals of the past century. Hundreds of tricks from card tricks and simple sleight of hand to elaborate illusions involving construction of complicated machinery. 332 illustrations. 563pp. 5⅜ x 8½. 23623-4 Pa. $6.00

MADAME PRUNIER'S FISH COOKERY BOOK, Mme. S. B. Prunier. More than 1000 recipes from world famous Prunier's of Paris and London, specially adapted here for American kitchen. Grilled tournedos with anchovy butter, Lobster a la Bordelaise, Prunier's prized desserts, more. Glossary. 340pp. 5⅜ x 8½. (Available in U.S. only) 22679-4 Pa. $3.00

FRENCH COUNTRY COOKING FOR AMERICANS, Louis Diat. 500 easy-to-make, authentic provincial recipes compiled by former head chef at New York's Fitz-Carlton Hotel: onion soup, lamb stew, potato pie, more. 309pp. 5⅜ x 8½. 23665-X Pa. $3.95

SAUCES, FRENCH AND FAMOUS, Louis Diat. Complete book gives over 200 specific recipes: bechamel, Bordelaise, hollandaise, Cumberland, apri-cot, etc. Author was one of this century's finest chefs, originator of vichyssoise and many other dishes. Index. 156pp. 5⅜ x 8. 23663-3 Pa. $2.75

TOLL HOUSE TRIED AND TRUE RECIPES, Ruth Graves Wakefield. Authentic recipes from the famous Mass. restaurant: popovers, veal and ham loaf, Toll House baked beans, chocolate cake crumb pudding, much more. Many helpful hints. Nearly 700 recipes. Index. 376pp. 5⅜ x 8½. 23560-2 Pa. $4.50

ART FORMS IN NATURE, Ernst Haeckel. Multitude of strangely beautiful natural forms: Radiolaria, Foraminifera, jellyfishes, fungi, turtles, bats, etc. All 100 plates of the 19th-century evolutionist's *Kunstformen der Natur* (1904). 100pp. 9⅜ x 12¼.   22987-4 Pa. $5.00

CHILDREN: A PICTORIAL ARCHIVE FROM NINETEENTH-CENTURY SOURCES, edited by Carol Belanger Grafton. 242 rare, copyright-free wood engravings for artists and designers. Widest such selection available. All illustrations in line. 119pp. 8⅜ x 11¼.
23694-3 Pa. $4.00

WOMEN: A PICTORIAL ARCHIVE FROM NINETEENTH-CENTURY SOURCES, edited by Jim Harter. 391 copyright-free wood engravings for artists and designers selected from rare periodicals. Most extensive such collection available. All illustrations in line. 128pp. 9 x 12.
23703-6 Pa. $4.50

ARABIC ART IN COLOR, Prisse d'Avennes. From the greatest ornamentalists of all time—50 plates in color, rarely seen outside the Near East, rich in suggestion and stimulus. Includes 4 plates on covers. 46pp. 9⅜ x 12¼.   23658-7 Pa. $6.00

AUTHENTIC ALGERIAN CARPET DESIGNS AND MOTIFS, edited by June Beveridge. Algerian carpets are world famous. Dozens of geometrical motifs are charted on grids, color-coded, for weavers, needleworkers, craftsmen, designers. 53 illustrations plus 4 in color. 48pp. 8¼ x 11. (Available in U.S. only)   23650-1 Pa. $1.75

DICTIONARY OF AMERICAN PORTRAITS, edited by Hayward and Blanche Cirker. 4000 important Americans, earliest times to 1905, mostly in clear line. Politicians, writers, soldiers, scientists, inventors, industrialists, Indians, Blacks, women, outlaws, etc. Identificatory information. 756pp. 9¼ x 12¾.   21823-6 Clothbd. $40.00

HOW THE OTHER HALF LIVES, Jacob A. Riis. Journalistic record of filth, degradation, upward drive in New York immigrant slums, shops, around 1900. New edition includes 100 original Riis photos, monuments of early photography. 233pp. 10 x 7⅞.   22012-5 Pa. $7.00

NEW YORK IN THE THIRTIES, Berenice Abbott. Noted photographer's fascinating study of city shows new buildings that have become famous and old sights that have disappeared forever. Insightful commentary. 97 photographs. 97pp. 11⅜ x 10.   22967-X Pa. $5.00

MEN AT WORK, Lewis W. Hine. Famous photographic studies of construction workers, railroad men, factory workers and coal miners. New supplement of 18 photos on Empire State building construction. New introduction by Jonathan L. Doherty. Total of 69 photos. 63pp. 8 x 10¾.
23475-4 Pa. $3.00

CATALOGUE OF DOVER BOOKS

GEOMETRY, RELATIVITY AND THE FOURTH DIMENSION, Rudolf Rucker. Exposition of fourth dimension, means of visualization, concepts of relativity as Flatland characters continue adventures. Popular, easily followed yet accurate, profound. 141 illustrations. 133pp. 5⅜ x 8½.
23400-2 Pa. $2.75

THE ORIGIN OF LIFE, A. I. Oparin. Modern classic in biochemistry, the first rigorous examination of possible evolution of life from nitrocarbon compounds. Non-technical, easily followed. Total of 295pp. 5⅜ x 8½.
60213-3 Pa. $4.00

PLANETS, STARS AND GALAXIES, A. E. Fanning. Comprehensive introductory survey: the sun, solar system, stars, galaxies, universe, cosmology; quasars, radio stars, etc. 24pp. of photographs. 189pp. 5⅜ x 8½. (Available in U.S. only)
21680-2 Pa. $3.75

THE THIRTEEN BOOKS OF EUCLID'S ELEMENTS, translated with introduction and commentary by Sir Thomas L. Heath. Definitive edition. Textual and linguistic notes, mathematical analysis, 2500 years of critical commentary. Do not confuse with abridged school editions. Total of 1414pp. 5⅜ x 8½.
60088-2, 60089-0, 60090-4 Pa., Three-vol. set $18.50

Prices subject to change without notice.

Available at your book dealer or write for free catalogue to Dept. GI, Dover Publications, Inc., 180 Varick St., N.Y., N.Y. 10014. Dover publishes more than 175 books each year on science, elementary and advanced mathematics, biology, music, art, literary history, social sciences and other areas.